# Fighting Traffic

**Inside Technology**
edited by Wiebe E. Bijker, W. Bernard Carlson, and Trevor Pinch

For a list of the series, see page 379.

# Fighting Traffic

## The Dawn of the Motor Age in the American City

Peter D. Norton

The MIT Press
Cambridge, Massachusetts
London, England

First MIT Press paperback edition, 2011
© 2008 Massachusetts Institute of Technology

For information on quantity discounts, email special_sales@mitpress.mit.edu.

Set in Stone serif and Stone sans by SNP Best-set Typesetter Ltd., Hong Kong. Printed and bound in the United States of America.

Library of Congress Cataloging-in-Publication Data

Norton, Peter D.
Fighting traffic : the dawn of the motor age in the American city / Peter Norton.
p. cm.
Includes bibliographical references and index.
ISBN 978-0-262-14100-0 (hc. : alk. paper)—978-0-262-51612-9 (pb. : alk. paper)
1. Transportation, Automotive—Social aspects—United States. 2. Transportation, Automotive—United States—History—20th century. I. Title.

HE5623.N67 2008
388.3'21097309042—dc22

2007035522

10 9 8 7

for Debby

# Contents

**Conclusion: History, Technology, and the Dawn of the Motor Age    255**

# Acknowledgments

In 1988 I got a job sorting through a collection of photographic negatives for the Historical Society of Delaware. Working in downtown Wilmington at the Society's headquarters—an old bank building on Market Street—I picked through unending piles of cellulose nitrate eight-by-tens, holding each up to a light so that I could classify it by subject. They showed Wilmington subjects taken by a local photographer from about 1910 to about 1960. I worked at a small wooden table in the society's imperfectly lit basement, usually alone and with little to remind me of my own era. Deprived as I was of other sensations, the job offered little to remind me of the world outside the old bank building. In return the negatives offered unfamiliar sensations. The smell of the decaying celluloid made the job something like working in a vinegar factory. My visual world was Wilmington, several decades earlier. It was a dim, brown version, with reversed shadows. The photographer's subjects were diverse. Those I remember best were the street views, when the photograph captured people going about their business. At 5 o'clock I would leave the basement and return to 1988. As I walked to the bus stop on King Street I would see the 1988 versions of the same streets. The contrasts between the streets I saw in the negatives and those I saw on my journey home planted questions in my mind that this book seeks to answer.

I am grateful to the people who made this book possible, and who made it better. Special thanks go to Brian Balogh. As my dissertation advisor, Brian set an example for me of professional commitment to scholarship. Because he would not let mediocrity slip by, his encouragement, when it came, was truly encouraging. I admire his decency, and have benefited from it. I am very lucky to have been his apprentice, and very proud now to be his colleague. He has my profound gratitude.

W. Bernard Carlson, Nelson Lichtenstein, and Edmund Russell read the dissertation from which this book is derived, and all offered insightful and expert criticism that I used to make the book better. Three referees read portions of the book manuscript, and their valuable advice was also a substantial help. Some manuscript readers offered generous words of encouragement. Because of my great admiration for the work of Clay McShane, Gijs Mom, and Zachary M. Schrag, I am honored and grateful for their kind comments.

I thank the editors of The MIT Press's Inside Technology series for helping me to see the material I studied from new and illuminating perspectives. I am lucky indeed to have Bernie Carlson just two doors away. Through their written work and in their visits to Charlottesville in 2005 and 2007, Wiebe Bijker and Trevor Pinch have also been sources of intellectual inspiration.

I thank the editors at The MIT Press, particularly Sara Meirowitz, Marguerite Avery, and Paul Bethge, for their support, expertise, and professionalism—and for their patience.

Friendships, conversations, and expressions of interest from many sources have helped me as well. I cannot name them all, but I wish especially to thank Jameson M. Wetmore, Dimitry Anastakis, Steve Penfold, John Staudenmaier, David Lucsko, Betsy Mendelsohn, Joel Morine, William Keene, Malcolm Bell, and Mildred Robinson. I owe distinct and substantial debts to each of these good people. In this category there are also many students and some former teachers, and though space will not allow me to name them, they have my thanks.

The Herbert Hoover Presidential Library in West Branch, Iowa, and the Headquarters Library of the American Automobile Association in Heathrow, Florida, were rich sources of evidence for this book, and both opened their collections to me for days at a time. Also helpful were the Hagley Museum and Library in Wilmington, Delaware, and the National Archives in Silver Spring, Maryland. Practical support for research at the Hoover Presidential Library came from the Hoover Presidential Library Association through a fellowship provided by the Robert R. McCormick Tribune Foundation of Chicago. I presented some of my research at a conference in Toronto; funds for travel were provided by the Social Sciences and Humanities Research Council of Canada.

Archivists and other professionals at numerous institutions provided indispensable assistance. They include Dwight Miller, Dale Meyer, and Cindy Worrell of the Hoover Presidential Library; Stephanie Haimes, Melissa Phillips, and Patty Wolfe of the Headquarters Library of the American Automobile Association; and Christopher Baer of the Hagley Museum and Library. Others saved me the expense of travel by helping from a distance, including Danielle Green of the Harvard University Archives, Tracy Larkin of the Eno Transportation Foundation, Robert Denham of the Studebaker National Museum, Raymond Geselbracht of the Harry S. Truman Library, Alan Raucher of Wayne State University, and the entire staff of the Interlibrary Loans office of the University of Virginia Library. In obtaining picture permissions, I was helped by the following people and institutions: Anne Brugman of *Motor Age,* Lynn Catanese and Marjorie McNinch of the Hagley Museum and Library, Deborah Cribbs of the St. Louis Mercantile Library at the University of Missouri–St. Louis, Lamar Gable and Susan Maturo of the Barron Collier Company, Lee Gardner of the American Risk and Insurance Association, Chris Hunter of the Schenectady Museum and Suits-Bueche Planetarium, Heather Hunter of the American Automobile Association, Carol Paszamant of the New Jersey Department of Transportation, Beth Payne of the International City-County Management Association, Kay Peterson of the National Museum of American History, Suzanne Powills of the National Safety Council, and Edward Reis of the George Westinghouse Museum.

For some, no words can adequately express my gratitude. These include Frances F. McMurdo (1901–1998) and Sally M. Hotchkiss (1929–2006). I remember my parents, David L. Norton (1930–1995) and Joan Carter Norton, later Dershimer (1932–1972). To my extraordinary good fortune, Richard A. Dershimer (1926–2007) and Greta G. Dershimer became parents to me, and thereby made all possible.

For my children, Will and Paul, I offer this book, entrusting the past, with its stories and its lessons, to the future. To Deborah, I dedicate this book. She has blessed me by her faith in me and in this project.

# Introduction   What Are Streets For?

Streets are public property—not to be abused but to be used with convenience for the good of the greatest number.

—George H. Herrold, city planning engineer, St. Paul, Minnesota, 1927[1]

The obvious solution . . . lies only in a radical revision of our conception of what a city street is for.

—*Engineering News-Record*, 1922[2]

## The Social Reconstruction of the City Street

How did the American city become an automotive city? Why was much of the city physically destroyed and rebuilt to accommodate automobiles? The case presented in this book is that before the city could be physically reconstructed for the sake of motorists, its streets had to be socially reconstructed as places where motorists unquestionably belonged.

This social reconstruction was only one of several ways in which people tried to solve a new problem. New automobiles were incompatible with old street uses. Until the 1920s, under prevailing conceptions of the street, cars were at best uninvited guests. To many they were unruly intruders. They obstructed and endangered street uses of long-standing legitimacy. As a Providence newspaper editor expressed the problem in 1921, "it is impossible for all classes of modern traffic to occupy the same right of way at the same time in safety."[3]

## The Social Construction of Technology

The social reconstruction of the street, as documented in this book, confirms others' findings about the social construction of other artifacts. First, it shows the importance of examining alternative constructions of an artifact "symmetrically"—that is, without presupposing the correctness (or falsehood) of any one construction. Today we tend to regard streets as motor thoroughfares, and we tend to project this construction back to pre-automotive streets. In retrospect, therefore, the use of streets for children's play (for example) can seem obviously wrong, and thus the departure of children from streets with the arrival of automobiles can seem an obvious and simple necessity. Only when we can see the prevailing social construction of the street from the perspective of its own time can we also see the car as the intruder. Until we do, not only will we fail to understand the violent revolution in street use circa 1915–1930, we will not even see it. This is why the full scale of the wave of blood, grief, and anger in American city streets in the 1920s has eluded notice.[4]

Success in such historical investigations requires not merely *looking* back from where we stand today at the actors of times past, but *getting* back to them, so we can stand next to them and adopt their perspective. The result can transform our view. Trevor Pinch and Wiebe Bijker, for example, discovered that the synthetic plastic Bakelite "was at first hardly recognized as the marvelous synthetic resin that it later proved to be."[5] Similarly, for years automobiles were not widely recognized as a good means of urban passenger transportation.

By adopting the perspectives of various social groups, we can recover more than one perspective. Borrowing their perspectives, constructivist historians of technology have discovered the "interpretive flexibility" of artifacts. One object can be different things to different people. To some young men of the 1880s, for example, a high-wheeled bicycle was a means of displaying physical prowess—a "macho bicycle"; to others the same device could be a dangerous machine—an "unsafe bicycle."[6] Constructivists have shown that this flexibility tends to be greatest when an artifact is new. But the present study confirms some researchers' findings that, under some conditions, flexibility can be reintroduced into a once-stable system. Prevailing social constructions of the street, for example, were stable in 1900. The automobile destabilized them. Social groups, such as pedestrians, parents, police, and downtown business associations, organized to preserve streets as they knew them. But their actions threatened to limit the automobile's urban horizons. In the 1920s, automotive interests (or *motordom*, as they were sometimes called) proposed that customary social constructions of the street were outdated and that only a revolutionary change in perceptions of the street could ease congestion and prevent accidents.

## Relevant Social Groups

Before motordom could champion such a daring cause, it had to give up hope in peaceful change. It had to find common interests strong enough to overcome many particular differences of interest between the groups that composed it (especially auto clubs, dealers, and manufacturers). In the 1920s the reactions of other social groups to the growing problems of accidents and congestion did just this.

Building their theory on historical case studies, Wiebe Bijker and Trevor Pinch have proposed that the social construction of artifacts evolves through interplay between "relevant social groups"—users and non-users with something at stake in the result.[7] In the case of city streets, these groups became distinct through their competing ways of fighting traffic accidents and congestion. Even before automobiles, diverse street users disagreed about what streets are for. Nevertheless, only with the arrival of automobiles in quantity were many street users forced under pressure to commit their loyalties. As the numbers of cars in city streets grew, the relevant social groups grew increasingly distinct. By the 1920s the groups were recognizable as pedestrians, safety reformers, police, street railways, downtown business associations, traffic engineers, and motordom. The categories were not tidy, however. In practice, streetcar patrons could be indistinct from pedestrians, since they normally had to enter streets on foot to reach streetcar stops. Street railways, however, sometimes sought stricter pedestrian control. Parents and educators concerned for the safety of children were often—but not always—in agreement with pedestrians about the dangers that automobiles posed. Small merchants often opposed the traffic platform of chambers of commerce dominated by bigger businesses.

But with time the relentless pressure of traffic tended to make social groups more cohesive. Groups more often acquired distinct names. More city people who wrote letters to the editor signed themselves "A Pedestrian." As improvised police traffic duties grew routine, some police became "traffic cops" or "cornermen." Chambers of commerce alarmed by congestion formed "traffic commissions." The municipal engineers they hired became "traffic engineers." And by the mid 1920s, organized automotive interest groups began calling themselves "motordom."

Traffic pressures also inspired rival groups to name each other. While older constructions of the streets prevailed, new motorists very easily became "joy riders," "road hogs," or "speed demons." Their machines were "juggernauts," "death cars," or "the modern Moloch." As motorists appropriated streets for new uses, respectable pedestrians became "jaywalkers" and streetcars became traffic obstructions. But not without a fight.

## Technological Frames

Each group comprised diverse people with diverse views. Nevertheless, the groups grew recognizable to themselves and to each other for some shared interests, habits of mind, and perspectives. Members of a relevant social group thus shared an approach to traffic problems. Bijker called such a shared approach a "technological frame."[8] Angry pedestrians tended to retain inherited notions about what streets are and what they are for. Parents, worried for their children's safety, tended to look at traffic safety in moral terms. They looked for the guilty and the innocent, assuming the innocence of child pedestrians. In a word, their technological frame was *justice*. Police, with conservative habits conditioned by long experience with other problems, tended to protect old street customs and to perceive their fundamental enemy not as congestion or accidents, but as disorder. We can call their technological frame *order*. To street railways, chambers of commerce, and the engineers they hired, congestion was indeed a frightening enemy threatening financial ruin. Before the mid 1920s, automotive interests often joined these groups to fight accidents and congestion. Their rallying cry was *efficiency*. But thereafter, automotive interest groups (especially auto clubs, dealers, and manufacturers) developed their own technological frame, at first defining it in opposition to all the others. Soon, however, they developed a positive case for new ways to fight traffic accidents and congestion, coinciding with their new self-identification as "motordom." Often they presented their position clothed in a rhetoric of *freedom*.[9]

Motordom was ultimately the most successful combatant. Yet the details of its struggle for the street are messy and show the extent of the power all street users could wield. Motorists had the advantage of horsepower, and with it they drove many pedestrians unwillingly off the pavements—even at crossings. Pedestrians had advantages of their own, in numbers and agility. A bold (or foolish) pedestrian could even win a fight for street access by calling the bluff of an oncoming motorist. Motorists mindful of children's poor judgment were sometimes forced to drive slowly. And the grim stories of those who were hit became powerful newspaper stories with an anti-automobile moral. Where signal timings did not suit their needs, pedestrians defied them. Los Angeles learned from Chicago that if pedestrians were to be controlled then signals could not ignore pedestrians' needs. Recalcitrant pedestrians preserved informal access to streets wherever traffic allowed.

Nevertheless, the prevailing social construction of the street changed. By 1930 most street users agreed that most streets were chiefly motor thoroughfares. Social constructivists call such declines in interpretive flexibility *closure*, which is followed by *stabilization*, when one interpretation prevails. Objections persist, but they typically do so within the frame imposed by the prevailing interpretation. For example, even most jaywalkers after 1930 would agree that they were jaywalking (that is, using the street in an unconventional way), though in 1920 most would have objected to the term. After closure, problems (such as casualties and congestion) can remain (or even worsen), but solutions are sought within the prevailing framework. Closure can "obscure alternatives," Thomas Misa explains, "and hence appear to render the particular artifact, system, or network as necessary or logical."[10] Thus, in the motor age, the solutions to casualties were pedestrian control, school safety instruction, penalties against reckless drivers, and "foolproof highways," and the solution to congestion was ample motor highways. Since we still live in the motor age, the apparent inevitability of motor age ways conceals the alternatives that prevailed before it.

### Closure Mechanisms

Constructivists have proposed various mechanisms by which closure is accomplished, particularly *rhetorical closure* and *closure by redefinition of the problem*.[11] Both "closure mechanisms" were at work in city streets. In rhetorical closure, problems (such as congestion or accidents) persist, but promotional language is used to assert the success of the new way, much as advertising promotes a product. Such promotional rhetoric for motor age methods grew common in the 1920s. "Motor age" was itself a promotional term, for it carried a built-in justification for overturning established custom. It combined rhetorical closure and problem redefinition, just as similar phrases have been used in more recent years to justify workplace smoking bans, cleaner fuels, and tightened security at airports.

Street railways and safety reformers attempted alternative rhetorical efforts, but these were dimmed by the shadows of a mammoth campaign to sell the motor age city. By 1930 the American Automobile Association had overtaken safety councils for leadership in school safety. In 1939 motordom's work culminated in one of the most monumental works of promotional showmanship in the history of technology: the Futurama model depicting the motorized city of 1960, displayed in General Motors' "Highways and Horizons" pavilion at the New York World's Fair. It was a motor age dream city, entirely dependent on automobiles but entirely free of accidents and congestion.

The closure of circa 1930 also followed a redefinition of the problem. When they were new, automobiles almost automatically strained the limits of street customs. The minor nuisances they caused were treated as violations of fairness. (Why should a motorist use his horn to drive a pedestrian out of his way?) More serious problems were injustices (perhaps legal, but certainly moral). Thus the prevailing problem definition was "What is just?" Justice, in turn, stemmed in part from custom, to which many appealed. Many safety reformers promoted their answers to the question through a rhetoric of innocence versus guilt, appeals to pity, and expressions of outrage. Police more often used a rhetoric of order.

Traffic engineers, influenced by experience in municipal engineering and by the needs of their clients (downtown business associations) for accessibility, defined the problem differently. They asked "What is efficient?" Some safety reformers joined engineers in this problem definition, decrying accidents as wasteful. A rhetoric of efficiency was ready to hand in the 1920s. Applied to traffic problems, the loss of street capacity to curb-parked cars became "the parking evil."

By the mid 1920s motordom had found that it could no longer work within existing problem definitions. It found an alternative stance in the problem "What is free?" By casting the problem in terms of political freedom and market freedom, motordom found that it could sidestep difficult questions of justice, order, and efficiency. Through this problem definition, it could characterize low speed limits as oppressive—an impediment to freedom. Overzealous do-gooders were "hog-tying the automobile," as an Ohio auto club put it.[12] Engineers who discriminated between modes of transportation on the basis of their spatial efficiency were violating free-market principles. Why should experts favor one mode over another? Let the market decide! As an ally to this rhetoric of freedom, motordom turned to a rhetoric of modernity. It was used to thwart appeals to custom, which could become "outmoded." Macabre safety publicity could look old-fashioned next to "modern" advertising, with its relentless good cheer. A new era demanded new ways. Motordom declared that a new era was dawning and named it "the motor age."

In the streets, rhetorical closure and closure by redefinition of the problem were accompanied by a third, closely related mechanism. We can call it "closure by control of use and misuse." The constant struggles to define use and misuse are seldom noticed as such. When a park bench acquires a central arm rail, those who define sleeping on a bench as a misuse have seized the high ground. Similar struggles to define the use and misuse of streets were at their hottest in the 1910s and the 1920s. When automobiles were new, many city people regarded them as a misuse of streets. By obstructing and endangering other street users of unquestioned legitimacy, cars violated prevailing notions of what a street is for. As long as defenders of automobiles fought their cause without questioning these notions, they were fighting on their adversaries' terms. By the mid 1920s, however, motordom knew its enemy. From then on it expressly challenged old ideas about what streets are for. It proposed that street uses that impeded automobiles were misuses of the street. Even as accidents and congestion continued, restriction of cars was no longer the only way to fight them. After all, cars *belonged* in streets. At first this claim was a difficult one to make, but by 1930 motordom was on the road to success.

## Whose Street?

Motorists arrived in American city streets as intruders, and had to fight to win a rightful place there. They and their allies fought their battles in legislatures, courtrooms, newspapers' editorial pages, engineering offices, school classrooms, and the streets themselves. Motorists who ventured into city streets in the first quarter of the twentieth century were expected to conform to the street as it was: a place chiefly for pedestrians, horse-drawn vehicles, and streetcars. But in the 1920s, motorists threw off such constraints and fought for a new kind of city street—a place chiefly for motor vehicles. With their success came a new kind of city—a city that conforms to the needs of motorists. Though most city families still did not own a car, manufacturers were confident they could make room for motor traffic in cities. The car had already cleaned up its once bloody reputation in cities, less by killing fewer people than by enlisting others to share the responsibility for the carnage. Engineers said they could rebuild cities to accommodate cars, and they were already breaking ground. In the following four decades, urban transportation problems were treated as tasks for highway engineers, and until the 1960s, among all urban transportation needs, state and federal policy recognized urban highway projects almost alone as a public responsibility.

The result was the automotive city—a city that made room for private automobiles. It was a city lacking good transportation choices. Those who argued for accommodating motorists often claimed that urban highways would let city people themselves choose the mode they preferred, since many would prefer automobiles. Yet transportation is a system of interdependent parts, and efforts to accommodate motorists degraded other modes. Since the middle of the twentieth century, people traveling in American cities have had few options. "The basic characteristic of the automobile-dominated city," observes a transportation economist, "is that, when one looks for an alternative to the private car, there is little or nothing there."[13]

In the 1960s and the 1970s, engineers and government began to encourage alternatives. Public funding began to benefit other urban transportation modes. In belated recognition of the interdependence of transportation modes, highway engineers renamed their profession *transportation engineering*. Since the 1970s, transportation engineers have striven to devise ways to lure motorists out of their cars and into other modes. They rejected the highway engineering orthodoxy of the middle four decades of the twentieth century.

Transportation engineers' recent aversion to automobiles in cities is not new. In the 1920s, traffic engineers also sought to limit the urban sphere of the car. Together, downtown business leaders and a popular safety movement strengthened the engineers' hand. The future of the automobile in city streets was the prize in a protracted and sometimes bitter contest. It was a clash not merely of methods but of first principles, as the conflicting views expressed in the epigraphs on page 1 attest.

These differences made the participants see the same problems in entirely different ways. All agreed that traffic jams were bad and that traffic accidents were intolerable. But was a traffic jam a symptom of wasted street space? Or was excessive urban concentration to blame? Or inadequate streets? If a motorist struck a child in the street, was the child responsible? Or was the newcomer to the street—the motorist—more to blame? The answers depended on who was asked, and the prevailing answers changed with time.

## The Origins of the Automotive City in America

How did the automotive city begin? Did it evolve, or was it made? The rise of the automotive city was most obviously a transition of prevailing transportation modes, and much of the historical scholarship begins with this observation. Thus simplified, the problem of the origins of the automotive city is a question of vehicles: How did the automobile displace the streetcar as the principle mode of urban passenger transportation?

## Mass Demand or Elite Imposition?

To some observers, the transition to automotive transportation was a kind of Darwinian evolution by technological selection: the fitter automobile drove the outmoded streetcar to extinction. Instead of Nature, city people selected (more or less rationally) the winning species in the struggle for survival. Others see a deliberate promotion of the automobile at the expense of the streetcar by corporate or professional elites, especially auto manufacturers and city planners.

**Evolution by Technological Selection: The Consumer-Demand School**   In the 1920s American automobile manufacturers made and sold millions of cars each year. Americans bought cars because cars served their transportation needs well. The automobile also suited Americans' taste for independence and individualism. It freed city people from subservience to timetables and from crowded and uncomfortable streetcars, let them live in green suburbs remote from the railway lines, and gave them Sundays in the country. According to this evolutionary interpretation, city people therefore bought cars, and gradually cities adjusted to this mass preference. As city people drove more and rode streetcars less, street railways lost money and ultimately failed. Because of their "love affair" with the car, Americans rebuilt their cities to accommodate it.[14]

In his 1987 study of Los Angeles, Scott Bottles concluded that "the public decided several *decades* ago that it would facilitate automobile usage as an alternative to mass transportation."[15] Bottles gave motorists—and the automobile itself—an autonomous, leading, and almost heroic role. He showed clearly that perceived and real corruption, combined with poor service, earned Los Angeles street railways the traveling public's resentment. For the people of Los Angeles, therefore, the automobile was an attractive alternative. Yet Bottles also contended that people simply preferred to travel by automobile. According to this interpretation, urban transportation has evolved in response to consumer preferences, much like other consumer goods in a free-market economy. As Bottles argued, Americans preferred to drive, and, therefore, "in a society that celebrates individual choice and free-market economics," city transportation must be primarily automotive.[16]

**The Automotive Elite versus City People**   Others have seen the problem differently. In crowded cities, mass demand for automobiles could not automatically transform transportation; cities would have to be rebuilt to accommodate cars. The promoters of such reconstruction were elites, they say, not a mass of transportation consumers. The elite promoters of the motor city pulled up streetcar rails and planned the deconcentration of urban populations. When there were no more streetcars to ride and when cities were replanned around motor transportation, city people rode buses or bought cars. Mass preferences were relatively unimportant. Some in this school have seen behind the automotive city not the free market but its deliberate subversion. Until the mid 1930s (the claim goes), street railways served city people well. To find new customers, however, automotive interest groups, led by General Motors, conspired to foil the free market by acquiring street railways, scrapping them, and substituting buses and, ultimately, urban highways. Automotive interests acted in concert, secretively and sometimes illegally.[17]

Others have seen more benign elites at work. Mark Foster argued that professional city planners promoted the automobile as a means of deconcentrating overcrowded city centers. To many city planners, deconcentration was the answer to various urban ills, and the automobile was the best instrument for accomplishing this end. Foster found that "the majority of planners enthusiastically endorsed both automobility and the suburban movement out of conviction."[18] He saw planners' work as fairly successful in relieving cities of the evils of overcrowding. Foster believed that elites were important, but he found nothing sinister in their work.[19]

Clay McShane documented an early affinity of interest between automotive groups and city planners. Auto interests wanted to rebuild cities for cars; planners saw cars as the basis of the new and better city they would design.[20] McShane dated the reconstruction of the American city for the sake of the automobile extremely early. Like Foster, he devoted much of his attention to the great masters of city planning. Daniel Burnham's famous 1909 Plan of Chicago was a blueprint for "providing roads for automobility," for example. Burnham and the other leading city planners were "fantasists" who "in effect declared traditional cities obsolete by calling for rebuilding downtown around the car." Though the fantasists were irrational and visionary "car-loving males," their designs were the foundation for the American automotive city.[21]

## The Limits of the Existing Explanations of the Automotive City

So far, researchers have concentrated their attention on *vehicles*—or, less often, on urban design. Contemporaries, however, usually regarded urban transportation as a problem of *existing streets*. They debated the proper function of the street, who belonged in it and who did not, and how to make the best use of it.

## What About Safety?

Traffic safety has not received the attention it deserves. In the 1920s the automobile's bloody reputation darkened its future in the city. Customarily, pedestrians were entitled to the whole street, and motorists and their cars were held responsible for injuring pedestrians almost as a matter of course. This condition seriously impeded motorists' use of the street. Before the automotive city could begin, pedestrians had to be regulated and they had to share responsibility for their own safety. This transformation was largely accomplished by 1930, and it is a foundation of the automotive city.[22]

## The Limits of the Consumer-Demand School

Mass demand for automobiles cannot alone explain the automotive city. Even in the United States there is little evidence in cities in the 1920s of a "love affair" with the automobile. In 1920 the very small minority of motorists was remarkably effective in obstructing streets. The result was an expert consensus, backed by most downtown business leaders, that the automobile should be restricted for the sake of efficient urban traffic flow.[23] Few suggested any extensive program for accommodating automobiles with special facilities. In such a climate there was little reason to expect an automotive future for the American city. Motordom, far from leaving the future of city transportation to the natural consequences of mass demand for automobiles, fought a strenuous campaign to defend the motoring minority's legitimacy and to redefine traffic problems.

Non-expert, popular opinion in cities was also generally unsympathetic to automobiles. With the sudden arrival of the automobile came a new kind of mass death. Most of the dead were city people. Most of the car's urban victims were pedestrians, and most of the pedestrian victims were children and youths. Early observers rarely blamed the pedestrians who strolled into the roadway wherever they chose, or the parents who let their children play in the street. Instead, most city people blamed the automobile. City newspaper headlines, editorials, letters, and cartoons depicted the automobile as a destructive juggernaut. Funereal parades and public ceremonies of grief in dozens of cities drew attention to the grim toll and spurred demands for mechanical limitation of cars' speed. These events did not promise a bright future for the car in the city. Such a future, at least, would not come unassisted. It would require a deliberate redefinition of the safety problem and a redistribution of responsibility for it.

Motordom did not trust the growing popularity of the car alone to lead to an automotive city. It saw in popular attitudes more reason for anxiety than for optimism. Early perceptions of the passenger automobile are reflected in a popular name for it: "pleasure car." When auto interests perceived the modifier 'pleasure' as a limitation, they worked to remove it.[24] Many city people also perceived the car as inherently dangerous and as a rural mode of transportation ill suited to city streets.

The consumer-demand school supposes that the automotive city was the result of a kind of economic majority rule. "America's present urban transportation system largely reflects choices made by the public itself," the claim goes.[25] Once most people traveled by car in cities, cities followed the lead of the majority and converted to automotive transportation.

The truth is the other way round. In the 1920s, motordom defended motorists as a persecuted minority suffering under a majority tyranny. Automakers feared that most city people would never buy a car. The prevailing school of city traffic engineering sometimes defined engineers' task as corralling the intrusions of the minority of motorists for the sake of the non-motoring majority. Motordom therefore appealed to the rights of the motorist minority, which, they claimed, were violated by regulations that favored more spatially efficient modes and by the absence of expensive facilities for motor vehicles. By the time motorists were indeed the majority (which, in city streets, was not until about the late 1940s, depending on the city), they had long since won the earlier struggle for principles.

Though in cities motorists and their passengers were a small minority in the 1920s, many of the passengers standing in crowded streetcars surely thought of themselves as future motorists. Bottles, indeed, showed that many streetcar patrons bought an automobile as soon as they could afford one. Yet the fight for an automotive future in cities began amid a popular outcry against the dangers of cars to city people. The fight was motivated by the *lack* of demand for cars in cities. A sudden drop in sales led manufacturers and other auto interests to reformulate the problem of urban transportation. In 1923 and 1924 city people were not buying cars, and the auto industry suspected that existing urban transportation principles were poison to motor transportation in cities. In response, auto interests were among the first to conceive of and promote an automotive city.

Finally, the consumer-demand explanation equates demands for street uses with demands in a free market. It supposes that in streets, as in other markets, the sum of individual preferences is the preference of the whole. Most market commodities work this way, but streets do not. All could enter; none paid for each use. In these circumstances, the individual interest of each user was to get as much of his or her share of the free street capacity as was possible. Yet the interest of the whole was to allocate street capacity equitably and efficiently, and to restrict uses which impeded others' use of the street. The best use of the street therefore required regulation. When, for example, curb space was free, why would a motorist not take it for a whole day—or a week? After 1935 the parking meter changed parking behavior radically by making parking space a commodity in trade. But there was no practical way to charge other street users for each use of the street. For example, a motorist paid less for 100 square feet of street space than a streetcar patron paid for 10. By 1930, in the city street, motorists had all of the benefits of the free market (virtually unrestricted access to street capacity), but they had few of its costs (they did not pay a market price for the street capacity they used). Individuals took advantage of this as soon as they could, but the cost to the whole was paid in the resulting inefficiencies. The real question, then, is not what city people preferred, for in this environment the car was a rational choice. We must ask instead how this peculiar environment arose.[26]

The automotive city was not simply the product of mass demand for automobiles. In 1920 there was a free market for automobiles but not for the use of city streets. The street was understood not as a marketplace for transportation demands but as a public service, subject to official regulation (however imperfect) in the name of the public interest. In the 1920s, motordom proposed that street uses be treated like demands in a free market, then fought for this new model. The contest was fierce. It was a struggle for the future of the American city.

### The Limits of the Elite-Imposition School
Consumer-demand interpretations do not give due weight to the divisions of interest at work and cannot explain why the automobile's promoters fought hard to win legitimacy for the car in the city. Historians who have ascribed the automotive city to deliberate efforts of elites have recognized these limitations, yet the existing models of elite control are inadequate.

In some explanations, the elites were a united, coherent, and durable foe of the masses (especially transit riders). Yet in the 1920s the most powerful local elites—police, chambers of commerce, and traffic engineers—often led efforts to *restrict* motor traffic. Local shopkeepers, whose claim to elite status was far weaker, usually opposed restriction of automobiles. Elite roles changed with time. In 1925 local auto clubs denounced the same traffic regulations that they had demanded in 1920. National transportation interest groups were also divided and changeable. Until the early 1920s, street railways were better represented by national interest groups than automotive industries were. There was no simple or lasting divide between the people and the interests.

Auto manufacturers' subversion of street railways did not create the automotive city. By the time the tracks were pulled up, the street railways had long been failing. The reasons are complex, but among them was a new tendency to treat the city street as a free market for transportation. Street railways suffered in this redefinition. In part this was because the redefinition was imperfectly applied. As intrusive restrictions on automobiles were eased, regulations still squeezed street railways from all sides. Their fares were limited by law, for example, and they remained closely regulated and heavily taxed. And free-market principles simply did not suit street railways well. By their very nature, street railways could not compete with each other (because they were natural monopolies), nor could they compete with automobiles (because in crowded city streets even a small minority of motorists obstructed them). Freedom for automobiles was in itself a restriction of streetcars.

Others assign the lead elite role to city planners. Foster found them enthusiastic promoters of cars in cities, and McShane followed Foster's lead. Both concentrated their attention on the most prestigious city planners, the more visionary of whom were McShane's "fantasists." In McShane's account, the fantasists began to convert cities to the motor age soon after 1900. McShane demonstrates that "fantasists manifestly affected city planning in the early twentieth century," but he is unable to show that city planners (fantasists or otherwise) shaped the cities themselves. Practical city planners did influence urban form, notably through planned civic centers and zoning ordinances, but fantasists achieved little. Referring to one such fantasist in 1926, a more practical-minded city planner wrote: "How idle the splendid plans recently prepared by Mr. Harvey Corbett and his committee of architects for New York!"[27] Visionary planners built a few broad boulevards, but these were influenced by the example of Baron Haussmann's work in pre-automotive Paris, and they were ill suited to motor traffic.[28] Motor age plans that were actually implemented were not on drawing boards until the mid 1920s and bore little resemblance to earlier visionary design. When promoters of the motor age city needed designers to plan it, they ignored the fantasists in favor of a new generation of highway engineers.

## The Struggle for the Street

In a study of the rise of the automobile in Chicago, Paul Barrett suggested another way to look at the problem. Barrett found that the prevailing definitions of transportation modes influenced the fate of transportation in American cities. Street railways were defined early as profit-seeking private enterprises, which should pay for themselves and return dividends to their investors. This idea took hold when street railways were profitable and investors were eager to lay tracks and get lines running. When profits disappeared, street railways soon followed.[29]

Roads and streets, however, were a public responsibility. Since street users could not be charged for each use, there was no practical way to make the street a private enterprise. Yet Barrett argued that urban transportation is properly understood as a "single, integrated phenomenon." By putting automobiles and streetcars in "artificially separate categories" public policy "did cripple mass transit and confuse public thinking on local transportation," because "facilities for the automobile were publicly subsidized while mass transit was regulated and taxed."[30]

Barrett found that the problem in Chicago before 1930 was not one of the superiority of one mode over another, or of popular preferences, but one of a disappearance of good choices. "A different local transit policy in Chicago would not have prevented the rise of the automobile," he concluded, but "it might . . . have provided alternatives for the urban commuter." Without good choices, automobiles soon became "the rational, practical alternative for those who could afford it."[31]

In a study of Hartford, Peter Baldwin found changing definitions of city streets. Early in the twentieth century, diverse reformers segregated street functions and "domesticated" streets.[32] Some of these changes belong in any attempt to explain the origins of the motor age city.

### Contested Definitions

Barrett and Baldwin have shown that cultural classifications and matters of economics and problem definition shaped city streets and their uses. Diverse city people and interest groups fought over these problems, and the future of the city was at stake.[33] The findings of Barrett and Baldwin apply far beyond the geographical limits of their case studies. Society shapes technological systems. Different people see evolving technologies differently. They fight for their vision of the technology's future. And if one vision prevails, those who fought for it portray the technology's history as an evolution by technological selection—a survival of the fittest technology. Such Whiggish distortion can be corrected by returning to the sources and recovering the indeterminacy of technologies when they were new.[34] Histories of streets and traffic in American cities, whether celebratory or critical, begin from the perspective of the triumphant automobile. But to travel to 1920 would be to travel to a time when such a future was unimaginable. What is a street for? The arrival of the automobile in quantity pried the question wide open. Anything seemed more probable than the motor age that was to come. Motorists in 1920 were more likely to defend their claim to the street as an oppressed minority than as champions of the technology of the future.

The struggle for the future of urban transportation was less a contest between vehicles than a competition for their urban medium: the street. There were far more participants in this contest than motorists, auto manufacturers, street railways, and city planners. The sudden arrival of large numbers of cars forced cities to face new questions about the street. The answers were of urgent interest to many, including professionals and business people, local and national interest groups, motorists, and pedestrians. Must cities find or make room in their streets for the private cars of all the motorists who choose to drive there? Can people walk in the streets wherever they choose? Will congestion deprive downtown merchants of their custom? Can a motorist claim curb space freely? Should children play in the streets? Must a few people in private vehicles impede the many in public vehicles? Must regulations deprive the motorist of the advantages of the car? Can traffic experts rightly decide who belongs in the street and who does not? These questions pitted street users against one another.

Centuries-old cultural and legal legacies led to answers unfavorable to automobiles in cities. In 1920 the city street was considered a public amenity for uses considered public, such as street railway service and walking. As a public good, the street was to be regulated by experts in the name of the public interest. Automobiles were individual, private property. Motorists were tolerated when they did not impede or endanger other users, but wherever congestion or accidents took their toll the automobile bore most of the legal responsibility and most of the popular blame. In the city street of 1920 the automobile was a nuisance, even an intruder. Automobiles were extravagant in their use of scarce space, they were dangerous (especially to non-motorists), they had to be parked, and they served only a small minority of city people. Cities, using police power delegated to them by the states, strictly regulated motorists on the grounds that automobiles were newcomers that moved few people at a heavy cost to street capacity.

The automotive city arose in part from an attack on the old customs of street use and an effort to let individual liberty and free markets rule there too. From American ideals of political and economic freedom, motordom fashioned the rhetorical lever it needed. In these terms, motorists, though a minority, had rights that protected their choice of mode from intrusive restrictions. Their driving also constituted a demand for street space, which, like other demands in a free market, was not a matter for expert scrutiny.

The struggle was difficult and sometimes fierce. In motordom's way were street railways, city people afraid for the safety of their children in the streets, and most of the established traffic engineering principles of the 1920s. Motordom, however, had effective rhetorical weapons, growing national organization, a favorable political climate, substantial wealth, and the sympathy of a growing minority of city motorists. By 1930, with these assets, motordom had redefined the city street.

In the new model, some users of once unquestioned legitimacy (notably pedestrians) were restricted. Traffic engineers no longer burdened motorists with the responsibility for congestion; their goal now was to ease the flow of motor vehicles, either by restricting other users or by rebuilding city thoroughfares for cars. New urban roads were treated as consumer commodities bought and paid for by their users and to be supplied as demanded. On this basis, over the following four decades, the city was transformed to accommodate automobiles.

## Overview

The book is divided into three parts, named for the perspectives or technological frames of leading social groups. Perspectives on safety and legitimate access to the streets are featured in part I. Part II turns to congestion. Part III is about safety and congestion in the new forms they assumed for the motor age. A strictly chronological approach would obscure the argument, which depends more on the temporally overlapping perspectives of relevant social groups than on the sequence of events. The frames are not to be understood as consecutive, except in the very loose sense in which (for example) the bronze and iron ages were consecutive.

Part I, titled Justice, examines street accidents and safety from the perspective of social groups that shared a perception that the arrival of automobiles threatened established customs of street use. It begins with a morally charged attack on the automobile's invasion (as many perceived it) of the city streets and the resulting bloodletting. Featured are pedestrians, especially parents and children, who often cast motorists as the perpetrators of injustice in a moral drama of good children and evil automobiles. Police and their attempt to make automobiles conform to older definitions of street order are then considered. Finally, the clash for street access between pedestrians and automobiles is examined as a struggle for legitimacy, culminating in a new effort by automotive interest groups to question pedestrians' customary rights to the streets.

Part II, titled Efficiency, turns from accidents to traffic congestion. In a new discipline called "traffic control," traffic engineers and the business associations that hired them constructed congestion as inefficiency and proposed to achieve traffic efficiency through regulation. This part ends with the emergence of an alternative construction of congestion by motordom.

Part III, titled Freedom, considers congestion and accidents from the point of view of automotive interest groups, which developed a new model of city traffic in response to dangers they perceived from those holding the perspectives of justice and efficiency. To legitimize the perspective of these interest groups beyond their own circle, motordom developed a case that appealed to American traditions of economic and political freedom.

Finally, a word about vocabulary. The word "automobile" is most often used here as it was used early in the twentieth century, as a generic term for motor vehicles, including passenger cars, trucks, and buses. While there are important distinctions between these vehicles, what they have in common is usually more important here. "The automobile industry" (or "the industry") is used broadly to encompass diverse groups financially interested in automotive transportation, including manufacturers, dealers, and auto clubs. These interests were sometimes collectively called motordom, especially to suggest their greater cohesion after 1923. Similarly, the word "motordom" is used here when a cohesive collection of diverse automotive interests is meant.

# I  Justice

It is impossible for all classes of modern traffic to occupy the same right of way at the same time in safety.

—"Jay Walker Problem," *Providence Sunday Journal*, June 26, 1921

Old street uses plus new automobiles equaled disaster. This fact transformed the city street between 1910 and 1930, but in ways few participants would have predicted. The three chapters that follow sketch reactions to this disaster from five social groups: parents, pedestrians, educators, various motoring interests (clubs, dealers and manufacturers), and police.

Distinctions between pedestrians, parents, and educators are difficult to draw, so their voices are often mixed in chapters 1 and 2. These chapters are instead distinguished by a sequence of action and reaction. In chapter 1, pedestrians and parents attack motorists as destructive intruders in the street. In chapter 3, automotive interests develop a counterattack, consisting of an attempt to redirect responsibility back toward pedestrians, including children. They did this through the invention of jaywalking and a reinvention of child safety education. Caught between these groups were city police, the subject of chapter 2. Police tended to favor long-standing street uses over innovations but also attempted to mediate between other groups.

All groups agreed that some of the effects of new cars in old streets were disastrous. This early agreement fostered hopes that common-sense solutions could be found. If all groups did their bit, perhaps street problems would be solved. This hope found expression in the slogan Safety First, borrowed from the young industrial safety movement. The National Safety Council, enjoying a high reputation for its recent successes in industrial safety, soon took advantage of new opportunities in the realm of traffic safety. Admitting membership from all social groups, local affiliates of the

NSC hoped to do for traffic safety what they had done for industrial safety.

In organizing local safety campaigns, however, local affiliates of the NSC soon found that there was no true common-sense solution, because no sense of the traffic problem was truly common to all social groups. Parents and pedestrians tended to blame motorists, especially for their speed. They usually regarded pedestrian casualties as innocent victims of irresponsible motorists. Parents also often saw safety campaigns as opportunities to recognize child traffic casualties as public losses, for example by honoring grieving parents officially. Educators tended to be more willing to give children responsibility for their own safety through traffic safety education. Automotive interests, faced with definitions of the safety problem that threatened their future in the city, organized as "motordom" to promote a new construction of the city street as a place chiefly for motorists, largely by discouraging free use of the streets by adult pedestrians and by children.

Between these groups lay the police, who regarded themselves as mediators. Police acted like peacekeepers, seeking to minimize conflicts between rival street users. Applying a perspective they had developed to manage other social problems, police sought order in the streets, even at the cost of efficiency. But as long as the various groups remained so much at odds, there was little the police could do.

# 1 Blood, Grief, and Anger

Under present conditions there is a deadly competition between pedestrian and motorist for the use of those strips of territory we call streets—a conflict deadly to the wayfarer, with the victory to the motorist.

—W. Bruce Cobb, magistrate, New York City Traffic Court, 1924[1]

On a Saturday afternoon in the spring of 1920, Leon Wartell was on a sidewalk, playing a ball game of some kind. The 9-year-old and his friends were in a residential Philadelphia neighborhood of row houses at Fifty-Third and Spruce Streets. A swerving car jumped the curb and rolled on top of Leon. In the back of an ambulance, on the way to Misericordia Hospital, Leon died.[2]

In time, the grieving parents, Mr. and Mrs. Barnett Wartell, found some consolation in their dead son's older brother, Howard. Howard grew into "an exceptionally gifted lad": an athlete, a fine student, and a promising violinist. Two years after his brother's death, Howard graduated early from Philadelphia's Central High School. He aspired to music school; for the short term, he took a job installing telephones. After work, late on an autumn afternoon in 1924, the 18-year-old rode his bicycle home to his parents along a quiet side street. A light truck hit Howard, throwing him and his smashed bicycle into a gutter. The panicked young driver fled the scene, leaving Howard to bleed to death.[3]

In the 1920s, motor vehicle accidents in the United States killed more than 200,000 people.[4] The deaths of the Wartell boys were instances of the most typical variety of motor calamity The dead were city people, they were not in motor vehicles, and they were young. Although automobile ownership rates were lower within cities, the risk of automobile-caused death was much greater there than in the nation at large.[5] In cities with populations exceeding 25,000, pedestrians accounted for more than two-thirds of the dead in 1925.[6] In large cities the proportions were still higher.

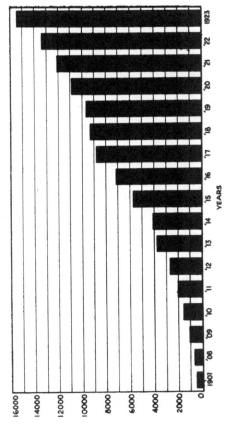

GROWTH IN NUMBER OF AUTOMOBILE
FATALITIES IN THE UNITED STATES

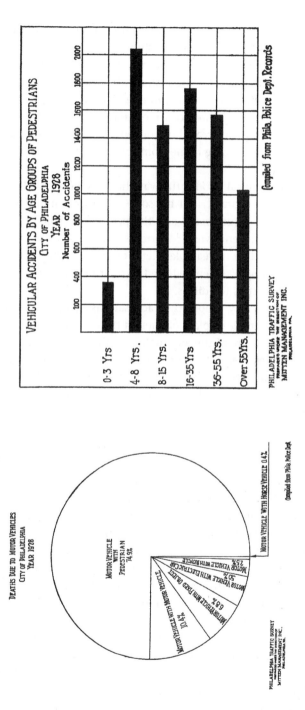

**Figure 1.1**

The sharp rise in total auto fatalities and the predominance of pedestrians and children among street accident victims contributed to widespread hostility toward automobiles and motorists among city people. Sources: vertical bar chart from "Public Accidents: America Leading the World in Accidental Deaths," *Journal of American Insurance* 1 (November 1924), p. 23, data from National Safety Council, courtesy American Risk and Insurance Association; pie chart and horizontal bar chart from Mitten Management, Inc., Accidents and the Street Traffic Situation (Philadelphia Traffic Survey Report no. 4), July 1929, 10, 18–19, Philadelphia Rapid Transit Company box 1, John F. Tucker Collection, accession 2046, Hagley Museum and Library, Wilmington, Delaware, courtesy Hagley Museum and Library.

In the Wartells' home city of Philadelphia in 1928, three-fourths of those killed by automobiles were pedestrians. In that single year 8,246 pedestrians were struck by motor vehicles in Philadelphia, and about half were under age 16.[7] In 1925 in the United States, cars and trucks killed about 7,000 children—about one-third of the total motor death toll.[8] Two months after Howard Wartell's death, *The Outlook* observed that "to the mother of young children" the "new problem created by the automobile" amounted to "the question, 'Will my child come home from school to-day alive and whole?' "[9]

The Wartells' losses were typical also for the way the boys' father understood them. Barnett Wartell characterized his sons' deaths as "murder." The motorists involved were "murderers," and the pedestrians killed (his sons and all others) were "innocent lives."[10] Wartell's is the view we would expect of a parent embittered by unbearable grief. Yet in the 1920s it was a typical position for city people, a view reflected in their newspapers and directly expressed in their letters to editors.

The fear and anger of many city people, their accusations against motorists and their cars, and their public displays of grief for the innocent victims constituted a threat to the future of the automobile in the American city. As long as the motor vehicle death toll consisted so largely of innocent children, and as long as motorists and their automobiles were saddled with sole responsibility for pedestrian casualties, the automobile's future in the American city was clouded.

## The Changing Face of Safety

Even more than most social phenomena, the problem of safety looked different to various observers at various times. All agreed that a new menace lurked in America's cities. As cholera and typhoid fever succumbed to better sanitation and as bacteriologists were winning their fight with diphtheria, motor vehicle accidents threatened to erase the gains. Most agreed also that, as in other public health efforts, fighting the motor death toll would require not a cure for the stricken but prevention, including controversial regulations. But what was to be regulated?

The answer lay in a combination of how the accident problem was defined and where responsibility for it lay, and here there was plenty of room for disagreement. Deaths like those of the Wartell boys could be laid to various deficiencies. Fault could be found in the presence of powerful motor vehicles in crowded cities, or in the presence of pedestrians and cyclists in the roadway. The motorists were blameworthy for their speed and carelessness. Yet pedestrians and bicyclists could be careless too; the driver of the truck that killed Howard Wartell blamed the accident on the victim's reckless cycling.[11] Suspicion could lead even to the design of the street: Leon Wartell met his death on a sidewalk that kept pedestrians close to motorists. While nearly all accidents could be explained by any of several faults, certain explanations were favored at the expense of others. The prevailing explanations changed—and continue to change—with time.[12] Until the mid 1920s, the dominant perspective was shaped by those most alarmed by the new danger—people who would have agreed with Barnett Wartell.

## Innocent Victims, Angry Pedestrians

After World War I, the scale of death and dismemberment on roads and streets in America grew fast. In the first four years after Armistice Day more Americans were killed in automobile accidents than had died in battle in France. This fact was widely publicized, and the news was greeted with shock.[13] The carnage inspired a popular safety movement unlike any before, a movement that touched everyone, old and young, in cities large and small throughout America. In its original, popular form, the movement did not last long; by the mid 1920s it was already ebbing. Yet it pressured government and business leaders to establish the traffic safety institutions that, though much evolved, are still with us today. The principles of this second, institutional traffic safety movement were devised largely as an answer to, and a deliberate rejection of, the popular notions of traffic safety that prevailed in the earlier safety movement.[14]

In the prevailing construction of the traffic safety problem before the mid 1920s, cities were not at fault for failing to provide safe accommodation for motorists. To frightened parents and pedestrians the problem was far simpler: they blamed automobiles and their drivers, regardless of the circumstances. City people were angry. Their anger is shown in mob attacks on reckless motorists, and in newspapers that played up automobile accident stories when the victim was easy to represent as innocent (a child, a young woman, an old person), the victim of an unambiguous "villain" (the motorist who leaves the victim bleeding in the street, the "speed maniac," the fleeing criminal, the drunk).

Prevailing perceptions of the problem are also clear in popular fiction. *Manslaughter*, a sensational novel by Alice Duer Miller, portrayed the death of a dutiful police officer at the wheels of a spoiled, speed-crazed socialite; the book juxtaposed the virtuous victim's nobility and the dissolute

"Oh! That's Different!"

**Figure 1.2**
Cartoon by "Red," *National Safety News* 6 (August 1922), p. 26.

motorist's depravity. The novel sold well and was made into a movie in 1922.[15] In 1928 one of the most important films of the silent era depicted child death in the form most distinctive of the 1920s. *The Crowd*, directed by King Vidor, included as its most poignant scene the death of a small New York City child struck down in the street under the wheels of a truck.[16] Like the daily press stories too familiar to city newspaper readers, *The Crowd* depicted the wasting of a pure innocent by a motor vehicle running its mundane errand. The dead in these stories were virtuous (the police officer in *Manslaughter* was a veteran "with a heroic record"[17]) or innocent (the little girl in *The Crowd* was perhaps two years old), and the motorists in both cases were speeding.[18] To the newspapers the calamity in the streets was a sensational moral drama pitting "death drivers" against innocents; to many of their readers, however, it was also nonfiction of the most serious sort.[19]

Enraged mobs attacked motorists who struck pedestrians. In 1920 the Philadelphia *Public Ledger* found that "any time an automobile collides with a post, a pedestrian, or other obstacle the crowd that gathers always displays prejudice against the driver . . . tho the latter's very middle name may be Caution." The newspaper advised readers who wished to be "in the height of fashion" that if a pedestrian is "hurt or annoyed" in an encounter with a car, "don't ask whether the victim was wholly or in part to blame. Suggest that the driver of the motor-car be lynched."[20] In 1923 car thieves speeding away in their stolen machine ran down and killed a woman waiting for a streetcar. The victim was just 20 years old and a popular church choir singer. A "menacing crowd," estimated by a reporter to number 2,000, surrounded the car's occupants and shouted threats to lynch the men. To rescue them, a policeman was forced to wave his gun at the throng and to call in reinforcements.[21] In separate incidents on a single spring day in 1927, eight children were killed by motor vehicles in and near New York City. One truck driver involved was attacked by a mob; police struggled to extricate him.[22]

Others confined their revenge to pen and ink. Editorial cartoons portrayed the accident problem as a matter of innocent pedestrians (overwhelmingly children) and motorist villains. A motoring Grim Reaper rivaled Uncle Sam for dominance of cartooning's iconography. Many city people wrote letters to their newspapers complaining of the new motorized scourge, and particularly of its invasion of the rights of pedestrians. They overwhelmingly outnumbered letter writers who defended the automobile or who faulted the pedestrian.[23] One pedestrian, nearly struck down at a streetcar stop, published a threat: "I am now ready for this brute, and if he ever makes a similar move where I am concerned it will be his last one, and there will be no court costs either."[24] Another letter writer's threats were more direct. After reading advice in the newspaper to look both ways when crossing streets, the writer sent a letter signed "Sic Semper Tyrranis": "When you get to the crossing, look to your left, pull out your automatic from the holster, step into the street and level the gun at the chauffeur coming. When in the middle of the street level it at the nearest chauffeur coming the other way."[25]

### The Automotive Juggernaut

The particular evils to which city people compared automobiles reveal how they understood the problem of traffic casualties. The more intellectual critics often called the automobile a pagan idol demanding sacrifice. The car was, for example, a juggernaut: a wheeled object of idolatry which crushed lives out under its wheels. The metaphor reflected the perception that the car's threat was to those afoot, not those in vehicles.

The comparison also showed the car not as a useful instrument subject to abuse by the irresponsible but as a needless and inherently dangerous machine. Driver and vehicle were not clearly distinguished; neither were responsible and irresponsible drivers. In 1916 an authority on the automobile industry wrote: "In the view of some of the press, the automobile is to-day a juggernaut, a motoring speed-monster, intent on killing and maiming all who stand in its way."[26] In 1920 a Milwaukee educator described the car as "the juggernaut of modern civilization."[27] An insurance executive writing later that year described the automobile as a "juggernaut . . . put in motion under the guise of economic necessity" that had left its record "in the blood of men, women and little children."[28]

In a similar vein, a cartoonist for the *St. Louis Star* portrayed the automobile as a machine age Moloch to which motorists sacrificed generous offerings of child victims.[29] A Cincinnati newspaper made the same comparison.[30] Papers carried a steady stream of editorial page cartoons showing the Grim Reaper behind the wheel of a speeding automobile harvesting a bumper crop of child victims with his scythe. The *Washington Star* commented in 1924 that newspapers often published "pen and word-pictures which reveal the motor car as a death-dealing monster."[31]

**Figure 1.3**
Cartoon by "James," *St. Louis Star*, November 6, 1923, p. 14.

Many joined Barnett Wartell in classifying pedestrians killed by automobiles as the victims of murder. In 1923 the *St. Louis Post-Dispatch* editorialized that, even in the case of "a child darting into the street" in "the excitement of play," the "plea of unavoidable accident in such cases is the perjury of a murderer."[32] The *St. Louis Star* called motorists involved in pedestrian fatalities "killers."[33] The City Club of New York published and distributed large "municipal murder maps" showing where children had been killed in traffic accidents.[34] A letter to a newspaper declared: "People are being murdered in the streets of our city every day."[35] Another letter writer, a motorist, was driven by the abundant accusations to plead "We are not a bunch of murderers and cutthroats."[36] The murderers, said others, should be punished as such. "Make them do time," one St. Louisan demanded.[37] Send them to the penitentiary, another advised.[38]

## Mothers and Children

In the safety publicity of the early 1920s the overwhelming majority of victims were children, and a large share of the rest were young women. Child death gave the traffic safety movement a feminine coloring. In safety publicity, young women victims often stood for pedestrians' innocence in general. Even more, mothers stood for the horror of child death and parental bereavement. When personified, the enemy was death (portrayed either as sexless or as male) or a reckless driver (nearly always male). Women sometimes represented safety or child protection. In a 1919 safety parade in Cleveland, a float carried a real woman representing the "Goddess of Safety"; she stood "with her arms extended over two children in an attitude of protection against the menacing figure of death."[39]

Women, especially mothers, were prominent among safety reformers in the early and mid 1920s. Thanks largely to the traffic safety problem, women joined local safety councils in quantity in the 1910s and the 1920s. They transformed these affiliates of the National Safety Council from business-dominated industrial safety bodies into something resembling an organized social movement. In 1918, when the National Safety Council began traffic safety work, it formed a Women's Section.[40] In 1922 St. Louis women formed a "women's division" of their local safety council. Its members lobbied city leaders for traffic reforms and went from door to door, canvassing other women for safety council memberships and appealing particularly to mothers.[41] For its symbol, the leader of the women's division took "the White Cross of Prevention." She claimed such work as "woman's chance to fulfill her ideals." "It is a woman's sphere," she said, "to carry the

White Cross far and wide."[42] A month later, women in Washington organized a memorial service to children killed in accidents there.[43]

In Philadelphia in 1927, women formed their own parallel, cooperating safety council. The founding core consisted of about 200 women who had lost children to traffic accidents.[44] As safety reformers, many women worked for more than accident prevention. For most of them, safety campaigns were expressions of public loss. Especially in the early 1920s, safety campaigns featured bereft mothers and crowds paid ceremonial respects to their children. Like fallen soldiers, children were publicly memorialized; their mothers, like the mothers of war dead, were honored as "white star mothers" or "gold star mothers."[45]

## Speed, Danger, and Motor Tyranny

In the early 1920s most city people found the blame for pedestrian casualties very easy to deduce. According to prevailing opinions, when a motor vehicle was involved in the injury or death of a pedestrian, responsibility lay entirely with the motorist. Few found it necessary to state, as one columnist did, that "the delinquencies of drivers are responsible for nearly all accidents."[46] Fewer tried to prove this proposition; letters to the editors of city newspapers, for example, treated it as self-evident. After all, by custom pedestrians held unrestricted rights to the streets, and therefore questions about where a pedestrian crossed the street or how alert he or she was to traffic conditions were immaterial. Motorists, as the newcomers, were under the greater burden to take such precautions. Even a motorist admitted that "the burden of not hitting pedestrians" belongs "entirely on the drivers."[47]

Implicit in all these attacks, both popular and official, was the assumption that the automobile was not a necessity. As long as automobiles were "pleasure cars" there were few grounds for tolerating the injury and death they caused. Walking, however, was obviously necessary, and pedestrians were virtually incapable of causing accidental injury to others. Motor vehicles were fast and therefore inherently dangerous to pedestrians, and the motor vehicle was a necessary condition of the pedestrians' injury or death.

### Speed and Inherent Danger

Angered by the terrible death toll on New York streets, an official from the city's medical examiner's office told the press "An automobile is a dangerous weapon."[48] Many city people shared his opinion.

Above all, the early critics of the automobile blamed its speed. Not until the mid 1920s was the inherent danger of speed widely questioned. In 1925 *The Outlook*, a popular magazine of news commentary, reported of traffic accidents that "the chief danger, it is now generally conceded, is not speeding, but bad and careless driving by immature or reckless drivers."[49] Today the distinction is not obvious. Speeding *is* "bad" or "careless" driving, indulged in by the "immature or reckless." Yet our definition of speeding has changed. Fifty or 60 miles per hour may be prudent driving, or even so slow as to disrupt the flow of traffic. We do not call this speeding; for us, speeding is unusual. Before the mid 1920s, however, speeding was what an automobile was made to do. When we speak today of a "speeding bullet," we mean a bullet merely doing what it was made to do. In the first quarter of the twentieth century, this sense of "speeding" applied to automobiles as well. The car distinguished itself from streetcars, horse-drawn vehicles, bicycles, and pedestrians by going faster—by "speeding." It was inherently a speeding machine.[50] The car's capacity to speed was its chief advantage over other modes.

Until the mid 1920s, however, nearly all agreed that speed was indeed the "chief danger." "Safety and care in driving," complained an automotive engineer, "are almost entirely translated, in the public mind and in the law, into miles per hour."[51] There was sense to this presumption. When the overwhelming majority of traffic accident casualties were pedestrians, speed alone made the difference between life and death. To a pedestrian, a horse-drawn wagon and a Model T of equal weight were not equally dangerous, because the car was likely to be traveling three, four, or five times faster, even in the hands of a prudent motorist. To many, therefore, the automobile was, like the gun, inherently dangerous, especially in cities.

Letters to the *St. Louis Star* reflect typical sentiments: "Most accidents are due to fast, careless driving."[52] "If accidents are to be reduced, speed must be reduced also; that is, unless speed is more essential than human life."[53] "The only way to prevent so many accidents" is to "force" motorists "by severe punishment to drive slow."[54] One writer proposed eliminating the trait that made cars dangerous: "This intolerable condition . . . can be easily prevented by restricting automobiles . . . to a speed not exceeding that of the horse-drawn carriage, before the day of the death-dealing speeding automobile."[55]

The popular press often agreed. The *Star* repeatedly blamed pedestrian fatalities on "speed maniacs" and "speed hogs" without so much as considering pedestrians' contribution to their own demise.[56] Speed helped critics explain how motorists were so willing to jeopardize their own lives and the lives of others. The *Baltimore Sun* explained that "motor-car psychology" turned even normally prudent people into deadly menaces; others diagnosed "gasoline madness" or "gasoline rabies."[57] *Illustrated World* blamed "speed-mad automobile drivers" for their "assaults" on "old people and children."[58]

### Saving Money by Saving Lives: From Industrial Safety to Traffic Safety

A Minnesota city manager concluded as early as 1919 that the remedy for traffic accidents "lay in education rather than in police regulation."[59] In safety matters, education in 1919 meant little more than publicity, especially in the form of brief public safety campaigns. Such campaigns for traffic safety were frequent from the late 1910s to the mid 1920s.

There was a precedent for such safety publicity, and the chambers of commerce drew from it. Losses in accidents had prodded insurance companies to nurture industrial safety from a frail shoot, stemming from heavy industry, into a mature, organized movement, complete with its own professionals ("safety engineers"). Beginning in 1911, state workmen's compensation laws raised both the success rate of insurance claims and their individual amounts.[60] Accident-prone heavy industries found their insurance costs soaring. When Illinois imposed its workmen's compensation law, in 1911, one manufacturer's accident insurance premium increased eightfold.[61] As insurers faced mounting claims and as industries faced rising premiums, both took a new interest in preventing accidents.

Their response quickly evolved into an industrial safety movement. While the movement's backers introduced safer machines, they kept much of the burden of accident prevention on the worker. Industry hoped to induce workers to take fewer chances by showing them the gravity of the risks they were running. If workers were shown safer methods, accident records would improve without expensive new capital investment. Industry used posters to show workers the bloody consequences of inattention and chance taking. Uniting such publicity was a new slogan for a new movement: Safety First.[62]

The First Cooperative Safety Congress, held in Milwaukee in 1912, soon evolved into the permanent National Safety Council (NSC).[63] Spurred on by workmen's compensation laws, employers of labor and their insurers worked together in the National Safety Council to prevent industrial accidents.

But traffic and other public safety problems were different. Workmen's compensation laws did not apply to public accidents. Liability ruled there, and insurers generally found a better return for their money by fighting claims in court than through safety publicity.[64] Thus there was at first no well-organized public safety movement of the scale of the NSC's industrial safety work. Yet institutions and methods devised for industrial safety were easily adapted to public safety campaigns, and in the late 1910s the transition began. Safety First spread beyond plant walls, becoming an admonition not only to industrial workers but also to motorists, streetcar patrons, and pedestrians.

The National Safety Council was decentralized. Local councils were affiliates of the national body, but autonomous and self-funded. They depended upon local sponsors. In time chambers of commerce joined industries and insurers as the patrons of locals. By 1915, four of the twenty locals were directly sponsored by the local chambers, and over the next five years such arrangements became common as the NSC courted local chambers for sponsorship.[65] St. Louis illustrates the close collaboration between chambers and councils; there the chamber supplied the council with offices and a stenographer, and defrayed incidental expenses.[66]

By 1924, 60 cities had local safety councils.[67] The spectrum of interests backing the St. Louis Safety Council in 1922 was typical. It included "the Chamber of Commerce, the business interests, and the schools."[68] Chicago's arrangement was also common; there the local safety council was both an affiliate of the national body and a department of the chamber of commerce.[69] These sponsors supplied the funds, while the council lent the prestige of its good name and expertise born of experience.[70] The sponsors' money and the councils' expertise together established and nurtured a new safety movement, distinct in function and character from the NSC's original industrial safety work. It was called "public safety."

## Public Safety

In Syracuse, New York, in the fall of 1913, a woman stepped off a streetcar. Rebecca Latimer, the wife a local grocer, was 55 years old. She began to cross the street when a motorist trying to pass the streetcar struck and killed her. This fatal accident and others like it in Syracuse inspired the first of many efforts to adapt industrial safety practices to city public safety campaigns. The Chamber of Commerce, the auto club, the Rotary Club, Boy Scouts, police, schools, churches, and newspapers in Syracuse made

December "Safety First Month."[71] For a dozen years thereafter, and especially from 1918 to 1924, such public safety campaigns were undertaken in cities large and small from coast to coast. Nearly all of them were one week long, and from the beginning the industrial safety movement served the public safety reformers as a model.[72] The NSC urged its locals to sponsor a safety week once a year, and by 1923 more than fifty cities had organized such campaigns.[73]

Safety weeks were, above all, publicity efforts. "Safety is essentially an advertising and selling proposition," explained a representative of the safety council of Niagara Falls, New York, in 1919; "I think publicity is the word every time."[74] At first, cities attacked public accidents of all kinds. In time, however, traffic safety grew to be the major and often the exclusive field of local safety campaigns. In 1914 the chamber of commerce in Rome, New York, sponsored a Safety First campaign. Like Syracuse, Rome borrowed the slogan direct from the industrial safety movement.[75]

### Charles Price and the Public Safety Opportunity

The transition from industrial to public safety is personified in the career of Charles Price. Price was an Iowa shoe salesman who went to work for International Harvester in Chicago. When the company saw workmen's compensation laws coming, it put Price to work to find ways to prevent accidents. By 1912 his success won him a reputation, and on it he built a new career as a safety expert. Price organized Wisconsin factories to comply with new state standards. A few years later, a reporter called him "the country's foremost life-saver."[76]

In 1916 Price brought his talents to the National Safety Council. The NSC recruited Price to act as a field secretary, to organize local councils. Over the next seven years he visited many cities, especially in the Northeast and the upper Midwest. To promote the new local councils, Price orchestrated conspicuous safety weeks. He was extending the industrial safety model, hoping that those with a financial stake in public accident prevention would join the movement. Price began with a 1916 visit to Rochester, New York, where he organized a public safety campaign directed at fire prevention.[77] In 1918 Rochester's local safety council launched a much broader six-month public safety campaign.[78]

Influenced by the 1918 Rochester campaign, and probably also by the earlier, independent public safety campaigns in Syracuse and Rome, Price soon saw greater opportunities for the local safety council movement in traffic accident prevention. Teaming up with St. Louis's Chamber of Commerce, he planned a big "Safety Week" in St. Louis that September to coincide with the National Safety Council's annual convention there. The organizer of Rochester's recent safety campaign traveled to St. Louis to present his city's methods. This time, Price's mission was to promote public safety in all areas, including traffic safety.[79]

Price's efforts won him election as general manager of the National Safety Council in 1919. By then, through his intimate connection with many locals, Price was noticing a dramatic increase in local interest in public safety. He saw opportunity for the NSC, and hoped to build an alliance of business interests (such as auto clubs and chambers of commerce) interested in practical dollars-and-cents accident prevention and city people angered by the automobile's depredations. He commented in 1919 that "very rapid development of public safety" was "bound to come in the next two or three years, because of the large increase of street accidents." Price admitted that safety "is a bigger job than we had thought when we had considered this only as a possible service to these industrial concerns."[80] He continued to travel, organizing local safety councils and promoting safety weeks. Within a year, ten cities, with help from the National Safety Council, had put paid executives in charge of public safety committees, and more were on the way.[81] In 1920 one of Price's colleagues in the NSC agreed that "the automobile problem is rapidly forcing the growth of the council."[82]

Price steered the NSC into the favorable winds. In September 1921, he announced the arrival of a "new and giant hazard," calling motor accidents "a problem more alarming and more far-reaching than any other in the history of the safety movement."[83] Then, at the NSC's conference in Boston, Price told delegates that "rapidly spreading alarm over the increase of deaths from motor vehicles" was fostering a "demand on the part of the people for some organized effort to control this hazard." The implications for the council were profound. "The effort to cope with this problem," Price explained, "will furnish a great stimulus to the organization of new local councils."[84]

Price appealed to angry pedestrians. "Each year," he explained, "it becomes more and more dangerous for a person to walk the streets." Like most of his contemporaries Price put the responsibility on the automobile and its driver, and credited their deadliness to their speed. "The obvious remedy," therefore, was "to improve constantly the traffic regulations." Price urged cities to "make their traffic regulations more and more rigid till they can point to low death rates from automobile accidents."[85] The first goal was to protect pedestrians. Price recommended more pedestrian crossings (not just at corners but also "in the middle of blocks") and "safety islands" for pedestrians in wide streets. He attacked "the tendency of some writers to exonerate automobile drivers and to place the blame of accidents upon pedestrians," which to Price revealed "lack of a full comprehension of the problems involved."[86]

### The Transformation of Local Safety Councils

The National Safety Council and its local affiliates were the leading force in traffic accident prevention from World War I to the mid 1920s. In turn, their growing role in public safety changed the councils.

As industrial safety bodies, the councils had been small and inconspicuous consultants for manufacturers. Workers—some of them under pressure from their foremen—could join the safety council (typically for $1), but the real impetus came from higher up.[87] A safety engineer for a meat packing house summarized the industrial safety problem: "Accidents are costly misfortunes, both in human suffering and cold cash."[88] Business had promoted industrial safety for the practical purpose of preventing compensation claims—that is, for the cash.

Public safety campaigns were different. Cash was on the line. Operators of truck, taxi, and bus companies could not afford accidents. Auto clubs wanted safer streets without undue restriction of motorists. For other businesses, however, the financial stakes in street traffic were lower. There was no equivalent of workmen's compensation laws to free up big money for public safety.

Nevertheless, there was a mass constituency for traffic safety. Those alarmed by traffic casualties in cities, especially parents and schools, joined local councils' public safety campaigns. Their finances were limited, but their numbers were great. Thus, if memberships were cheap enough, mass membership promised to compensate for the limited funds of each member. With $1 individual memberships, the St. Louis Safety Council was already 25,000 strong in 1922. That October it launched a two-week drive to enroll another 25,000.[89] The public safety movement transformed local safety councils in many other cities too, especially those north and east of St. Louis. When the council in Erie, Pennsylvania, began public safety work in 1921, city people quickly transformed the small, little-known organization into a city-wide movement.[90]

Unlike the business-minded promoters of industrial safety, however, rank-and-file members of public safety campaigns drew their inspiration from human suffering. As members of local safety councils, they strengthened the councils' tendency to portray frankly the grim consequences of accidents, both as a practical warning to others and as public, emotional expressions of the toll of human suffering in accidents.

Influenced by such lay participation, local safety councils tended to reflect lay assumptions about the causes of accidents. They often blamed motorists exclusively and linked danger directly to speed. At the NSC's 1920 "Safety Congress," for example, "the automobile as a death dealing instrument was unanimously decided upon as the greatest present day menace to public safety."[91] As the Milwaukee chamber was organizing a local safety council, it bought a full-page newspaper advertisement blaming accidents simply on a "never-ending call for speed."[92]

The citizen members of local safety councils profoundly influenced the character of public safety campaigns. From them, councils solicited posters, plays, and pageants that reflected parents' and pedestrians' grief, fear, and anger. These lay participants in the public safety movement gave safety campaigns a less purely utilitarian role, a role that was entirely new for safety councils. Industrial safety campaigns were cool, practical efforts to cut losses from accident claims. To public safety campaigns, however, city people added a moral dimension. With public participation, safety drives became heated expressions of grief and anger, such as one would expect at the burial of the victims of a massacre. After all, to many city people, dead pedestrians, and especially dead children, were the innocent victims of murder.

### Safety Weeks: Practical Accident Prevention and Public Mourning

Public safety campaigns bore a strong family resemblance to their forebears in industrial safety. Manufacturers fostered competition within or between plants to keep mishaps down, and often these efforts evolved into periods in which workers tried to avoid accidents altogether. Local safety councils grafted such "No Accident Weeks" or "Safety Weeks" onto city streets. It was a deliberate effort to duplicate the success of the industrial safety movement. The chairman of the city of Washington's 1922 safety week committee hoped to apply "the same safety first principles which captains of industry have inaugurated . . . to the every-day life of the general population."[93]

Applied to city streets, the councils' safety methods combined the lingering influence of their industrial origins and the new elements of grief and anger in public safety. To work, public safety campaigns had to be visible; their audience was the whole public, not just workers inside plant walls.

And since traffic accidents evoked intense emotions, city people were not going to be a passive audience. They joined in shaping the public safety movement.

The difference is clear in the iconography of safety campaigns on shop floors and in city streets. Industrial safety publicity made extensive use of simplified representations and personifications of more complex phenomena. For example, posters in factories caricatured the inattentive or careless worker as a foolish bumbler. Workers seeing such a fool would presumably avoid resembling him and thus become more cautious.

### From Otto Nobetter to Auto Demons

One oafish character became a mainstay of industrial safety posters. Otto Nobetter stumbled over tools, smoked near explosives, and generally flaunted his ignorance of elementary safety precautions. He invariably paid dearly for his mistakes.[94] Compensation laws gave industry responsibility for workplace accidents, but Otto Nobetter gave it back to the workers.

The public safety cousins of Otto Nobetter appeared in the safety weeks organized in cities in the 1920s. Like Otto Nobetter, they warned those who saw them. They fell largely into two classes, one the creation of the business-minded elements in the safety councils. This class included various personifications of the "jaywalker," who ridiculed careless pedestrians much as Otto Nobetter ridiculed careless workers. Angry parents and pedestrians and editorial cartoonists developed a much more sinister class of caricatures. For example, in a parade for Milwaukee's safety week in the fall of 1920, a streetcar pulled a flatbed trailer through the city. The trailer displayed a wrecked automobile driven by a likeness of Satan.[95]

In the public safety movement, Satan or the Grim Reaper joined bumblers as personifications of the accident problem. The demonic images clearly originated outside the original constituency of the safety councils, for the councils had used nothing like them in their industrial safety work. They thrived in realms beyond the reach of safety experts, especially in city newspapers. Otto Nobetter had put the burden of responsibility for accidents on the group that suffered from them—in this case, workers. Caricatures of jaywalkers did the same for pedestrians. The demonic images of traffic safety campaigns, however, exonerated the leading victims (pedestrians, including children) by putting the caricature in the driver's seat of an automobile, and often by making the vehicle itself the demon. The victims were portrayed not as pedestrians in general, but specifically as children or other innocents who could not be held responsible for their own demise. The inventive minds behind sinister displays such as

Milwaukee's float were not only urging safer driving; they were accusing motorists—and even automobiles—of moral failure. The intent was only partly to advance a practical remedy. Such publicity was an expression of grief and anger.

In Milwaukee as elsewhere, posters for the safety week were drawn by ordinary citizens without charge, not by commercial artists in agencies contracted by the safety council. The winning posters were grim. Milwaukee's first-prize winner shows a grief-crazed mother holding the corpse of her small son. Her hand covers his bloody face, as if to stop the bleeding. There is blood all over the little boy's summer playsuit and on his bare legs. Just behind them is the wheel of an outsized truck.[96] Second place went to a poster showing a woman sobbing into a handkerchief. Her little daughter clutches her mother's hand. She asks "Was Daddy hurt much?"[97] Another winner shows Death represented as a skeleton in a shroud. He stands atop a mountain of skulls, clutching the corpse of a woman. Vultures circle above. Amid the skulls lies a wrecked automobile.[98] Another winning poster personifies the temptation to take risks as the Devil in disguise.[99]

Participation in safety weeks was open; the safety councils did not try to control it. Thus, for safety week, the *Milwaukee Journal* sponsored a safety poster contest of its own, this one for grade school children. The newspaper awarded first prize to a poster drawn by an 11-year-old boy showing an automobile and a locomotive speeding into the vacant eye sockets of a human skull. Beneath this image, stretcher bearers carry casualties.[100] Many local safety councils favored images of grieving parents clutching their children's corpses. "The pity of it!" exclaims the caption of a Memphis Safety Council poster, as a man (a father or a remorseful motorist) supports the tiny, limp body of a struck-down little girl.[101] Such posters were directed not to pedestrians but to motorists, and their messages were often blunt. In 1922 the Safety Commission of Oak Park, Illinois, posted fifty large signs throughout the city warning motorists "DON'T KILL A CHILD."[102]

## Grief

For several years after the Armistice of 1918, Americans in cities large and small dedicated memorials to those who never returned from France. In these years, Americans also publicly grieved the children killed in street traffic. Early city traffic safety campaigns were in large part memorials to the dead. Funereal ceremonies and monuments were these campaigns' most conspicuous features, setting them profoundly apart from the tidier safety publicity that would prevail in later years.

**Figure 1.4**
Poster by George Starkey, reproduced in "Winning Safety Poster," *Milwaukee Journal*,
September 25, 1920.

The Detroit Safety Council's Safety First campaign of 1919 drew special
attention to street fatalities with the tolling of bells. At City Hall, at a
church, at a fire station, and at every school in the city, twice a day bells
slowly tolled eight times on any day in which a life was lost to a traffic
accident. Teachers or police officers announced the names of the dead and
the manner of their deaths to the school children.[103]

Detroit thus paid its respects, as at a funeral, to all the dead in traffic
accidents. In most cities, however, safety campaigns concentrated their
attention on child deaths, and in so doing these campaigns put the burden
of responsibility squarely on the shoulders of motorists and their cars. The
guiltless children were pedestrians, and their injuries and deaths came

**Figure 1.5**
Poster by Harry De Bauffer, reproduced in "Poster Wins Second Prize," *Milwaukee Journal*, September 28, 1920.

nearly always from motorists, and the implication of the drivers' guilt was almost automatic. Adults on foot benefitted by their association with children in the larger class of pedestrians; motorists therefore could not share the burden of guilt with other adults.

During its safety week in June 1922, Baltimore introduced a new feature: a monument of war memorial proportions. The 25-foot wood and plaster obelisk was carefully built to resemble a permanent stone memorial. It was conspicuously located and dedicated publicly by the mayor himself. The marker memorialized the 130 children killed in all accidents in Baltimore in 1921, but its inscriptions singled out traffic accidents. Several hundred school children, Girl Scouts and Boy Scouts, and a delegation from the Women's Civic League attended. A church choir sang, and a minister led the assemblage in prayer.[104]

**Figure 1.6**
Baltimore's memorial to child accident victims during its 1922 dedication by Mayor
William Broening. Source: "Baltimore Puts Over Successful No-Accident Week,"
*National Safety News* 6 (August 1922), p. 38.

With help from the National Safety Council, Baltimore's safety week was
influential. In New York's safety week in 1922 a procession of 10,000 children was "the most spectacular feature of the week." Among the marchers
was a "Memorial Division" of 1,054 children, each representing one of the
1,054 children killed in accidents (most them in traffic) in the city in 1921.
The cortège was led by Boy Scouts carrying crepe bunting and a papier-
mâché tombstone for the child victims. The young marchers passed a
reviewing stand, where, among other dignitaries, there were about fifty
"white star mothers"—mothers of those killed. In front of a crowd of

thousands, at least 200 "white star mothers" attended the dedication of the stage prop tombstone. Flowers lay at the monument's base as speakers eulogized the dead; one speaker prayed: "May the memory of these children cause us to devise methods to save their companions and others who will bless us in the days to come."[105] Three days later, in the safety week's main event, a procession of open cars carried children maimed and disabled in accidents.[106]

"What Baltimore can do, Pittsburgh can do," pledged Pittsburgh's mayor, William Magee, a week after New York's safety campaign. Magee was speaking at the dedication of Pittsburgh's own memorial to children killed in accidents. Children brought flowers to the dedication ceremony, one for each of the 286 children killed in Pittsburgh in 1921. More than 5,000 people attended the ceremony, including families who had lost children to traffic accidents. The city's safety parade repeated features innovated elsewhere: a local auto club carried injured children in open cars and a red devil represented danger. One float displayed "a little girl, crushed between two colliding automobiles."[107]

**Figure 1.7**
Children dedicated Pittsburgh's monument to child accident victims by holding up a flower for each child killed in the city in 1921. Source: "Deaths Drop 60 Percent During Pittsburgh's Safety Week," *National Safety News* 6 (December 1922), p. 16.

Six weeks later, the people of Washington mourned their dead children. More than a thousand people attended a solemn dedication of a temporary monument to the 97 children killed in accidents in Washington in 1921. The Marine Band played Chopin's funeral march as children stepped into formation around the memorial. A clergyman delivered a mass eulogy, then the gathering sang "Rock of Ages." Church choirs sang hymns. Ninety-seven children dressed in white—one for each of the dead—dropped 97 flowers at the monument's base. A Boy Scout played "Taps." Children then left wreaths at the monument.[108]

Washington's safety week closed six days later with a procession of 2,000 marchers and 80 floats along Pennsylvania Avenue. Employees of federal departments marched in the parade, and their home-made signs and floats reflected popular constructions of the public safety problem, including easy references to death and a predilection to blame speed. Treasury Department marchers carried signs discouraging haste: "Get Up Sooner!—You Won't Have to Hurry!—You Won't Get Hurt!," read one banner. Another read: "What's Your Hurry?—Take Your Time!—Don't Get Hurt!" The Agriculture Department's float included mock tombstones inscribed with snappy references to the common causes of accidental death. The War Department paraded four floats which together told a story: a wrecked car, doctors' hopeless efforts in surgery, a last trip in a hearse, and a tombstone. An unidentified group marched as corpses in shrouds.[109]

In June 1923, club women dedicated a similar public monument to the dead in Louisville. A platoon of little children dressed in white attended, each representing a child accident victim of 1922. They left white chrysanthemums at the monument's base as a band played "Nearer My God To Thee."[110] Days later, Louisville's Safety Week parade was a spectacle of "coffins and skeletons, with wrecked automobiles and hospitals, . . . and small children."[111]

In the autumn of 1923, St. Louis staged perhaps the last of the grander safety weeks in this popular public safety movement. On the eve of the safety week—Sunday, November 18—clergymen preached sermons on safety. The next day, in the heart of the city, the St. Louis Safety Council unveiled an impressive monument, a broken column atop a "huge" pedestal inscribed "In Memory of Child Life Sacrificed on the Altar of Haste and Recklessness." Hundreds gathered around the memorial for its solemn dedication to the 32 children who had already been killed that year on the city's streets. The innocence of the child victims was represented by four carved cherubs seated at the base of the broken column. During the dedication ceremony, a local band played a dirge as seven children covered the

**Figure 1.8**
St. Louis's monument to child accident victims, erected in 1923. Source: "Monument to Child Auto Victims and Speedometer to Show Deaths," *St. Louis Globe-Democrat*, November 20, 1923, p. 8. Courtesy *St. Louis Globe-Democrat* Archives of St. Louis Mercantile Library at University of Missouri-St. Louis.

base of the monument with flowers. The crew of an airship dropped more flowers around the monument. The mayor told the crowd that "drivers of automobiles must be taught to adhere to the doctrines of the St. Louis Safety Council." The president of the council addressed the crowd himself, but as befitted a funeral service, clergymen spoke too. Among them were the archbishop of St. Louis and the president of the protestant church federation. One of the two rabbis who also spoke asked the audience to help protect the right of the city's children to live until adulthood.[112]

In another expression of public grief, in 1925 the Memphis Safety Council began displaying a black flag of mourning at sites in the city where children had been killed. The council also posted signs reading "A Child Was Killed Here—WHY?" The chamber urged the people of Memphis to "support this movement by your full cooperation as well as with your purse!"[113] The Toledo Safety Council later also flew a "mourning flag" from its office window in any month in which a child was killed.[114]

The blood, grief, and anger in America cities were not due entirely to high casualties. They showed the persistence, well into the twentieth century, of a traditional perception of the city street. To safety reformers, to pedestrians angry at motorists, and to grieving parents, the street was *their* space—a place to alight from a streetcar, a place to walk, a place to play. In this traditional construction of the city street, motorists could never escape suspicion as dangerous intruders. While this perception prevailed, the motor age could not come to the American city.

# 2 Police Traffic Regulation: Ex Chao Ordo

Ex Chao Ordo [order from chaos]

—motto of the Eno Foundation for Highway Traffic Regulation, 1921

A sane application of common sense will point the way to a solution of traffic problems. . . . The immediate problem of every municipality is to do all that can be done to fit traffic to present streets.

—Guy Kelcey, traffic device manufacturer, 1926[1]

The blood of street traffic casualties was the most shocking effect of the new motor traffic in cities, but it was not the most common. Day in and day out, the new abundance of cars in cities meant traffic chaos. In the automobile's first decade or two in the city, police struggled to restore order.

By "chaos" most police meant something related to but distinct from congestion. Traffic engineers would later make term "traffic congestion" common, to refer to a kind of inefficiency characterized by high traffic density and poor traffic flow, with low average speeds. Police seldom spoke of congestion or its synonyms because few conceived of it as their enemy. What later generations would consider low speeds had always before been normal speeds, to police and everyone else. Police instead saw their problem as a new kind of disorder caused by a new kind of vehicle—the automobile. Their conception of the solution followed from their conception of the problem. If the motorcar, thanks to its new capacities, disrupted the old street order, the answer was to make automobiles conform to the old street order. Police traffic regulation was thus fundamentally conservative.

Looking back at the first quarter of the twentieth century, a government expert concluded that "the change in traffic demands and the loads on our highways have amounted to a revolution."[2] Until about 1910, American cities did little to cope with the sudden and dramatic increase in motor traffic. Soon, however, cities had to devise new means of managing the new kind of traffic. Until this crisis, street traffic had been a problem managed informally, by custom. In the thick of the motor traffic crisis, cities formalized the regulation of street traffic. To bring order to the streets, police departments codified custom and supplemented it with common-sense regulations. To police, the benchmark of "common sense" was an orderly pre-automotive street.

Before 1903 no city had a traffic code worthy of the name. Cities managed heavy vehicular traffic much as hospitals, shopping malls, and other large institutions manage heavy pedestrian traffic today. Order depended on the customary practice (followed with varying degrees of conscientiousness) of keeping to the right. Signs did not regulate; they only indicated the location of popular destinations. The occasional official directed the lost, and traffic conflicts were resolved informally. Remembering her school days near the beginning of the twentieth century in an Indiana county seat, one woman recalled that at intersections "it was every man for himself."[3] A New Yorker remembered that before 1900 "the only rule, if such it might be called, then in existence, was that if you met another vehicle, you were supposed to keep to the right."[4]

### The Automobile and Traffic Disorder

Few blamed street design, heedless pedestrians, or slow streetcars for the traffic problems of the first two decades of the twentieth century. More blamed tall buildings. By far the most commonly accused suspect, however, was the automobile. "Everyone blames the automobile," the architect Harvey Corbett observed.[5] One engineer argued that the car's arrival in the city constituted an "emergency."[6] Another held that the auto caused traffic conditions to undergo a "complete change" after about 1912.[7] In 1915 city planners in Newark, New Jersey, pressed the mayor and the city council to take steps to respond to the "most remarkable change in the character of vehicular traffic."[8]

The prudence of Newark's city planners stemmed from sobering facts. Between 1912 and 1915 the number of automobiles in the city more than doubled; motor trucks more than trebled. Planners expected the quantity of motor vehicles to increase by "at least" 20 percent a year "for some time to come."[9] Milwaukee enumerators counted 39 motor vehicles crossing a checkpoint over a two-day period in 1911. Three years later they counted 1,373 crossing the same point in a single day.[10] Motor vehicle registrations in Chicago increased tenfold between 1911 and 1921, despite the World

War.[11] In St. Louis there were 16,000 registered motor vehicles in 1916; by 1923 there were more than 100,000.[12]

"Traffic" and "traffic congestion" were not even distinct categories of thought in city administration until about 1915, when motor vehicles made them so. In its first five years, from 1909 to 1913, *The American City*—soon to be the leading national journal for city administrators—did not devote one major article to street traffic. Other professional and trade journals gave the problem similarly scant consideration, as did the proceedings of professional bodies such as the American Society of Civil Engineers. "Traffic" was simply not yet a topic of city administration. Articles about "traffic congestion" before 1915 were more likely to be about crowding on streetcars or inadequate harbor facilities than about the crowding of streets with vehicles and pedestrians.[13] Around 1915, however, the term "traffic congestion" was transformed. Thenceforward it nearly always meant the crowding of streets with motor vehicles. With, for example, the number of Chicagoans per automobile falling from 106 in 1913 to 52 in 1916, city officials simply could no longer treat the problem in the customary way.[14]

The character of traffic was changing too. The decline of the horse in favor of the gasoline engine increased the need for formal traffic regulation, because automobiles were faster (and therefore more dangerous) than horse-drawn vehicles, and because motor vehicles lacked the horse's ability to resist the usual consequences of reckless driving. Although the human population of eight major cities grew a combined 19.9 percent between 1910 and 1920, the number of horses fell 57.4 percent.[15] New York's Health Department found that 30 percent of the horses stabled in the city in 1917 were gone just two years later.[16] By the mid 1920s, large cities reported that only 3–6 percent of the vehicles in their streets were drawn by horses.[17] Traffic was changing fast.

## Police Traffic Regulation: The Search for Order

In most large American cities, police traffic regulation had appeared by about 1910; it came earlier in New York and later in smaller cities. In 1909 few would have disagreed with the reporter Ira Judson, who remarked that the automobile had "not yet been adjusted to its proper place."[18] The pioneers who implemented these traffic measures seldom articulated the principles on which these measures were based, or their goals. They practiced largely as individuals or as isolated teams, and they followed the unwritten code of common sense, shaped by a pre-automotive conception of the street.

To Robert Wiebe, the late nineteenth century and the early twentieth century in America were characterized by a "search for order," and although he did not examine city police departments, their case supports Wiebe's view.[19] Erik Monkkonen illustrated the construction of one dimension of urban order in his study of American police. He demonstrated that late in the nineteenth century police departments grew detached from the political networks that had formerly constrained them. By 1900 police departments' customary social welfare functions were ebbing. The change gave them more freedom to defend urban order from the "dangerous class."[20] Although he did not examine traffic police, their techniques corroborate Monkkonen's interpretation. The motor revolution prodded police to go beyond their long-standing role as the arbiters of individual traffic conflicts. They became also the formulators and enforcers of codified traffic rules for the management of crowded streets. Police worked to bring order out of chaos.

## Between Custom and System

In 1903 New York instituted the first traffic code of the coming motor age. While many cities used New York's "Rules" as a model, many more devised their own codes. By World War I, however, cities had begun to institute more innovative measures to attack traffic problems at particular points, with, for example, signals at intersections and signs indicating traffic rules.

Police traffic regulation filled an interval between custom and system in city traffic management. Beginning in 1903 with New York's "Rules for Driving" and culminating in the early 1920s with the great diversity of local traffic measures and devices, police department tinkerers tried anything to fix the traffic problem. Tinkerers offered what custom by its very nature could not—innovation. Lacking tested theory or empirical research, their laboratory was the street itself, their method was trial and error.

## Eno

One of these tinkerers was a New York aristocrat turned gentleman traffic reformer. William Phelps Eno became an example to police traffic regulators throughout the nation. He did not admire unguided police methods and thus was not a typical exponent of police traffic regulation. He valued vehicular speed more than the police did, and was less attached to the pre-automotive street order. He recollected that city streets before 1900 never failed to make him "astonished at the stupidity of drivers, pedestrians and police."[21] But any consideration of police traffic regulation in this period must begin with Eno, because his energy prodded cities into action. He was not a police official himself, but he worked through New York's police department to make his vision real. Neither Eno nor the police officials with whom he worked had formal training in traffic regulation. Their recommendations for remedial steps had little theory and less research behind them. They were tinkerers; they applied common sense to traffic problems.[22] Eno inherited his father's fortune in 1899, and at the age of 41 he left his family's real estate business to begin a career in traffic regulation.[23]

Eno was an eccentric who worked at the periphery of the public sector. At age 80 he recalled that his father's office in New York "must have been somewhere below the City Hall and near P. T. Barnum's Museum."[24] The description places his own career as accurately as his father's address. Eno was a genteel monomaniac who saw the police as the natural implementers of the reforms he advocated. He shared with police officials in cities throughout the country a faith that order could be restored in city streets, and that sound police regulations could achieve it.

Eno was also a minor prophet, warning nineteenth-century people of twentieth-century problems. Declaring "It is time something should be done," he issued his first manifesto of traffic reform in January 1900, when automobiles were still of no significance in the street traffic of any city in the world. "Properly understood and regulated," Eno claimed, "several times the present traffic in our streets could go on with less delay, more safety and more comfort than what there is now."[25]

To fight traffic, Eno and other reformers declared, cities would have to bring order where there had been chaos. Indeed, Eno made "Ex Chao Ordo" the motto of his traffic foundation. He described his work as an effort "to substitute order where chaos now reigns almost supreme."[26] His mission was a search for order in city streets; to pursue it, Eno sought first to make traffic regulation "scientific." Although a journalist termed Eno's work "The Science of Street Traffic,"[27] Eno was no scientist. His academic record was little better than embarrassing. Though he could cite reasons of health to explain his early departure from Yale, it is likely he had other compelling grounds for dropping out.[28] He did not base his traffic recommendations on empirical research; he offered common sense dressed up as science.

Eno formulated his "Rules for Driving" for New York in 1903. The rules were a modest beginning; most of them were no more than a codification of custom. Drivers were directed to keep their vehicles to the right, to pass on the left, to use hand signals, to yield to emergency vehicles, and to travel at a "safe and proper" speed. The only noteworthy innovation was the "wide" or "outside" left turn: drivers were to keep the center point of the intersection to their left as they turned left. Yet police traffic regulation began here, in written codifications of custom, with gradual accretions of untested common-sense ideas.[29] Later Eno added higher speed limits and circular traffic flow at intersections to his proposals.[30]

With the adoption of the "Rules," New York was well ahead of other American cities. Ten years later police traffic regulation was still inconspicuous everywhere else. New York needed police regulation first because it was biggest, but it was also favored by chance. It had an eccentric millionaire obsessed with traffic reform. Eno was a true reformer; he sought no personal gain from his crusade. He paid the entire bill for the 100,000 copies of "Rules for Driving," merely on the condition that the city adopt the rules and distribute the four-page guide.[31] He accepted no paid work as a traffic expert until 1913.[32]

For a dozen years after New York adopted "Rules for Driving," most cities took few or no steps to manage traffic with anything more systematic than custom. By 1915, however, custom was clearly unequal to new conditions. Nationwide in 1910 there was still only one auto per 200 persons, but by the end of 1915 the rate exceeded one per 40, with the number of autos increasing at the rate of 40 percent a year.[33] American cities large and small soon enlisted police rules and devices in the new traffic fight.

Eno's rules were influential, though most cities rejected elements of his system. Eno saw himself as a progressive, welcoming a fundamental change in traffic. In contrast, most police were conservative in outlook. Automobiles, as the dangerous newcomers, would have to conform. Some cities gave horse-drawn vehicles the right of way over motor vehicles.[34] "Motor vehicles in the business quarter," *Municipal Journal and Engineer* held, "should have no precedence over others, as it leads to confusion and on occasion causes accidents."[35] Above all, few police shared Eno's high opinion of speed. Police departments borrowed what they wished from Eno and introduced some ideas of their own. The outside left turn rule was adopted almost everywhere.[36] Without Eno's help, police began the practice of painting lines on streets cities to guide traffic about 1915.[37] They also introduced one-way traffic.[38] Few shared Eno's enthusiasm for traffic circles. Eno had proposed them as a means of keeping traffic moving, but many police departments betrayed a common-sense conviction that speed was part of the problem, not part of the answer.

Unlike Eno, police tended to see speed and safety as mutually exclusive ends, and to reduce accidents they frequently restricted traffic flow. Police usually saw traffic congestion not as an enemy but as a helpful ally in their effort to slow vehicles. In traffic management as in law enforcement, police applied an "adversary model" of regulation in which speed was inconsistent with safety.[39] In a survey, only 3 percent of police chiefs chose congestion as the most important factor in accidents. Though the questionnaire did not prompt them to do so, 24 percent of the chiefs volunteered their opinion that incompetent drivers were another major factor, and 39 percent added an objection to speed.[40] In selecting Safety First as its motto, the International Traffic Officers Association (whose members were confined largely to the United States and Canada) indicated its relative disinterest in the expeditiousness of street traffic.[41] Sharing his police department's views, a Philadelphia judge made it his mission to fight speed. Abandoning judicial circumspection, he declared: "This mad desire for speed must be checked and I am going to do everything in my power to stop it."[42] To an Atlanta judge, the most urgent goal in traffic safety was "the slowing-down of men who by their speed are slaughtering these men, women and little children."[43]

Police-imposed speed limits were always low, sometimes extremely so. The median state-designated speed limit in cities in 1906 was 10 miles per hour, and local authorities usually could (and often did) set still lower limits.[44] States were slow to raise limits; Indiana's limit in cities, 8 miles per hour in 1906, had been raised only to 10 miles per hour by 1919.[45] Police usually enjoyed the support of city newspapers for their low speed limits.[46] The limits were impossible to enforce; indeed some were so low that motorists risked stalling if they observed them.[47] But police worked around this problem through other regulations that incidentally slowed traffic.

Police officials also tended to defend the customary rights of those on foot, and to expect automobiles to defer to them. Since most traffic casualties were pedestrians struck by motor vehicles, the safety problem, in this view, was not to make the streets safe *for* motorists (relatively few of whom were getting injured), but to make them safe *from* motorists.[48] Police authorities and ordinary "cornermen" (traffic police) tended to blame the motor vehicle and its driver for street casualties. In a survey of 480 city police chiefs, 72 percent chose "carelessness of the driver" as the leading cause of street casualties. Only 16 percent blamed pedestrians.[49] To police,

speed was the factor that made automotive traffic deadly. To protect pedestrians and to make motorists conform to long-standing street customs,
police departments tried to limit automobiles to pre-automotive speeds.

### Silent Policemen: Affordable Order

Street intersections were the epicenter of the traffic crisis, where danger
and disorder were concentrated. The National Safety Council estimated in
1923 that 70 percent of auto accidents occurred at street crossings, and a
study in St. Paul in 1924 and 1925 found that 86 percent of downtown
auto accidents occurred at intersections.[50] State laws and city ordinances
therefore often designated lower speed limits at city intersections. Although
motorists in South Bend, Indiana, were limited to 10 miles per hour in
1919, police ordered them to slow down still more at intersections.[51]

Eno's most enduring legacy was the outside left turn. From the beginning Eno had identified intersections as the crux of the traffic problem. He
was attached to "rotary traffic control" (traffic circles), but where this was
impracticable, he favored the wide left turn.[52] The rule required a turn at
something close to a right angle. The rule could be enforced by a policeman with a shrill whistle, but there were less expensive (and quieter)
alternatives. In 1904 New York City, at Eno's suggestion, installed posts at
the centers of some intersections to help drivers "understand the necessity
of passing around the central point."[53] As a consequence of the mobilization for World War I, a road transportation committee of the Council of
National Defense drafted recommendations for traffic rules for adoption
by cities nationwide. Eno served on the advisory panel of the transportation committee, and seems to have had no difficulty in having the rule
included among the resulting recommendations.[54] By then the most visible
reminder of traffic regulation in American cities of all sizes was the intersection center-point marker, usually marked "keep to the right," so that
motorists turning left would go around it. These markers went by numerous names, but the most common was "silent policeman." This humble
traffic device marked the victory of common-sense traffic reform where
custom alone had proved inadequate. For the first time, many cities
installed similar physical objects to control traffic between the curbs. No
longer was the entire burden left to the motorist's knowledge of the traffic
code. The device reveals the prevailing traffic regulation principles of police
departments. Police sought to bring order to chaotic streets. Intersections
were in the middle of the search for order from chaos that characterized
the approach of Eno and city police officials nationwide. To police officials

of the 1910s, order was the foundation all other traffic progress would be built upon. Guy Kelcey, a promoter of silent policemen, recommended that the "ultimate solution" for traffic problems at intersections lay in "the elimination of disorderly movements."[55] The silent policeman stood for this ideal.

The "disorderly movements" silent policemen controlled were those of "corner cutters," motorists who endangered pedestrians by approaching them at speed from behind. One pedestrian complained about the effects to his city newspaper: "If you obey the 'go' signal, a vehicle speeding up from behind and around the corner" will threaten you.[56] By compelling motorists to take outside left turns, silent policemen and the traffic police they replaced slowed automobiles down and gave them more predictable paths. In effect, the left-turn rule made automobiles conform to pre-automotive street customs. It slowed them to pre-automotive speeds and gave them a path of travel as confining as a streetcar's tracks. It reduced automobiles' opportunities for collisions with pedestrians while increasing their points of conflict with other cars. Since they now crossed pedestrians' paths at right angles, left turners were less likely to surprise them.[57] The rule and the device that enforced it thus indicated a value for pedestrian

**Figure 2.1**
A depiction of corner cutting. Source: George Kelcey, "Traffic and Parking Regulations as They Affect Public Safety and the Business Man," *City Manager Magazine* 8 (September 1926), p. 22. Courtesy International City-County Management Association, Washington, D.C.

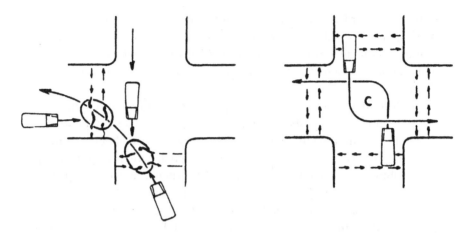

**Figure 2.2**
Intersection center posts, often called "silent policemen," discouraged corner cutting (illustrated in the left diagram). Source: G. G. Kelcey, "Traffic Study in Relation to Accident Prevention," *National Safety News* 8 (Oct. 1923), p. 26. Courtesy National Safety Council.

safety and street order, without regard for speed. To police, the element of danger was entirely the product of automobiles' speed, and congestion, as a suppressor of speed, was to many an ally in the struggle against accidents. Eno intended the silent policeman to expedite traffic, but it did not—and to many police departments, this was its chief virtue. Guy Kelcey, a supplier of silent policemen, guessed that low traffic speeds made the congested business district safer.[58] Kelcey held that the outside left turn, by slowing traffic down, prevented accidents. "The object of traffic control," he maintained, "is to *check speed* so that a vehicle is under easy control, and to compel right-angle intersections."[59] Manufacturers before 1920 almost never promoted a device as a facilitator of traffic flow; indeed, they more often promoted their products' beauty than their efficacy at speeding up traffic.

The silent policeman also reflected a need for economy. Eno faulted the numerous fanciful proposals for elaborate traffic facilities as "tremendously expensive," proposing instead simple, regulatory measures.[60] All police departments sought to limit their dependence on costly, overburdened and exposed traffic police ("cornermen"). Traffic regulation innovators worked to find ways to limit the need for traffic police. To one engineer, the expense of the cornerman was "perhaps the most important consideration" behind the search for alternatives.[61]

To provide an intersection with a cornerman for 8 hours a day typically cost cities almost $2,000 a year. The expense was onerous. For example, the Michigan city of Grand Rapids found itself spending more than 36 percent of its police budget on traffic regulation at intersections.[62] Cities simply could not afford to hire enough cornermen to keep up with the growing crowds of automobiles. Chicago, for example, kept the size of its Traffic Division practically flat from 1916. Indeed, the force was four officers smaller in 1925 than it had been nine years earlier; meanwhile, the number of motor vehicles had tripled.[63] The silent policeman was a simple, inexpensive substitute for the traffic policeman. Advertisements for such devices proclaim their affordability prominently. In one, an illustration shows one of its standards casting a shadow shaped like a cornerman directing traffic, likening the lifeless post to "a policeman on perpetual duty."[64]

**Figure 2.3**
Advertisement by George Cutter Company, *American City* 20 (January 1919), inside back cover.

The left-turn rule also slowed and congested motor traffic by increasing the points of conflict in the intersection. In La Grande, Oregon, silent policemen were deliberately made solid enough to damage the cars of hot-headed motorists. City Manager George Garrett reported "complete satisfaction" with the concrete markers, adding enthusiastically that "a number of machines have collided with them," causing sufficient injury to the vehicles so that "no machine has yet hit a standard a second time."[65] A manufacturer of silent policemen urged cities to put its devices in the center of their intersections, "near to or in the path of travel," since the sturdy post could withstand the collisions with "wild and careless drivers."[66]

In their diversity, silent policemen reflect the improvisational inventiveness of police traffic regulation. While the outside-left-turn rule was perhaps as close to universal as any police traffic regulation method, among motorists who frequented more than one city the physical variety of devices and rules was notorious. Cities, lacking a body of traffic professionals, were left to their own resources for most of their traffic management techniques. "If we were to collect specimens of traffic control devices from those cities where traffic regulation is active," a committee of investigators remarked in 1923, "we could fill a museum with signs and signals, no two of which would be alike in color, shape, size, or marking."[67] Cities often fashioned their own silent policemen.[68] Others turned to manufacturers of the devices, who promoted their products directly to city officials, without expert intermediaries. Professional traffic engineers were scarce, so representatives of such companies combined the roles of marketer and expert, supplying "recommendations as to signal location, type of signal to be used," and "plan of operation."[69] Guy Kelcey of American Gas Accumulator saw his articles on traffic regulation, which endorsed his company's silent policemen, published in prestigious national journals consulted by city officials.[70]

Finally, silent policemen exemplify police traffic regulators' informal approach to problem solving. In 1917 Eno claimed that traffic regulation had "grown to be an almost exact science," but neither he nor any other practitioner of traffic regulation had yet gone beyond the careful application of common sense.[71] The outside-left-turn rule was based on no preliminary traffic surveys or articulated theoretical models. It seems to have begun with Eno's 1899 hunch that the rule would bring order to chaotic intersections.[72] Kelcey developed a common-sense case that the "short left turn" (turning before the center of the intersection) increased the opportunity for an accident, especially to pedestrians.[73] The rule must have seemed a self-evident improvement over the disorder of unregulated intersections.

## Street Corner Experimentation

In their pursuit of affordable order, traffic regulation entrepreneurs devised many other money-saving devices.[74] Most supplemented or replaced the cornerman. The earliest and simplest innovations for intersection traffic regulation merely improved the visibility or the clarity of the cornerman's signals. Before World War I, at many intersections a policeman rotated a sign resembling a weathervane to show "go" to two points of the compass and "stop" to the other two. With other devices a single policeman could operate signals at several intersections simultaneously. Manufacturers marketed traffic devices directly to cities, without extensive field trials. Officials installed devices without preliminary surveys of traffic volume or accidents. Each of the numerous manufacturers produced products of its own (often patented) design; almost none specialized exclusively in traffic devices.[75]

The ultimate extreme of local trial-and-error methods is found in the workshops of city police and street departments. These turned out makeshift devices for trial in local streets. Detroit's street department improvised a machine to paint lines on pavement from a tennis court line maker, and other cities imitated the device.[76] Six years later a Yonkers entrepreneur was marketing it as the Line-O-Graph.[77] A later innovation in Detroit was an intersection center post with a light on top to attract attention. A small windmill was attached to a rotating shield that covered the light on one side so that "whenever the wind blows, which is nearly all the time, the light is made to flash."[78]

Many early traffic regulation devices were adaptations of railroad signals. City officials widely accepted *The American City*'s view: "What signaling has done for railroads it can do also for towns and cities."[79] At least one railroad signal company, Crouse-Hinds, developed a line of street traffic signals.[80] Those known as "traffic semaphores" closely resembled their counterparts along the rails, except that nearly all required human operators.[81]

Because of its greater clarity and visibility, the traffic light was a considerable improvement on the semaphore. In principle, however, it was not much different. In Cleveland on August 5, 1914, a local traffic device firm installed the first practical traffic lights at a busy intersection. A policeman in a booth at the corner operated the signals manually, but the city no longer had to post more officers there at rush hours.[82] Traffic lights, however, remained unusual and were found only on major streets in the largest cities before the mid 1920s.

The traffic tower (Eno insisted the correct term was "crow's nest") was another way to supplement the cornerman's efforts at major intersections. Eno characterized towers as a development on the silent policeman.[83] He reported that the first appeared in Detroit in 1917,[84] and by the early 1920s towers could be found in many cities.

City officials liked traffic towers because they gave cornermen sufficient visibility to coordinate their signals with those at other intersections. The opportunities for impressive orderliness in traffic regulation could justify elaborate arrangements. Beginning about 1918, cornermen in Detroit's traffic towers matched their signals to those of a master tower; cornermen on the ground were to follow their lead.[85] Other cities imitated the plan. New York's police department installed coordinated signals between towers on Fifth Avenue in 1920. The main tower was visible from Washington Square to Thirty-Fourth Street, and between them cornermen on the ground matched their signals accordingly. Twenty-six blocks were thus under simultaneous control.[86] In Philadelphia, city officials took the idea to an extreme. They put a "master light" in the tower of City Hall, the tallest building in the city. The many cornermen who could see the light were to synchronize their intersections to signals from the master light. At other intersections, Philadelphia, like other cities, used timers and relays to synchronize signals.[87]

The common-sense assumption behind this plan was that simultaneous signaling is orderly signaling. Deputy Police Commissioner John Harriss described the resulting "unison" in New York's traffic as "virtually clock-like—and certainly efficient," and the traffic authority Miller McClintock agreed that simultaneous signaling made an "impression of great neatness and system."[88] Traffic towers, however, soon fell into disfavor. They began vanishing in the mid 1920s. While police administrators like Commissioner Harriss of New York sometimes defended towers as "successful beyond a doubt," others more interested in efficient traffic flow than street order considered them a failure. These critics explained to city officials that simultaneous signaling actually rewarded speeding while reducing vehicular capacity. The simultaneous starting of scores or hundreds of streetcars also elevated expensive peaks in power demand.[89]

## The Milwaukee Mushroom: Entrepreneurs as Experts

Even silent policemen earned widespread hostility. As City Manager Garrett of La Grande, Oregon, had discovered, motorists struck the posts, damaging their vehicles. What was worse (at least to Garrett), vehicles often destroyed the post. One Pittsburgh motorist was noticed at night "making a terrific racket," dragging an elaborate silent policeman he had struck but never seen.[90] In 18 months, at a single intersection in Milwaukee, the city suffered $1,658 in losses through the repeated destruction of silent policemen.[91] Besides replacement costs, such incidents raised the specter of legal liability, a danger that materialized in 1923 when an Indiana judge ruled that a silent policeman could constitute a street hazard.[92] By then Milwaukee had abandoned silent policemen altogether after city officials calculated that they were losing them at the rate of 400 a year.[93]

For a solution, cities turned not to experts but to entrepreneurs. In a 1923 magazine advertisement, an enterprising paint manufacturer instructed city officials in the proper marking of an intersection. The company, in effect, sold not just a particular product, but a traffic regulation policy.[94] Between device manufacturer and purchaser lay no intermediary with technical expertise, theoretical training, survey results, or the resources of a professional society. Instead, manufacturers promoted new traffic contrivances directly to city officials, who in turn kept the companies abreast of their demands. Manufacturers urged cities to consult them directly, asking officials to "let us submit designs and quotations on your needs,"[95] or to "bring your traffic problems to us."[96] At commercial fairs, such as Chicago's Good Roads Show of January 1922, city officials could evaluate examples of the numerous traffic regulation devices by common-sense standards, consulting manufacturers directly for advice. Beginning in 1918, annual conventions of the International Traffic Officers Association gave manufacturers the chance to get specifications from the police officials and to win recognition for their products through its awards.

The Chicago company that introduced the Milwaukee mushroom received the Association's "Award of Merit" in 1921.[97] A product of non-expert, entrepreneurial policy advice, it was sold as a superior variation on the silent policeman. To police officials a mushroom was a cast iron object the size and shape of a salad bowl, turned upside down and attached to the pavement. If struck in traffic it would jolt the driver without damaging device or vehicle.

In 1915 a New York transit expert reported the use of a "mushroom-shaped base of iron" on the streets of Detroit, where it served to keep motorists out of "safety zones" (streetcar landings).[98] Four years later a device manufacturer suggested replacing the bulky silent policeman with the low-profile mushroom in the center of intersections, and illuminating it for visibility.[99] In August 1920 Milwaukee's street lighting department introduced the "Milwaukee mushroom type," which was hollow and

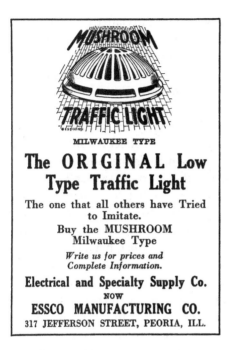

MILWAUKEE TYPE

## The ORIGINAL Low Type Traffic Light

The one that all others have Tried
to Imitate.

Buy the MUSHROOM
Milwaukee Type

*Write us for prices and
Complete Information.*

### Electrical and Specialty Supply Co.

NOW

## ESSCO MANUFACTURING CO.

317 JEFFERSON STREET, PEORIA, ILL.

**Figure 2.4**
Advertisement by Essco Manufacturing Company, *American City* 27 (July 1922),
advertising page 68.

perforated like a colander, but with larger holes. Inside was an electric light, visible through the holes to motorists. In the early and mid 1920s this new design proved immensely popular, and within two years several manufacturers were supplying the device to cities throughout the country.[100]

The traffic mushroom's clear advantages over the silent policeman made it easy to sell. Like the silent policeman, it could substitute for a living cornerman. A manufacturer of a variation on the traffic mushroom appealed to city officials facing "the expense of detailing traffic officers," which had become "very serious."[101] Unlike the silent policeman, however, mushrooms were "indestructible" and "accident-proof."[102] A mushroom manufacturer reminded city officials that "the ordinary upright street intersection or safety zone sign is often the cause of serious accidents by collision with the sign itself."[103] Another company advertised a variation on the device that would "never...involve your city in law suits." Its mushroom was mounted on a spring and would retract into the pavement when struck.[104]

Through trial and error cities abandoned one device for another, while the principle upon which both were based—the outside-left-turn rule—remained entirely unexamined. Thus the two devices had much in common. Like the silent policeman, the lighted mushroom was intended to bring order to left turns. The original manufacturer promoted it as a way to "control the most habitual 'corner cutter.' "[105] Until 1924 no traffic mushroom advertisement claimed that such markers ease congestion or expedite traffic flow.[106] Some traffic mushroom manufacturers sold their product as a safety device.[107] But manufacturers most often sold mushrooms as a means of imposing order and as an economical substitute for cornermen. Soon, however, the reason for being of both devices would itself be challenged. Both presupposed the value of the outside-left-turn rule and the apparent order it conferred, and both protected pedestrians crossing streets. Other social groups in the city came to doubt that these benefits were worth the cost to traffic efficiency.

## The Limits of Common Sense

Cities treated the arrival of the automobile as they might any other emergency. Through existing agencies, they drafted rules for managing the problem. The public, however, knew little of the proceedings, and police lacked the means to inform them. Police tended to rely on voluntary compliance to codes too complex to remember. A traffic professional would later advise city officials that "all of the provisions" of a traffic code "cannot be carried in mind by the users of the streets."[108] Throughout the country, cities followed New York's lead by drafting "Rules for Driving," relying on motorists' knowledge of and adherence to these rules for most of the traffic regulation burden. Public street signs indicating anything besides directions and distances were almost unheard of. In 1921 motorists were still complaining that traffic rules varied from intersection to intersection, and that "there is in no case a sign informing the driver on these points." Instead, the motorist's "first intimation that he is not playing the game according to the rules usually consists in a raucous bawling out from the nearest officer."[109] Unsympathetic traffic court judges referred unwitting violators to codes "published in pamphlet form," which were "available for your inspection and study."[110]

Well-intended police traffic methods could cause accidents and worsen congestion. Common-sense expedients such as silent policemen and traffic towers were untested experiments, despite their promoters' claims of scientific rigor. Motorists learned to doubt the value of such innovations. In

1927 the engineer Burton Marsh reported that "a considerable proportion of traffic signals are now generally regarded with disfavor."[111] A more systematic method was needed. By 1927, the Detroit Police Department's own director of traffic was lamenting "ill-considered or unstudied minor experiments in widely scattered localities."[112] In such circumstances, the 1926 protest of one engineer that traffic regulation is "an engineering problem, which regular police officers have neither the time nor the technical training to handle with the greatest efficiency" found ready assent.[113]

Police efforts showed a bias in favor of conditions in the streets before the arrival of the automobile. The message of their methods was clear: automobiles would have to conform to cities as they were; cities would not conform to the needs of automobiles. "The immediate problem of every municipality," wrote Guy Kelcey, "is to do all that can be done to fit traffic to present streets."[114] If the speed of motor vehicles was their greatest threat, then they must be confined to speeds typical of horses and streetcars. Police departments guarded pedestrians' traditional rights to the street from motorists' demands for more street space. City officials defended both the outside-left-turn rule and low speed limits largely in the name of pedestrian safety. In these years they generally ascribed pedestrian casualties to impatient motorists, not to insouciant pedestrians. Police relegated the auto, like the slaughterhouse, to the status of a tolerated nuisance.

Police traffic regulation was not popular with motorists. Police found themselves unable to force drivers to conform to their exacting regulations. Kelcey admitted that nearly two out of three failed to execute the outside left turn properly when they encountered silent policemen.[115] Police complained that the complexity of traffic codes made them "difficult to enforce" and they bemoaned motorists' "extremely defiant attitude."[116]

Though police administrators generally agreed with the Safety First movement, safety crusaders betrayed their lack of faith in the police by working outside of the departments. Some found the police methods unscientific in theory and arbitrary in application. To others, police traffic regulation's most conspicuous failure was that it did not keep traffic moving. "It sometimes seems as if the first object of the policeman were to hold things up," New York City's police commissioner admitted in 1916.[117] In the 1920s, groups seeking to prevent accidents or to fight congestion looked elsewhere for answers.

# 3 Whose Street? Joyriders versus Jaywalkers

Are streets for commercial and pleasure traffic alone?
—Bessie Buckley, Milwaukee, 1920[1]

If safety is not realized, there will come a time when the public will demand drastic action—no matter how hard it may hit the pocketbooks of automobile manufacturers, dealers and owners.
—*Milwaukee Leader*, September 25, 1920[2]

This dreadful slaughter must be stopped. If necessary, regulations severe and searching enough to do it must be adopted and enforced. . . . If reasonable safety of life and limb can only be had by impairing the motor car's efficiency the motor car will have to pay that price.
—*St. Louis Post-Dispatch*, November 19, 1923[3]

Beneath the grief and anger of many safety reformers lay an old assumption: city streets, like city parks, were public spaces. Anyone could use them provided they did not unduly annoy or endanger others. Under this construction of the city street, even children at play could be legitimate street users, and even careful motorists were under suspicion. In the 1920s, however, the pressure of traffic casualties divided old allies. Some renewed their resolve to compel motorists to conform to the customs of the street as it had been, especially by limiting their speed. Others, more pragmatic, wanted to save lives by giving pedestrians more responsibility for their own safety. Finally, some newcomers proposed a more radical social reconstruction of the street as a motor thoroughfare, confining pedestrians to crossings and sidewalks.

With these perspectives went diverging answers to the question "who belongs in the street"? To protect pedestrians, some proposed restricting their use of street space. Pedestrians angry at the automobile's intrusion resented such ideas, and demanded restriction of the car instead. As these disputes grew hotter, automotive interests began to appreciate their own stake in the result. They lost confidence that safety councils could prevent accidents without curtailing the car's urban future. By late 1923, therefore, motordom joined the safety fight as an independent player. It struggled to stop definitions of the safety problem that threatened the automobile's place in the city. This chapter traces this parting of ways, following it up to the full mobilization of those promoting a new social construction of the street for the motor age.

## Pedestrians' Rights

Today it is a commonplace that the automobile represents freedom. But to many city people in the 1920s, the car and its driver were tyrants that deprived others of their freedom. Before other auto promoters, Charles Hayes saw that industry leaders had to reshape the traffic safety debate. As president of the Chicago Motor Club, Hayes warned his friends that bad publicity over traffic casualties could soon lead to "legislation that will hedge the operation of automobiles with almost unbearable restrictions." The solution was to persuade city people that "the streets are made for vehicles to run upon."[4]

Pedestrians would have to assume more responsibility for their own safety. But how? Where they had been tried, legal regulations alone had been ineffective. From 1915 (and especially after 1920), cities tried marking crosswalks with painted lines, but most pedestrians ignored them.[5] A Kansas City safety expert reported that when police tried to keep them out of the roadway, "pedestrians, many of them women" would "demand that police stand aside." In one case, he reported, "women used their parasols on the policemen." Police relaxed enforcement.[6]

City people saw the car not just as a menace to life and limb, but also as an aggressor upon their time-honored rights to city streets. "The pedestrian," explained a Brooklyn man, "as an American citizen, naturally resents any intrusion upon his prior constitutional rights."[7] Custom and the Anglo-American legal tradition confirmed pedestrians' inalienable right to the street.[8] In Chicago in 1926, as in most cities, "nothing" in the law "prohibits a pedestrian from using any part of the roadway of any street or highway, at any time or at any place as he may desire."[9] The most restrictive interpretation of pedestrians' rights was that "All travelers have equal rights on the highway."[10]

Conversely, the motorist's claim to rights in the street at the expense of pedestrians was very hard to make. By law and by custom, all had a right

**Figure 3.1**
Woodward Avenue at Monroe Avenue, Detriot, 1917. Source: *Detroit News* photo archive.

to the street, and none could use it to the detriment of others' rights. In 1913 the New York Court of Appeals observed that it was "common knowledge" that the "great size and weight" of automobiles could make them "a most serious danger," and so the responsibility for preserving the safety of the streets lay overwhelmingly with motorists.[11] In New York City's traffic court in 1923, a judge explained that "Nobody has any inherent right to run an automobile at all." Rather, "the courts have held that the right to operate a motor vehicle is a privilege given by the state, not a right, and that privilege may be hedged about with whatever limitations the state feels to be necessary, or it may be withdrawn entirely."[12] The law would not deprive pedestrians of their customary rights so that motorists could roam at will in cities.

By custom the street had always been free to all; the law had intervened only in the names of safety and equity. Equity could demand, for example, that no one obstruct the roadway with standing vehicles.[13] Yet fast and dangerous automobiles imperiled pedestrians' traditional right. It did not make sense to most city people to protect the pedestrian majority by

**Figure 3.2**
Eleventh and F Streets NW, Washington, ca. 1915. Source: Prints and Photographs
Division, Library of Congress.

curtailing this right and turning the pavement over to motorists. A Philadelphia newspaper editor reproached motorists for usurping "pedestrians'
rights" by passing standing streetcars and preventing pedestrians from
crossing streets.[14]

Readers' letters to the *St. Louis Star* express pedestrians' indignation at
motorists' intrusion upon their rights. One letter, signed "Pedestrian," complained that "the pedestrian is forced to submit to the tyranny of the automobilist."[15] Other letter writers urged pedestrians to organize to defend
their claim to the streets. "It might be necessary to organize an anti-
automobile league," wrote one.[16] "The time is ripe for the common people
and the pedestrian to organize," wrote another.[17] "We must all pull together,"
wrote a third, and "insist on our rights to use the streets" until the
"auto-hogs . . . wake up to the fact that they cannot do as they please and
monopolize the streets."[18]

Local police tended to blame motorists for pedestrian traffic casualties.
With their traditional mission of defending custom and seeking equity,

police were unwilling to abridge pedestrians' rights to the free use of city streets. New York police magistrate Bruce Cobb in 1919 defended the "legal right to the highway" of the "foot passenger," arguing that "if pedestrians were at their peril confined to street corners or certain designated crossings, it might tend to give selfish drivers too great a sense of proprietorship in the highway." He assigned the responsibility for the safety of the pedestrian—even one who "darts obliquely across a crowded thorofare"—to drivers.[19] Most American police would have agreed with the Ontario authorities who regarded pedestrians as victims of "an unfortunate attitude of mind which belongs to some drivers and which assumes that the pedestrian should get out of the way of the vehicle."[20]

Police and judicial authorities recognized pedestrians' traditional rights to the streets. "The streets of Chicago belong to the city," one judge explained, "not to the automobilists."[21] Some even defended children's right to the roadway. Instead of urging parents to keep their children out of the streets, a Philadelphia judge attacked motorists for usurping children's rights to them. He lectured drivers in his courtroom. "It won't be long before children won't have any rights at all in the streets," he complained. As the usurper, the motorist, not the child, should be restricted: "Something drastic must be done to end this menace to pedestrians and to children in particular."[22]

As might be expected, judges tended to defend customs of street access. A Philadelphia newspaper declared that as a general rule, "indignant judges tell the average driver who is called before them that 'they and their contraptions should be driven from the streets.' "[23] In 1921 an Illinois judge struck down Joliet's requirement that pedestrians cross streets at right angles or on crosswalks, and that pedestrians follow other traffic rules.[24] In 1926 a Detroit judge admitted that in accident cases "his sympathies were always with the pedestrian, and that a driver of a motor vehicle which had caused injury to a pedestrian, coming before him in his court could expect as severe treatment as the law would permit him to hand out."[25] Another magistrate lectured an errant motorist for threatening to make America a "race of cripples." Upon convicting a truck driver of manslaughter he declared: "Such persons as you are a disgrace to humanity."[26] Even the ordinary traffic police won a reputation for hostility to the motorist. The director of police in Philadelphia was forced to remind his officers publicly that well-intentioned drivers who overlooked one of the city's many motoring rules "should not be treated as speed maniacs or criminals."[27]

Juries tended to favor pedestrians as well. "Juries in accident cases involving a motorist and a pedestrian almost invariably give the pedestrian the benefit of the doubt," a safety expert explained in 1923; "the policy of the average juryman is to make the automobile owner pay, irrespective of responsibility for the particular accident."[28]

More often than not, the press took much the same view. The leading city paper in Syracuse, New York, argued that the burden of safety lay properly with motorists. "The public, for the most part, is not so greatly in need of constant warnings against the dangers of the streets."[29] The *New York Times* claimed in 1920 that pedestrians' rights to the streets were so extensive that "as a matter of both law and morals they are under no obligation" to exercise "all possible care." The greater share of responsibility (moral and legal) lay with the motorist: "drivers justly are held to a greater care than pedestrians," the paper contended. Pedestrians "have rights in the streets, even tho they choose to cross elsewhere than at the appointed places."[30] *The Outlook* agreed that a higher order of justice was at work than the merely legal. Motorists have a "moral responsibility" on the road, a responsibility too few were fulfilling.[31]

Before the American city could become a largely automotive city, the automobile had to win a superior right to most of the street's surface. Unless it succeeded in this claim, in crowded towns those motorists who were unwilling to run down pedestrians would be forced to a virtual standstill. Yet before 1920 American pedestrians crossed streets wherever they wished, walked in them, and let their children play in them. The extent of these practices was such that in one of the first organized street safety campaigns in 1914, the Chamber of Commerce, in Rome, New York, had to ask pedestrians not to "visit in the street" and not to "manicure your nails on the street car tracks"—with limited success.[32] Under these circumstances, an automotive city seemed a dim prospect.

### Safety Weeks: Internal Tensions

The National Safety Council was a diffuse organization with porous boundaries. Especially in its public safety work, it never stood for any distinct, coherent principle. Council members working in public safety included insurance executives, lawyers, police, auto club secretaries, street railway people, truck operators, teachers, and individual members, and NSC membership therefore could stand for little more than a general interest in accident prevention. A 1924 safety week in New York's Westchester County was backed by a typical spectrum of local organizations, constituting "a group of the most influential and most respected citizens of the county," including representatives of the Chamber of Commerce, local government, schools, women's clubs and other civic organizations, and churches.[33] Auto clubs often backed safety council campaigns as well, at least until 1923. Because local safety councils financed themselves, the national headquarters in Chicago could not impose a party line on them. Local safety reformers could be dues-paying members of local NSC affiliates without joining the NSC. They often held views quite distinct from those of local council leaders and at the Chicago national headquarters.

Safety First was based on the hope that the accident problem could be solved if all did their part. "The safety first doctrine," the *Chicago Tribune* explained, "asks each person to guard himself."[34] It was a common-sense doctrine, but time soon demonstrated that there was no sense of the accident problem common to all categories of street user. The interests backing city safety campaigns shared no common motive. At one extreme were practical-minded motor fleet operators with insurance premiums to pay and auto clubs with publicity problems. At the other were committed reformers, including parents and teachers. Some were pragmatic people who saw pedestrian control and child responsibility as unfortunate necessities. Others—"crusaders," as one newspaper editor called them—were less willing to compromise.[35] They regarded motor vehicles in city streets much as St. George regarded the dragon.

Appeals to caution were ubiquitous, but the disparate participants urged different kinds of caution. Local safety councils unified safety weeks only superficially; the threads that tied their constituent groups together would hold only in fair weather. Auto clubs urged their members to drive carefully. Schools taught children not to play in streets. Boy Scouts asked pedestrians to cross streets at corners. For a time these various recommendations all seemed quite harmonious. But securing safety meant placing responsibility, intruding upon rights (real or perceived), and sacrificing convenience. Under the pressure of worsening street casualties and offended senses of right, disagreements among the diverse promoters of safety grew divisive. In 1923 an officer of the Brooklyn Safety Council reported to the NSC "This subject is loaded to the muzzle."[36]

### The Discovery and Reinvention of Jaywalking

One explosive ingredient in the gunpowder was rhetorical. Words betrayed their users' prejudices, and sometimes partisans redefined old terms or devised new ones to fight traffic.[37]

The interjections that pass between pedestrians and motorists can gauge street users' status. Who, for example, is entitled to shout "You don't belong in the street!"? Such an exchange is preserved in fiction. In 1910 the pulp publisher Edward Stratemeyer introduced his Tom Swift series of stories for boys with *Tom Swift and His Motor-Cycle*. At the outset of his first adventure, young Swift is bicycling along a road. His archrival, the spoiled Andy Foger, comes speeding along in an automobile. A quarrel over road rights ensues. Foger makes the first move. By maintaining speed and sounding his horn, he makes a claim by right of conquest. But Swift cannot evade him, so the bluff is exposed. Foger swerves, landing the car in a ditch. The exchange continues in words. Foger blames the accident on Swift. Swift replies by painting motorists as usurpers and tyrants. "You automobilists take too much for granted!" he declares. "I guess I've got some rights on the road!" Foger questions the relevance of old ways in a new age: "Aw, go on!" he says. "Bicycles are a back number, anyhow."[38]

Real variations of this fictional battle were fought and fought again for the next two decades. Pedestrians (and bicyclists) claimed prior rights, but motorists' advantage in power tended to make pedestrians relinquish them. But motorists had rhetorical weapons too, and they found more over the years. Like Andy Foger, they claimed that old ways no longer suited the new motor age.

When motorists first intruded upon city streets, annoyed pedestrians found epithets for the more aggressive ones. Some called them "joy riders," others "speed maniacs."[39] Both terms connoted irresponsibility and a reckless disregard for the rights and safety of other street users. In 1909, for example, joy riders were motorists who abused their "power of life and death" over rightful but weaker occupants of streets: "pedestrians, . . . the aged and infirm, . . . children playing in the streets"; in 1912 they were "automobilists" whose "aggressions" intruded upon the rights of pedestrians "to the very great danger of children and aged people."[40] Both views reflected the unspoken assumptions of their time: that people on foot, including children at play, had a rightful claim to street space.

Motorists replied with epithets of their own. They hit upon the most effective one early: "jaywalker." A "jay" was a hayseed, out of place in the city; a jaywalker was someone who did not know how to walk in a city. Originally the term applied as much or more to pedestrians who obstructed the path of other pedestrians—by failing, for example, to keep to the right on the sidewalk.[41] As autos grew common on city streets, jaywalkers were more often pedestrians oblivious to the danger of city motor traffic. According one early, more general definition (1913), jaywalkers were "men so

THE MAGIC STROP OIL AT BROADWAY AND WALL STREET.

**Figure 3.3**
This cartoon, from a newspaper article titled "In Simple, Child-Like New York" (*Kansas City Star*, April 30, 1911), is the first known illustration of "jay walkers."

accustomed to cutting across fields and village lots that they zigzag across city streets, scorning to keep to the crossings, ignoring their own safety" and "impeding traffic."[42]

Overworked police "cornermen" soon applied the term to pedestrians who ignored their directions. By 1916 "jaywalker" was a feature of "police parlance."[43] Police use modified the word's meaning and sparked controversy. "Jaywalker" carried the sting of ridicule, and many objected to branding independent-minded pedestrians with the term. In 1915 New York's police commissioner, Arthur Woods, attempted to use it to describe anyone who crossed the street at mid-block. The *New York Times* objected, calling the word "highly opprobrious" and "a truly shocking name." Any attempt to arrest pedestrians would be "silly and intolerable."[44] The word's meaning had become an objective in the emerging traffic fight. In December 1916, "Don't Jay Walk" banners appeared in small public safety campaign in Washington, D.C.[45] According to a 1917 Boston definition, a jaywalker was "a pedestrian who crosses the streets in disregard of traffic signals."[46] In a 1918 public safety campaign in St. Louis, leaflets informed pedestrians that a "jay walker" was someone who "refuses to use the crossings and cuts the corners."[47]

Soon after war ended in France, however, a fight over "jaywalking" began at home. In the city safety week campaigns of those years, new allies joined police to promote the word and extend its application. To them it was a defensive weapon in the fight against traffic casualties. Others, however, saw it as calculated aggression against innocent pedestrians. On this fight hinged the future of the city street. In the coalitions behind the safety weeks, many agreed that pedestrian safety would cost pedestrians some of their rights to the street. Police, practical-minded safety reformers in local safety councils, and some school administrators and teachers joined auto industry people to urge pedestrian control. They resorted to the explosive epithet because pedestrians often resisted and resented pedestrian control. Few rules were made and fewer enforced. In 1921, Charles Price, easily the top public safety expert in the United States, was asked "What towns have ordinances against jay walking?" "I don't know of one," Price answered.[48] Los Angeles had attempted pedestrian control in 1919, with little discernible effect on pedestrians' behavior.[49] Police in Washington also dabbled in pedestrian control in the 1910s, getting nowhere. In 1921 an officer of the American Automobile Association invited a guest to observe rush-hour traffic from the top of the Riggs building there. "The streets there," he recalled, "were absolutely black with people. They made no attempt to come anywhere near the cross-walks. An automobilist went through the crowd at his own risk. A great many of those people were not particularly careful of how they crossed the streets, they just bowled across."[50]

In 1921 a National Safety Council member from Baltimore confessed to his colleagues that, at least in pedestrian control, "We haven't got public opinion with us today," because "You are affecting personal liberty when you keep people from crossing the streets at certain places."[51] Even within the leadership of the NSC some questioned the value of pedestrian control. At the 1921 NSC safety congress a delegate from Newark, New Jersey, warned of "a possibility of slopping over on this thing, of going a little bit too far in giving the automobile driver the idea that he is not obliged to look out for pedestrians at any other point than at the cross-walks. If we as an organization devoted to safety get that thought across even in the abstract, we will do wrong. Let's be very careful about that."[52]

Critics of jaywalking put little hope in outlawing the practice. Their problem was more a matter of custom than of law. They knew many would resent fines or arrests for a nearly universal practice, and that without pedestrian cooperation bans would be unenforceable. Change, said one safety reformer, would require "education instead of prosecution or perse-cution."[53] "We want to educate the people rather than arrest them," said a deputy police chief.[54] In city safety weeks, the leading feature of safety education for adults on foot was intense sloganeering against "jaywalking," directed at "reckless pedestrians."[55] Advocates of pedestrian control fought to classify as jaywalker any person who walked anywhere in the roadway, except in intersections at right angles to traffic. To one promoter of pedes-trian control, anything else was "crossing the street in the rube fashion."[56]

To work, the epithet "jaywalker" had to be introduced to the millions. In 1921 a collector of dialect found the term "not common."[57] That would have to change. In city safety campaigns, safety reformers found their opportunity. In Syracuse's pioneering safety campaign of December 1913, a man in a Santa Claus suit used a megaphone to denounce careless pedes-trians as "jay walkers." According to one safety reformer, those singled out for this treatment "never forgot it."[58] In St. Louis in 1918, at the first big public safety week organized by the National Safety Council, leaflets intro-duced pedestrians to the word. They defined a "jay walker" as "the man who refuses to use the crossings and cuts the corners."[59] Other cities went further. In a 1920 safety campaign, San Francisco pedestrians who thought they were minding their own business found themselves pulled into mocked-up outdoor courtrooms. In front of crowds of onlookers they were lectured on the perils jaywalking. The idea was to "kid the people into taking care of themselves"—but surely many defendants didn't appreciate the joke.[60] A year later, Boy Scouts in Providence, Rhode Island, summoned jaywalkers to a "school for careless pedestrians" for reeducation.[61]

In many cities, police or Boy Scouts distributed anti-jaywalking cards to pedestrians who crossed streets in disapproved ways.[62] The technique was quick and easy, and relatively unintrusive. It also introduced tens of thou-sands to the newer, official definitions of "jaywalking." During a 1921 safety week in Grand Rapids, Michigan, the safety council posted Boy Scouts to hand out cards to jaywalkers, informing them of the risk they were taking, and teaching them that they were "jay-walking." The cards made the case that the practice "was permissible when traffic was horse-drawn," but "today, it is dangerous—conditions have changed!"[63] Thus, in Grand Rapids, a new label for an old practice was introduced. Through the campaign, "thousands of people who never knew what jaywalking meant have learned the meaning of the word."[64]

In some cities' safety campaigns, actors were recruited to attract ridicule as conspicuous "jaywalkers." In a 1919 Cleveland parade "crowds of 'jay

**Figure 3.4**
In 1923, the Automobile Club of Southern California paid for signs notifying Los Angeles pedestrians that jaywalking was prohibited by order of the police department. Source: Robbins B. Stoeckel, "Our Rights on the Highway," *National Safety News*, December 1923, 19. Courtesy National Safety Council.

**Figure 3.5**
Boy Scouts handed out cards like this one to pedestrians in Hartford in 1921. Source: "Boy Scouts and Kiwanis Club of Hartford Put On Anti-Jay Walking Campaign," *National Safety News* 3 (February 7, 1921), p. 4.

walkers' " (recruited for the purpose) demonstrated for spectators what jaywalking was.[65] In a New York City safety parade, a boorishly dressed character allowed himself to be rear-ended repeatedly by a slow-moving Model T.[66] In a Philadelphia safety campaign, a pair of "Country Cousins" dressed "as rubes" distributed "Cautious Crossing Crosser" buttons to pedestrians.[67]

The cleverest anti-jaywalking publicity effort was in Detroit in 1922, where the Packard Motor Car Company exploited the new fashion for monuments to traffic fatalities. Packard built an oversized imitation tombstone that closely resembled the monument to the innocent child victims of accidents in Baltimore. But Packard's tombstone redirected blame to the victims. It was marked "Erected to the Memory of Mr. J. Walker: He Stepped from the Curb Without Looking." Packard entered the monument on a float in Detroit's safety week parade. The Detroit Automobile Club voted Packard's float the best in the parade, and awarded the company a silver cup.[68]

In Washington, the fight over jaywalking turned traffic chiefs against each other. Congressionally appointed district commissioners chose an American Automobile Club executive to serve as Washington's traffic director, despite controversy over the "the danger of filling the new post with a person who might see only the automobile driver's side of the question."[69] The new traffic chief, M. O. Eldridge, soon showed himself an aggressive opponent of stubborn pedestrians. His office elevated "jaywalker" from slang to a word dignified by edict. The new, official definition was broad. A jaywalker was "any pedestrian who undertakes to cross the stream of vehicular traffic on his own responsibility when there is a signal or a policeman at the corner to start and stop automobiles, or one who crosses in the middle of a block under any circumstances not involving an emergency."[70] Eldridge wanted police to move jaywalkers back to the curbs, hoping that "the ridicule of passing motorists and bystanders" would reform them.[71] But when Eldridge left Washington for a two-week vacation, his deputy, I. C. Moller, changed course. "I think it is an unfortunate mistake to call pedestrians 'jay walkers,' " Moller said. "Neither motorists, policemen, nor anyone else should ridicule pedestrians. Our job is to protect persons using the streets, not humiliate them. Traffic can be regulated without demeaning citizens." Moller did not consider the motorist's claim to the street superior to the pedestrian's. He feared that Eldridge's anti-jaywalking campaign would give pedestrians "the idea that we are trying to slow them down in order to speed up motor vehicle traffic at their expense."[72] When Eldridge returned from his vacation, he and his deputy compromised. Eldridge dropped official use of the term "jaywalker," but ordered arrests of jaywalkers. "I hope they land in jail," he told reporters.[73] Police soon arrested 83 people, of whom only 45 appeared in court. The judge found all of them guilty but did not jail them, on condition that they become charter members of a "Careful Walkers' Club" he formed on the spot. All 45 took a solemn oath of membership, promising to obey pedestrian regulations. The judge then released them.[74]

Many pedestrians resented and resisted anti-jaywalking campaigns. In Grand Rapids the local safety council chief admitted that "some of the people were very indignant over it, as they naturally are when their personal liberty is interfered with."[75] A young couple there took resistance almost to the point of a civil disobedience campaign, crossing streets repeatedly "in every place but the crosswalks," collecting "about 15 blue warning cards" from Boy Scout enforcers.[76] In Detroit a judge campaigned for reelection by appealing to such dissenters. He attacked the notion that pedestrians walking in the street could be "reckless." "The pedestrians have a right in the street, however much reckless drivers insist to the contrary," he said.[77]

### Jay Walkers versus Jay Drivers

A St. Louisan, defending pedestrians' traditional rights to the street, tried to turn the "jaywalking" label against those who promoted it. "We hear the shameful complaint of *jay walkers*, to console *jay drivers*," he wrote. "It is the self-conceited individual who thinks people are cattle and run upon them tooting a horn." "Make every machine stop and wait," he demanded, "until the road is clear, and give precedent to people who are walking. The streets belong to the people and not to any one class, and we have an equal right, in fact, more right than the automobile."[78] Nine months later the *Washington Post* argued that "the jay driver is even a greater menace to the public than the jay walker," and in 1925 Washington's deputy traffic director I. C. Moller endorsed the term.[79] Meanwhile, in Chicago, a stage sketch called "The Jay Driver" amused audiences at variety houses.[80]

But promoters of the epithet "jay driver" failed.[81] Critics of motorists could call them cold-hearted, tyrannical, or selfish, but a motorcar's power, modernity, and worldly sophistication made its owner anything but a jay. But what of pedestrians who risked their lives to cross busy streets, without the protection of a police cornerman or a signal? Some such risks could indeed be foolish, and the element of reckless exposure to danger became part of the first dictionary definitions of "jaywalker."[82] Thus motorists' most persuasive claim to first place among street users lay in the physical threat they represented to other users.

Challenged by an automobile, most people on foot conceded the roadway (including crosswalks) as a matter of practical necessity, leaving aside finer matters of custom, right, or equity. This change in habits lent support to those who claimed that pedestrians did not belong in the streets. "The custom which made the common law is likely sooner or later to change it," one journalist observed.[83]

In 1920, when the wave of public safety campaigns was just beginning, "jaywalker" was a rare and controversial term. Safety weeks, more than anything else, introduced the word to the millions. Frequent use wore down its sharp edge, and it passed into acceptable usage as a term for lawless pedestrians who would not concede their old rights to the street, even in the dawning motor age. In 1924, soon after the intense publicity of safety weeks, "jaywalker" first appeared in a standard American dictionary. The entry officially gave the word its new, motor age definition: "One who crosses a street without observing the traffic regulations for pedestrians."[84] Many pedestrians continued to cross streets wherever and however they pleased, but by 1930 most agreed that such persons, when they obstructed vehicles, were jaywalkers. Except at crosswalks, busy streets in 1930 were for vehicles only.[85]

### The Safety Salesmen versus the Grundys

Promotion of the word "jaywalker" was only one arm of a much larger publicity effort intended to limit pedestrians' customary rights of street access. In 1928, Stanley Resor, president of the J. Walter Thompson advertising agency, contended that traffic accidents could be "advertised out of existence."[86] By then, the National Safety Council's leaders agreed. The NSC had long compared its task in public safety publicity to that of an advertising agency. "Public safety is a product which has to be sold to the public," explained an early NSC leader in public safety.[87] Success would require the best salesmanship techniques of twentieth-century marketing.

Lay safety reformers' amateur publicity did not fit this bill. Reformers in local councils annoyed NSC leaders. Their safety publicity was macabre, and some NSC members turned against it. At the NSC's 1921 convention in Boston, Laura Roadifer urged her colleagues to resist those who disseminated grim publicity. Roadifer was a council delegate from Philadelphia's Rapid Transit Company, where she was known as "Miss Safety First." She called such safety reformers "Grundys" who gave their audience "a mental picture of themselves maimed, or blind, or even dead." Roadifer's perspective, true to the original motives of the safety council movement, was practical. Advertisers had already found that grim publicity does not work. "Haven't you noticed," she asked her fellow council members, "how modern advertising, which is the greatest medium of salesmanship, features the happy, prosperous and joyful side of life"? She compared this approach favorably against old insurance publicity depicting the "sorrow-stricken home with grim death hovering near." Why can't "the 'constructive' method," she asked, "be just as successfully employed in selling safety?"[88] Miss Safety First concluded: "Let us eject the Grundys and adopt the constructive, happy method of teaching or selling safety."[89]

Some local councils followed the national organization's advice. In 1927, for example, Toledo put away its black mourning flag and substituted a white banner reading "*No* Fatal Child Accident This Month," displaying it as long as the month's record went unblemished.[90] Yet the autonomy of local councils and their dependence upon local support left the National Council with little influence over local methods. Nevertheless, in the drafting of the traffic safety message, new groups would soon overshadow the councils. They were more interested in ridding the traffic safety message of Grundyism and more capable of doing so.

### Protecting the Children: Safety Education and Pedestrian Responsibility

The National Safety Council's industrial safety publicity counteracted compensation laws by giving workers much of the responsibility for their own safety. But in traffic safety, local safety councils tended to ascribe street casualties almost exclusively to motorists and their machines; pedestrians were considered innocent. The law could not restrict pedestrians much. They had recognized rights to the street and enforcement was a practical impossibility. Yet as a practical matter many council members reasoned that they had to persuade pedestrians to stay out of the path of moving vehicles. By the early 1920s, councils everywhere were attempting to do just this. Safety councils cooperated in this effort with other local institutions and used their usual technique: publicity. They directed most of this effort at school children, who clearly needed the protection and who were less likely to object to infringements on their rights as pedestrians. In effect they gave pedestrians, especially children, a share of responsibility for their own safety. By the mid 1920s, through such efforts, pedestrians' unconditional innocence was lost. In recognizing the practical necessity of training pedestrians to avoid accidents, safety councils helped to foster the principle of pedestrian responsibility.

Schools were among the councils' most important patrons in public safety, and together schools and the councils pioneered safety education. They picked up where purely local efforts left off. From about 1906, with help from insurance companies, some schools began requiring lessons in fire prevention.[91] In 1915, New York's schools began inviting police sergeants to the classrooms to instruct children in "safety-first principles," including traffic safety.[92] A year later, a Cleveland principal, Annie Salter, introduced daily safety instruction in her elementary school.[93] Then, in fall 1917, Cleveland lost twelve school children in two weeks to street accidents, prompting the Board of Education there to introduce traffic safety into schools citywide.[94]

In St. Louis, E. George Payne introduced broader and more formal safety instruction in city schools in 1918. He devised a program of safety instruction adopted in numerous city school systems.[95] Detroit's Board of Education introduced safety education in 1919, organizing committees of teachers under a part-time supervisor. In the same year, the NSC formally adopted Payne's safety education plan and launched a program to supply schools nationwide with materials for safety education.[96] By 1920 Detroit had hired a full-time supervisor, Harriet Beard. Payne and Beard were perhaps the first two professional safety educators. Beard had to improvise; there were no textbooks and no authorities to which to turn. Teachers devised safety games. Beard invited Detroit police into the schools to instruct the children in safety. Police records showed that accidental deaths among children there were halved between 1918 and 1920.[97]

Most safety educators were torn between a conviction that automobiles were to blame for child traffic casualties and the practical necessities of accident prevention. J. Wesley Brown, a patrolman who worked with Harriet Beard, explained the position of the police and city schools. When city coroners ascribed responsibility for two-thirds of accidental child deaths to the children themselves, Brown and his fellow safety reformers found the claim repugnant. "We could hardly concede that," said Brown; "the children were not the ones to blame." Brown credited the coroners' statement with spurring the police to join the schools in their safety work. Yet, Brown admitted, "it was not the driver's fault always."[98] To safety reformers like Brown, children were innocent, but drivers were not always guilty. Many safety reformers in schools therefore gave up placing responsibility to turn to the purely practical matter of accident prevention. As schools elsewhere in the country introduced safety education, Detroit's experience was widely repeated. Educators who in principle held motorists responsible and who considered children unconditionally innocent found, like Beard and Brown, that to save children's lives they had to give children some responsibility for their own safety. Schools taught children to stay out of streets except to cross them, and to cross them only at designated places.

On their own initiative, teachers and school administrators devised games to inculcate safety habits. In 1919 a Cleveland schoolgirl described one: "Every morning at school . . . half of us pretend we are automobiles and half pretend we are people crossing the street. Those crossing the street must look first to the left and then to the right. The automobiles must go slow at crossings and sound their horns. Whoever doesn't do this, whether he or she is an automobile or a person is told not to go about without a nurse."[99]

Like safety weeks, school safety education efforts exposed a tension between local safety reformers and the professionals in the NSC. Local reformers tended to add the expression of grief and anger to the practical goal of accident prevention, and their warnings often included vivid or macabre depictions of the bloody consequences of accidents. Images of death or severe injury were sometimes practical warnings and sometimes evocations of woe, but in either case they were most often non-professional creations of the local laity of safety reform. Safety council leaders preferred to keep the practical goal of accident prevention unencumbered by emotional appeals, and they preferred that safety publicity keep away from blood and corpses. Later in the decade, safety publicity most often depicted settings of positive good cheer, of the kind found on the faces adorning cereal boxes.

An early, non-professional attempt to extend Safety First from industrial safety to child safety education was Lillian Waldo's book *Safety First for Little Folks* (1918). Waldo evidently emulated L. Frank Baum's fabulously successful series of Oz books. Instead of Munchkinland and the Emerald City, however, Waldo's fanciful places included "Careless Town" and "Danger Land." Their inhabitants included a "crippled army" of maimed children (among them double amputees).[100] A teacher in Newark, New Jersey, teamed up with a principal to write *Safety First Stories*, which brought home to elementary school children the dangers of street play. As a local, non-professional effort, the book was susceptible to the charge of Grundyism. A posed photograph shows two boys in a street in the aftermath of risky play. One boy lies face down by the wheels of a truck. The photograph's caption reads "THE LOSER."[101] A school in Allentown,

Pennsylvania, worked safety messages into its arithmetic word problems. Among these, for example, pupils read of a motorist who stopped to get gasoline at night. "Thoughtlessly the driver struck a match to see how near the tank was filled. An explosion followed. . . ." Students calculated the driver's costs for a destroyed vehicle and three weeks of burn treatments.[102]

Concurrently, however, the NSC was beginning to introduce classroom traffic safety materials free of such grim imagery. In the winter of 1923–24 the council issued traffic safety posters in which—pointedly—"actual accidents are not depicted." It launched a poster contest for school children to "stimulate the expression of safety ideas in positive terms." All submissions had to reflect "a positive and not a negative conception of safety."[103]

### Playgrounds and Fences

Streets were the playgrounds for most city children, but in the 1920s street games were becoming a high-stakes gamble. Nearly half of the children struck down in city streets were on their home block, a fact indicating that unsupervised street play was probably a much bigger risk factor than journeys to school or stores.[104] Leon Wartell was among the first of at least 10,000 children to meet their end in street play during the 1920s. One contribution to a newspaper's safety slogan contest was "Kids in the Street Make Sausage Meat."[105]

Many parents sent their children to the streets, and did not welcome police efforts to curb street play. For a time, New York City police arrested "small boys who have recklessly defied the perils of crowded thoroughfares," but they soon desisted because "It frightened and shamed the child and angered his parents and guardians."[106] "Children must play," a St. Louisan wrote in 1918, "and even in the more exclusive residence sections it is difficult to always keep a child out of the street. In other and more crowded sections, it is practically impossible."[107] Despite the dreadful toll of death and injury, some safety reformers defended children's right to play there. They objected to efforts to bar children from using the streets except to cross them at designated points. Since most people held motorists responsible for accidents involving pedestrians, and since pedestrians had extensive rights to the streets, forcing children off the streets could seem like making the innocent pay for the crimes of the guilty. A Milwaukee educator defended street play, even at the expense of inconveniencing motorists, as consistent with "the widest enjoyment of our streets for the greatest number." "Are streets for commercial and pleasure

**MOTHERS!**

Autos will kill 220 children and injure 5854 in Massachusetts this year unless you help prevent it.

**Don't let your children play in the street!**

MASSACHUSETTS SAFETY COUNCIL.
SAFE ROADS FEDERATION

**Figure 3.6**
Poster by Massachusetts Safety Council, reproduced in Lewis E. MacBrayne, "Saving the Massachusetts Child," *National Safety News*, July 1923, p. 36. Courtesy National Safety Council.

[automobile] traffic alone?" she asked. Street play was an exercise of the child's "inherent rights." She recommended barring automobiles from some streets and turning them over to children, so as to allow "precious childhood to enjoy its legitimate and God-given desire, play."[108]

To other safety reformers, such high-minded principles no longer made sense. Though reformers of both kinds wanted to save children's lives, their perspectives were entirely different. Without shared definitions, they lacked a common language. The principled and the practical-minded could not always understand each other. For example, in 1922 a principled Detroit mother asked a deputy police commissioner "What are *you* going to do to make the streets safer?" The practical-minded commissioner replied "*You* can help make streets safer for your children by seeing that the children always use the crosswalks."[109]

Pragmatic safety reformers also worked to keep children from playing in the streets. "We cannot expect to limit accidents," explained Harriet Beard of Detroit, "unless we provide playgrounds for the children."[110] Early Progressive reformers had begun a playground movement for the moral and physical development of children even before automobiles came to cities, but the surge in accident casualties lent it a new, more practical justification.[111] By 1918, city planners were warning cities that "one of the chief causes of loss of life on public streets is due to the fact that many city children have no other place on which to play," and already local organizations were building such places off the street.[112]

About 1922 a sharp rise in child traffic casualties inspired a new playground movement still greater than its Progressive Era forebear. The Playground and Recreation Association of America organized local efforts; some chambers of commerce and rotary clubs funded them.[113] Detroit's 1922 safety parade featured a float with a real playground, with children playing on swings and slides.[114] Safety reformers taught children new traffic safety games on many of the new playgrounds. The Memphis Safety Council devised a game for its playgrounds, called "Traffic," which introduced children to the new idea that motorists were not usurpers of pedestrians' rights. Instead, the game assumed "the equal rights of motor vehicles and pedestrians in the highways," and taught children "to observe and obey traffic signals." Teams represented pedestrians and motorists; children competed to have the fewest "casualties."[115]

Suppliers of playground equipment encouraged the trend. "Have your empty lots made into playgrounds where the kiddies can have a safe place to exercise and play away from dangerous traffic," one equipment distributor suggested.[116] The Everwear company, a playground equipment manufacturer, distributed a free pamphlet to show civic groups how they could start local playground movements, and company advertisements sold equipment as a way to keep children out of traffic.[117] Fence manufacturers joined them by selling their product as a way to keep children out of the streets.[118] Auto clubs also recognized the need. When Youngstown's Playground Association arranged for the closing of some streets for the exclusive use of children, the local auto club backed the plan and alerted its members to the closings.[119] The new playgrounds facilitated children's withdrawal from the streets, slackening the child traffic casualty rate, and limiting the claim of a class of pedestrians to street access.

### Safety Patrols

Even if children could be coaxed into playing elsewhere, they still had to cross streets. By 1920 schools had begun to assume particular responsibility for the safe journey of children to and from school, either by taking on the job directly or by working through safety councils. Police could not protect children at every crossing near schools.[120] As Detroit schools organized for traffic safety in 1919, a first grade teacher arranged for parents to serve as crossing guards. Teachers devised some of the playground games that taught safe street-crossing practices.[121]

Schools introduced another new method of child protection. It saved lives by giving children more responsibility for their own safety. Its origins are detectable as early as 1913, when Brooklyn's schools and the local electric railway organized the Children's Safety Crusade. As part of the campaign, schools organized "safety patrols" to protect children on their journeys to and from school.[122] At about the same time, a spate of child traffic casualties in Rochester, New York, prompted similar steps there. Rochester schools had "boy's clubs" in the fifth and higher grades, and one of them formed a safety committee that stationed boys near the schools, like Brooklyn's safety patrols. In 1915 the local safety council (a committee of the local Chamber of Commerce, affiliated with the National Safety Council) extended the safety committees to boy's clubs throughout the city. These separate committees were then united into a city-wide "Junior Safety Council," based in the Chamber of Commerce's offices.[123]

After World War I, the safety patrol idea spread rapidly to cities of all sizes.[124] At first Boy Scout troops often cooperated with safety councils to supply schools with patrols. In the early 1920s, however, most patrols were pupils organized in junior safety councils affiliated with local safety councils, and these displaced the Scout patrols.[125] The junior safety councils usually had a formal affiliation with the local Chamber of Commerce. Some of their patrols acted on the prevailing view that pedestrian safety was a matter of controlling motorists. In Omaha, for example, where schools and the Chamber of Commerce organized boys into patrols, the boys reported offending drivers to the Chamber, which then sent letters to drivers notifying them of their offense. Upon a fourth infraction, a motorist was subject to arrest.[126] Some boy patrols were empowered to regulate motor traffic in the interest of child safety.[127] In Flint, Michigan, schoolboys were organized in "junior traffic squads" under police department direction. The boys could direct traffic in the street and report motorists to the police. Flint's police department promised to handle "any violation of traffic regulations" by motorists, and any failure to heed the boys' directions, "in the same manner as in the case of a regular traffic officer."[128] Elsewhere, however, patrol organizers emphasized children's responsibility for their own safety.[129] The boys and girls in Cincinnati's young "Safety Guards," for example, pledged to work for their own safety and for the safety of their schoolmates.[130]

Classroom safety education, playgrounds, fences, and safety patrols, together with safety weeks, helped prevent accidents to children. Already in the mid 1920s there was measurable progress, even as adult casualty numbers grew.[131] The trend was an indispensable foundation of the automotive city, for the number of child deaths had been too great (and the consequent controversy too intense) for the automobile win a secure place in city streets. As adults began to constitute a larger proportion of the dead and injured, the innocence of the automobile's victims grew less obvious. It grew harder to make the moral drama of a villain and its victim convincing. These trends did not erase the threat to cars in cities, or widespread resentment at their intrusion. Most safety publicity remained macabre. Funds for school safety education were scarce. Calls to restrict cars for the sake of safety remained common.

## Councils and Clubs: From Collegiality to Estrangement

Beginning with the first city-wide public safety campaign (Syracuse's, in 1913), local automobile clubs backed public safety campaigns.[132] In the late 1910s, auto clubs joined the coalitions backing local safety councils. As vilifiers of the automobile joined safety councils, auto clubs competed with them to shape the character of city safety campaigns. They worked to promote the idea that at the bottom of the accident problem lay the bad habits of a small minority of motorists—and of some pedestrians. By keeping these culprits off the streets, and with better pedestrian control, the clubs hoped to make streets safer, and remove a blot on their reputation.

In the 1910s and the early 1920s, auto clubs, safety councils, and chambers of commerce joined together to plug the gap between overtaxed police and the sudden influx of automobiles by forming their own supplementary enforcement bands. In the manner of other unpaid citizen police forces, the groups were often called "traffic vigilantes." Evidence of a proposal for such a group of supplementary traffic law enforcers goes back to 1913.[133] In 1916, to make Berkeley, California, a "safe city for women and children," the mayor deputized volunteer "Citizen Police" into a force 75 strong. They were empowered to track down "the reckless and ignorant automobile driver" and the "speed maniac" and do to them "what the Vigilantes did to the lawless gunman and thug in the early days of San Francisco." Though summary executions were strictly out, the Citizen Police could make arrests. Most of them were business or professional men, and most owned automobiles.[134] About the same time, the Chamber of

Commerce in Minneapolis gathered informal reports of traffic violations from citizens, and notified violators of their infractions. In 1919 the Minneapolis chamber stepped up this enforcement effort by joining with the police department to deputize unpaid "special traffic officers." In their first fourteen months the "specials" had more than 25,000 violators brought before the police, 3,365 of whom were convicted and subjected to penalties ranging from $1 fines to 90-day sentences in the workhouse. By 1920 the specials had cut traffic violations by half.[135] That year, the police chief of Newark, New Jersey, organized a "Citizens' Traffic Squad," which was much the same kind of organization as those in Berkeley and Minneapolis; similar volunteer traffic enforcers appeared in other cities.[136]

Chambers of commerce typically organized vigilantes through local safety councils.[137] This trend lent some national uniformity to the numerous local traffic vigilante organizations. By 1923 there were "vigilante traffic squads composed of business men organized under police direction" in "many cities."[138] Together, for safer streets, local chambers and the local safety councils narrowed the wide gap between traffic law and enforcement.

Until 1923, auto clubs often backed such vigilante work. Clubs were reluctant to leave enforcement to police alone. To the clubs, the police were at best biased against motorists, and clubs accused police of harassing motorists simply to raise money from fines. "Some of our cities live off the motorist," one auto club executive complained. "The fine is what they are interested in."[139] Police were, furthermore, attached to a notion that was very unpopular in the clubs: that to be safe in cities, automobiles must be confined to pre-automotive speed limits. The clubs saw their opportunity in the traffic vigilantes. If auto clubs backed traffic vigilantes where they existed and organized them where they did not, they could weed out reckless drivers, and thus improve the standing of the law-abiding majority.

Auto clubs and safety councils worked together to organize vigilantes. In Minneapolis the auto club was an important part of a coalition of support for the local vigilantes in 1919.[140] In Milwaukee the vigilantes enjoyed the "hearty endorsement" of the auto club.[141] In 1920, Robert Lee of the St. Louis Automobile Manufacturers and Dealers Association began a major vigilante effort in his city, designating himself "Chief Vigilante." Lee worked through the local safety council to organize about 325 vigilantes who patrolled city streets in their cars, incognito. They reported about 600 motorists a week for various offenses, concentrating their efforts on reckless drivers. Lee preferred to recruit businessmen to his vigilance committee, explaining that "the bigger a man is in the business world the more anxious he is to help make his city the safest to live in."[142] There was similar cooperation in Cincinnati, where auto dealers, through the safety council, furnished plainclothes police with automobiles. The resulting crowd of motorists summoned to police court won the safety council "a lot of favorable publicity."[143]

Until 1923, auto clubs also joined with safety councils and chambers of commerce to back safety weeks. In 1920, Milwaukee's auto club backed a chamber-council safety week, even though safety stickers used in the campaign alleged that motorists were "largely responsible" for deaths of children in the streets.[144] In New York City's 1922 safety week an auto club sponsored the motorized parade of maimed children.[145] A few auto clubs helped organize school safety patrols, but most patrols were organized under local safety councils as "junior safety councils." The National Automobile Chamber of Commerce, the voice of all major manufacturers except Ford, had no safety committee and very little to say on the subject. Change came suddenly, however; by the middle of the decade, motordom withdrew from its alliance with safety councils and rivaled them as one the two strongest voices in traffic safety.

## 1923–24: Bad Sales and Bad Publicity

Disastrous traffic casualty numbers were behind the change. In 1922–23, deaths rose 20 percent. Since 1918 no year has ever had a sharper increase in the traffic fatality rate.[146] At the same time, automotive interests were organizing rapidly, and thus their capacity for taking a coherent position on traffic safety was growing. Motordom was losing its enthusiasm for the existing traffic safety movement, both in its popular manifestations (for example, in the newspapers and in legislatures), and in its leadership (the NSC). It was also looking for an explanation for a sales slump in 1923–24.[147]

The auto industry blamed the slump on several factors, including growing sales of used cars and market saturation. But they also blamed the car's growing reputation as a ruthless killer. In 1922 a St. Louis auto dealer warned that "every automobile accident, whether serious or minor, causes a sales resistance."[148] Others agreed. Detroit's former police commissioner was well acquainted with the situation in the streets, and he saw what it portended. In the spring of 1923 he warned a convention of auto club representatives that "the loss of life and injury to persons and property is causing prejudice against the car owner that can only be mitigated by a reduction in the number of accidents."[149] In June the trade journal *Automotive Industries* worried that "to some extent traffic conditions and the numerous accidents are retarding sales."[150] Yet summer always promised bigger volume, and the industry took little action.

In December the industry heard the warning again. This time the audience was much larger, and the source was the most prominent public safety expert in the country. Sales had already soured, so the warning was heeded. Charles Price had left the National Safety Council and was working as a public safety consultant in New York. He had grown disenchanted with the local safety weeks he had systemized in 1918. By accepting safety reformers' lurid and emotional attacks on the motorist and the automobile, safety councils had failed to preserve the industrial safety movement's standards of dispassionate professionalism. Price objected to safety reformers' habit of demonizing the automobile, contending that "the machine itself is safe enough." Ahead of the auto industry, Price saw that traffic accidents were sullying popular perceptions of the automobile, and he saw an opportunity in the need to clean up its image.[151] One cause of the unexpected drought in the showroom, Price said, was the car's bloody reputation. "As a result of the accident situation a reaction against the purchase, use and toleration of motor vehicles has already set in." "There is a great danger," he added, "of large numbers of people beginning to look on the automobile as more of a menace than a blessing." The public was "cynical"—in part because safety week publicity reinforced anti-automobile prejudice, and people were therefore forcing "restrictive and 'half-baked' legislation and regulation" on the motorist. "Family men—and to a greater extent, women—are refraining from buying cars because of the fear of accidents," Price claimed. "The actual accident experiences of 365,000 persons a year" were creating formidable "sales and advertising resistance." Price even knew of a large trucking firm that was "displacing motor vehicles with horse teams for this same reason."[152] The persistence of the sales slump into 1924 seemed to confirm the truth of Price's warning that traffic accidents—and the consequent bad publicity—were keeping customers away. As if to confirm the claim, in April 1924 an Ohio Chevrolet dealer wrote to President Calvin Coolidge about the sales slump, letting him know that "this condition came about by unjust propaganda."[153]

Price offered a way out. The industry should lead the safety movement, he said, and redefine it so that automobiles and motorists themselves would no longer have to shoulder all the responsibility for the bloodshed. They must find a way to reduce accidents without regulating the automobile into uselessness. "The whole problem of the accident situation is still in the formative stage," he explained, "awaiting the leadership of some group of interests." For any who missed his drift, he then named the interests: "the automotive industries." They could give people "the right attitude toward the motor vehicle." If the industry could reverse "this unfavorable attitude" among the public, "an even greater number of cars, tires, parts and equipment could be sold each year."[154]

In short, Price was telling automotive interest groups to reconstruct the traffic safety problem. They listened. An automotive engineer accepted Price's case, warning "drivers and makers of automobiles" that they "must solve" the accident problem "or face increasing public opposition."[155] An industry journalist warned that "many persons who could afford to own a motor car . . . will decide that it is better to spend Sunday afternoon on the front porch rather than risk life and limb on the road—or run over a pedestrian."[156]

Auto clubs took the lead in attacking the popular moral drama in which cars and drivers were the villains. Zack Elkins of the Chicago Motor Club expressed a common complaint among his auto club colleagues. "Every time there is a fatal automobile accident newspapers play it up with scare headlines and frequently paint the motorist as a bloodthirsty maniac bent on destruction," he said. "Motorists are being listed as vampires who annually snuff out the lives of some twenty thousand citizens." The "good will" of the public "is being injured by this mass attack of the newspapers."[157] An Ohio auto club executive, Charles Janes, explained the cost of this public relations fiasco: "The average city council and city official will ride with the tide of popular sentiment as reflected in the newspapers." As a result, "popular hysteria" about the bloodshed was leading to stifling "ordinances and rules."[158]

## "The Safety Fight"

Meanwhile, auto clubs were growing disenchanted with the safety councils and their toleration of the auto vilifiers. In 1925, Charles Janes took his objections to a national convention of local and state club leaders hosted by the American Automobile Association, where he found a very sympathetic audience. "Like the clubs in many other cities," Janes told the gathering, "the Cleveland Club has found the traffic situation hopelessly muddled, due largely to the activities of the local safety council." "For years," Janes continued, "the Safety Council had been spreading its propaganda, taking always the attitude that nearly all accidents were caused by automobile drivers; that the majority of drivers were reckless, or drove too fast, or were careless, or incompetent. The pedestrian, of course, was held up as the innocent victim."[159]

During a 1923 safety week in Louisville, the local auto club also grew disgusted with the safety council. The club president accused safety reformers of attacking motorists "as the exclusive cause of our motor death rate." Following the example of Baltimore, Pittsburgh, and Washington, Louisville erected a monument to its child accident victims. The local safety council, the Louisville and Nashville Railroad, and city club women collaborated on the project. The monument's inscription honored the 25 child victims of *all* accidents in 1922, but the auto club president accused the monument's builders of permitting "the impression to prevail that the monument" was "dedicated to the city's death toll from automobiles" alone. When a newspaper headline called the memorial a "Monument to Auto's Youthful Toll," the auto club president's suspicions were confirmed.[160] Later, at the monument's dedication, a clergyman intoned in prayer against "the deadly automobile speeding up and down the streets."[161]

Some auto clubs resorted to organizing safety campaigns of their own, outside of safety council auspices. The councils did not welcome the help. Three auto clubs around Davenport, Iowa, united in such a campaign in 1921, when club-council relations were still generally excellent. The clubs' move, however, led a Davenport safety council member to complain "we don't want the system that we have devised jimmed up now by outsiders who assume they know as much as we know."[162]

A more immediate threat from safety councils also angered leaders of auto clubs. In the first half of the 1920s an inconsistent maze of local, regional, state, and national automobile clubs organized themselves into an orderly national hierarchy under the leadership of the American Automobile Association. By 1924 the AAA was calling the new movement "organized motordom," embodying "one national association, one national policy and one national service."[163] Yet even then, groups outside the AAA occasionally organized motorists. The AAA watched vigilantly for these renegade groups, and attacked them repeatedly as "gyp clubs."[164] Most of them were local bodies that competed with the AAA to serve and represent motorists. Gradually, however, one rival loomed above all the rest—a kind of national "gyp club" called the National Safety Council.

To influence motorists and to gather intelligence from them, safety councils in the early and mid 1920s went directly to them. This offended and threatened the AAA and its member clubs, since they considered themselves motorists' sole representatives. Fact gathering, for example through surveys of workers, was an elementary technique in all durable industrial safety campaigns. It showed safety councils where the dangers lay, and it gave them workers' own suggestions for remedies. In public safety campaigns, safety councils soon applied similar techniques, but in doing so the councils exposed fissures in the safety campaigns' backing.

During its 1923 safety week, the St. Louis Safety Council sent about 100,000 questionnaires to motorists in and near the city. Local government offices supplied street addresses, and the mayor signed the questionnaire's cover letter. To the auto clubs of St. Louis and Missouri, however, the St. Louis Safety Council was intruding on the clubs' territory. In assembling the list the council was taking the first steps toward organizing motorists, and thus directly competing with the clubs. The council's action was also a threat to the auto clubs' prestige. In bypassing the clubs, councils implied that auto clubs fix flats, but do not represent motorists' views or solve the larger social problems of the automobile. The council's organizing effort, moreover, carried the official sanction of local governments, giving them an advantage over the private clubs. Roy Britton, president of both the state and the city clubs, denounced the council's "interrogation" of motorists on "unrelated phases of traffic regulation" in the press, calling it "untimely and ill-advised."[165]

Still more alarming were local councils' efforts to recruit motorists as dues-paying members of "safe drivers' clubs." The threat caught auto clubs unawares. During Milwaukee's safety week in 1920 the safety council there (operating as a division of the Chamber of Commerce) organized 12,000 motorists in the city into such a club, with $1 memberships.[166] Some of the money went to a driving school and the traffic vigilantes.[167] The local auto club, far from objecting to another group organizing motorists, backed the move. The auto club's chief served as one of the new Safe Drivers' Club's organizers. The Milwaukee Automobile Dealers' Association even signed up all of its members, and a local motorist predicted "the hearty cooperation of all auto owners who have any regard for the lives of their neighbors."[168] St. Louis emulated Milwaukee's example. By 1921 the Safe Drivers' Club there had 5,000 members. It put some of its money into a "safe drivers' school" for commercial drivers.[169]

But traffic casualties, the bloody accident publicity of the safety councils' safety weeks, and the sales slump put an end to such teamwork. Meanwhile, local safety councils organized many more safe drivers' clubs. Regular auto clubs began to suspect the safety councils' motives. An officer of the NSC admitted that safe drivers' clubs "furnish a not inconsiderable addition to the local council's income."[170] The St. Louis Safety Council's Safe Drivers' Club gave members very little to show for their money. The

**Figure 3.7**
In 1922 this young woman recruited motorists to join the local safety council's Safe
Drivers' Club in Milwaukee. Source: C. M. Anderson, "How Organization Brought
Us 15,000 Safe Drivers," *National Safety News* 6, no. 2 (August 1922), p. 12.

council privately admitted that in its first year its club had "not been able
to keep up a personal contact" with members, who therefore knew nothing
about its work "except what they read in the press."[171] When the St. Louis
Safety Council nevertheless persisted in soliciting memberships in succeeding years, auto club leader Roy Britton wrote a letter of protest to the
council, releasing a copy to the press. Britton publicly accused the safety
council of forming the "so-called 'Safe Drivers' Club,' " with its $1 member-
ships, as a way "to raise funds for the Safety Council." Britton objected
especially to "the solicitation of the members of the Automobile Club of

Missouri," because "one of the functions of the [auto] club is to promote safety."[172]

Safety councils were also early leaders in driver education. Here, however, they tended to concentrate their efforts in training commercial drivers (especially truck drivers), causing less offense to the auto clubs.[173] Nevertheless, as auto clubs grew disenchanted with the safety councils over other matters, they began to compete with the councils to lead in driver education. Finally, in the 1930s, the American Automobile Association coordinated a national driver education program, supplying schools with books and materials.[174]

The battle over safe drivers' clubs, in St. Louis and elsewhere, eroded a once cohesive traffic safety movement, in which auto clubs had cooperated in a council-led coalition. At Milwaukee in 1920 the local auto club played on the council's team. As early as 1923, however, the auto club in St. Louis was refusing to follow the local council's lead, and was starting to act like a free agent. Elsewhere, turf wars between councils and clubs worsened the tensions. In Wilkes-Barre, Pennsylvania, the auto club's leader, Norman Johnstone, complained: "The National Safety Council is simply duplicating the work of the automobile clubs, and in some cases they are becoming a nuisance in regard to posting highways which are already being posted for safety by the motor clubs." Johnstone urged his colleagues at the AAA to pass a motion condemning the council. AAA executives squelched the idea. Instead they told member clubs to "win the safety fight" at the local level with effective safety campaigns of their own.[175]

Thus a growing rivalry between safety councils and auto clubs prodded motordom to take a more independent role in shaping the traffic safety debate. A greater threat, however, drove it to organize to reconstruct the safety problem, as Price had urged. It came not from the safety councils but from lay safety crusaders.

## The Speed Governor War

"If safety is not realized," a Milwaukee newspaper editorialized in 1920, "there will come a time when the public will demand drastic action—no matter how hard it may hit the pocketbooks of automobile manufacturers, dealers and owners."[176] If the great chorus of critics of the automobile was right—if, indeed, the pedestrian had unlimited rights to the street and the automobile was inherently dangerous, then dire restrictions of automobiles in cities were needed.

That speed made automobiles deadly was the prevailing popular view, and many in the safety councils accepted it. The industrial safety movement had formulated the precept that "speed must always give way to safety."[177] When Charles Price first guided the National Safety Council into national leadership in traffic safety, he declared that "automobile traffic must be slowed down and controlled until it becomes safe." At a time when official downtown speed limits averaged ten miles per hour, Price called for "reduction of the speed limit, especially at crossings."[178] In a full-page newspaper advertisement, the incipient Milwaukee Safety Council (then only a committee of the Chamber of Commerce) blamed accidents on a "never-ending call for speed."[179]

In the early 1920s, few remedies besides restriction of automobiles were even considered. In solidarity with Milwaukee's 1920 safety week, the local Yellow Cab company ordered its drivers not to exceed 20 miles per hour.[180] The editors of the *St. Louis Post-Dispatch* in 1923 were blunt in calling for regulations of this kind. "This dreadful slaughter must be stopped," the newspaper argued. "If necessary, regulations severe and searching enough to do it must be adopted and enforced." There was no sympathy for motorists. "If reasonable safety of life and limb can only be had by impairing the motor car's efficiency the motor car will have to pay that price."[181]

More specifically, the *Post-Dispatch* warned that "the automobile seems to be hurrying along to mechanical restrictions that will reduce its speed."[182] Many city people in this period demanded that motorists be required by law to equip their cars with speed governors. One letter writer to the *St. Louis Star* suggested "gear them down to fifteen or twenty miles per hour and quit joking about speed limit laws"; another suggested "equipping" cars with "some sort of governor" to limit them to "fifteen miles per hour."[183] Police departments were sympathetic to such proposals. In a survey, two of every three city police chiefs agreed that their cities should require governors on automobiles.[184]

Cincinnati had an unusually bad safety record. Like city people elsewhere, most Cincinnatians blamed motorists, and many were prepared to force them to limit their cars' speed mechanically. In 1923, 42,000 people—more than 10 percent of the city's total population—signed petitions for a city ordinance requiring local motorists to equip their cars with governors that would shut their engines off at 25 miles per hour.[185] But they ran into fierce resistance.

Though local, the Cincinnati speed governor war of 1923 brought home to motordom its danger more powerfully than any ink-and-paper warning could. It frightened an industry. It convinced it to give up hope in the

# VOTE "YES"
## On the Ordinance to Curb Speeding

*Which Shall*   A Limit of 25 Miles Per Hour and
*It Be---*         SAFETY
              ----- or -----
            No Limit and the Lurking Danger
                    of DEATH!

**Figure 3.8**

Advertisement in *Cincinnati Post,* November 1, 1923. Courtesy *Cincinnati Post.*

prevailing definition of traffic safety, and to do battle against those who advanced it. Motordom mobilized to fight the threat, and in so doing it formed new, well-funded safety institutions that reconstructed the safety problem.

The enemies of speed faced many practical difficulties. A speed governor would be easy to tamper with or to disconnect. Out-of-town motorists would have to be exempted. Cincinnati was hilly; to keep under the mechanically enforced speed limit, motorists coasting downhill could be forced to brake all the way. City motorists traveling outside Cincinnati would still be confined by their sealed governors, even on straight, rural, hard-surfaced roads. Police would have to find time and staff to inspect governors and punish offenders. Nevertheless the measure won enough backers to appear on the ballot.

Opponents of speed governors organized. Cincinnati's daily newspapers fiercely opposed the plan. The leading daily, the *Enquirer,* may have objected to the idea on its merits, or it may have been keeping the sponsors of its lavish Sunday Automobile Section content; in either case, editors repeatedly urged Cincinnatians to fight the initiative. The Automobile Section was the voice of Cincinnati motordom, and its columns show them evolving from isolated bands into a unified front as election day approached. Opponents ignored the local safety council; the Automobile Section proclaimed that "the automobile clubs" and "motor car dealers" were the "traffic experts of this city," then revealed to readers that these experts unanimously opposed a governor ordinance. According to the Automobile Section's transportation experts, the problem was that "every once in a while some whang-doodle of a charlatan jigs open his wide bazoo and crys aloud the damnation of the automobile."[186]

The Cincinnati Automobile Dealers Association organized a "General Citizens' Committee" of local auto interests, raising $10,000 for it. The Citizens' Committee was an early example of a trend in which groups

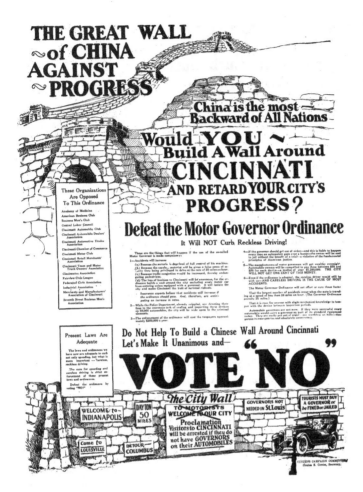

**Figure 3.9**
Advertisement by Citizens' Committee, *Cincinnati Post*, November 5, 1923. Courtesy *Cincinnati Post*.

organized by motordom gave themselves names that concealed their parentage.[187] The committee sent a letter to every motorist in the city denouncing the plan as "the most vicious Ordinance any community has ever been asked to vote upon." Besides urging readers to vote No, the letter asked them to send money to the Citizens' Committee.[188] The secretary of the Cincinnati Automobile Club, A. E. Mittendorf, mobilized precinct workers for the Citizens' Committee. He stationed 400 people (including "every girl employed by the Cincinnati Automobile Club") at the polls on election day to buttonhole voters and press them to vote No.[189] On November 6,

1923, the proposition was crushed by more than six to one. Although 42,000 had signed petitions to put it on the ballot, the plan won only 14,000 votes.[190]

Well before election day, the speed governor proposition was creating a stir in offices of motordom far from Cincinnati. *Motor Age*, a national organ for the automobile trade, treated the fight as a warning. Local auto interests throughout the United States had to unite and set forth their own plan for better traffic: "Do not wait until you are put on the defensive with 42,000 ignorant and angry people in a league against you."[191] Three weeks before election day, Alfred Reeves, general manager of the National Automobile Chamber of Commerce, visited the city to rally local auto interests. Reeves urged them to use their influence not only to fight the speed governor proposition, but to lead the traffic safety movement in a new direction.[192]

## The Battle for Leadership in Traffic Safety

Motordom mobilized to fight the speed governor war. After it won, it never returned to a peacetime footing. Institutions and cooperative arrangements formed during the fight persisted and grew. One day after the speed governor vote, the National Automobile Chamber of Commerce formed a Safety Committee.[193] Its chairman was George Graham, vice president of the Chandler Motor Car Company of Cleveland. Graham said the committee would seek "ways and means of making it easier for vehicles to operate on city streets and to make the streets safer for pedestrians."[194] In time, with growing funds, it evolved into the leading national institution of traffic safety.

The Cincinnati speed governor initiative showed motordom how easily people equated speed and danger. The formula was a threat. It kept the burden of responsibility for accidents on motorists, and it inspired campaigns to restrict automobiles' speed, depriving them of their chief advantage over other modes. A city judge, sympathetic to motorists laboring under a ten-mile-per-hour limit, put the industry's problem clearly. Such a low speed limit "stripped the automobile of all its efficiency. You might better return to the horse-drawn vehicle days and do away with all the hazard and all the danger than to try and drive automobiles or have automobile trucks driven for you at no greater rate of speed than 10 miles an hour."[195] An auto industry executive later explained that "the motor car was invented so that man could go faster" and that "the major inherent quality of the automobile is speed."[196]

Together, the safety councils' safe drivers' clubs and the Cincinnati speed governor initiative drove motordom to abandon the councils' traffic safety coalition and strike out on its own. During the speed governor war, many AAA leaders for the first time privately agreed that auto clubs would have to wrest leadership in traffic safety from the safety councils. In September 1923 the chief of Pennsylvania's auto club complained that "an organization of business houses" (such as a Chamber of Commerce, with its NSC affiliation) "cannot decide on how motor accidents are caused and what remedy can be used to reduce them." Instead, "It is the function of an automobile organization to work for safety matters."[197] Yet to challenge the safety councils' leadership openly was risky. AAA leaders, fearing an open break, worked to keep the ill feeling out of view.

In 1925 the AAA's general manager, Ernest Smith, still counseled caution to the impatient leaders of member clubs.[198] Yet those who looked could see the tension. The AAA members' magazine, *American Motorist*, published a thinly veiled hostile reference to "a certain civic organization" in Cleveland that was given to "diatribes against the motorist, with his 'juggernaut of death.' "[199] A year later, Secretary of Commerce Herbert Hoover warned Smith to keep the animosity under control, noting that "these jealousies amongst safety organizations are highly destructive."[200]

In 1927, Ernest Smith broke ranks with the NSC openly. Although the councils had been ahead of the clubs in traffic safety, Smith said the councils instigated the turf war. "They are invading our field," he complained. "We have told them frankly that they are treading in our field and that we propose to continue to expand our safety work, and that if they go into a town where our motor clubs are operating they may expect strenuous opposition." He urged club leaders to fight for leadership in traffic safety. "It is an obligation upon the motor clubs of the country to get into safety work in greater measure, particularly in view of the fact that the National Safety Council are treading upon our heels in their endeavor to take over that field of work." There was no time to spare, Smith said. "If our motor clubs which have not already adopted safety work do not do it very quickly, the National Safety Council will have their organization in there, and they do a very high grade work and are composed of fine men."[201]

Frightened into action by safety council rivalry and popular constructions of the safety problem, the National Automobile Chamber of Commerce, the American Automobile Association, and others in motordom mobilized to join the safety fight and to advance alternative models of traffic safety. The effort required coordination, money, and organization. None were forthcoming before 1923. Bloody safety publicity, poor sales, and the speed governor war changed everything. Within a year motordom had organized and its members were negotiating common principles. They fought to overcome the rhetoric of justice deployed against them, using rhetorical counterattacks that would put safety crusaders on the defensive. Motordom found ready allies in some of the more pragmatic safety reformers, who agreed to measures of pedestrian control and child safety education. But motordom soon developed new traffic safety principles as well.

Motordom's new self-confidence was also clear in its response to another kind of traffic problem. As it perceived threats in prevailing constructions of the accident problem, motordom discovered that engineers' construction of traffic congestion also limited the car's urban future. The origins of traffic engineering, the application of engineers' efficiency principles to city traffic, and motordom's growing alarm at the implications of the efficiency model are the subjects of part II.

# II  Efficiency

Scientific organization of traffic . . . could cut traffic congestion at once by half.

—J. Rowland Bibbins, "Traffic-Transportation Planning and Metropolitan Development," *Annals* (American Academy of Political and Social Science) 116 (November 1924), 205–214 (212)

Tourists gazing at Niagara Falls from the safety of the railing do not all see the same thing. Some watch the American falls, others the more curvilinear Canadian. Some look on as the dark gray-green water heads horizontally for the precipice, and only hear the thunder of its descent; others see the vertical falls through white spray. Yet these tourists all share a common experience as observers of the falls at a safe distance. Their perspectives have more in common with each other than with the point of view of tourists aboard the *Maid of the Mist.*

Like the tourists at the railing, pedestrians, parents, and police shared a common general perspective, even though they did not all see the traffic problem the same way. Most were, in a sense, conservatives: they compared street conditions as they found them against their experience of the street before automobiles crowded the pavement, and saw their mission as restoring pre-automotive order. Cars were like a herd of unruly children intruding upon a grownups' dinner party. The adult guests will regard the children as raucous interlopers and do what they can to restore decorum.

But there were other, quite different perspectives. Some in the city welcomed change as "progress," and called cities that failed to adapt to it "backward." "Progressive" cities sought no return to the nineteenth century; for twentieth-century problems they wanted twentieth-century solutions. In cities this perspective belonged, above all, to business associations that wanted a vibrant commercial future downtown. The rise and the decline of their construction of city traffic problems are the subjects of chapters 4–6.

Downtown business associations—which usually styled themselves "chambers of commerce"—came somewhat later to city traffic problems than pedestrians, parents, and police, in part because for a time they hoped police would rise to the challenge. In time, however, chambers of commerce discovered that police did not see the problem the way business did. Chambers were not attached to "order" for its own sake. To protect their investments downtown, they wanted traffic to keep moving. In their quest for order, police did little to speed traffic and often expressly sought to slow it down.

In response, business associations gradually articulated a different construction of the traffic problem. Either as allies or as members, diverse transportation interest groups, including street railways and auto clubs and dealers, generally backed the business associations' cause. Their common enemy was not so much accidents and disorder as congestion. Traffic jams threatened to strangle downtown commerce. Business wanted a way to keep traffic density high while keeping vehicles moving. What they wanted, in short, was traffic efficiency.

Efficiency experts were ready to hand. Most called themselves engineers. By 1920, when business was turning to them to keep traffic moving, engineers already had a long and impressive record achieving efficiency in diverse modern city services, and business was confident they could do the same for street traffic. Pooling their resources, downtown businesses turned to engineers for answers and funded their solutions. In the process, they together invented a new professional discipline: traffic engineering. The new traffic engineers called their traffic efficiency technique "traffic control." Chapter 4 traces the traditions traffic engineers drew upon to form their own, new approach to city traffic problems. Chapter 5 looks more closely at traffic control in practice, with particular attention to traffic signals and curb parking.

Just as the police quest for order gradually prodded business groups to join in the effort to shape the city traffic problem, engineers' quest for efficiency soon frightened other groups into joining in. Automotive interest groups, like downtown business associations, were unhappy with police regulation and wanted to fight traffic congestion. Thus they joined with business associations to back traffic control. But under the traffic engineers' efficiency model, automobiles were a poor match with cities. Traffic control measures could thus seem hostile to motorists. Confronted with traffic control, therefore, those who wanted a brighter future for cars in cities organized to propose another new construction of the city traffic problem. The origins of their model and its early competition with the dominant traffic control model are the subjects of chapter 6.

# 4 Streets as Public Utilities

We are trying to accommodate human nature to new physical facts.
—*Engineering News-Record*, 1924[1]

As they began their work in the 1910s and the 1920s, traffic engineers did not question the prevailing social construction of city streets as public space. Nevertheless, they attacked traffic problems from a new angle. Backed by downtown business associations, they first redefined their problem as a technical matter for experts. This was new; there was as yet no discipline or body of accepted knowledge on modern urban street traffic. Though they conceived of streets as public spaces, their model was less the busy city park than the overburdened city service or public utility. Drawing from their experience in other technical problems of municipal service, engineers seldom saw their enemy as injustice or disorder. Instead, they attacked inefficiency, and saw in the pursuit of efficiency the answer to other, less fundamental problems. They used the term "efficiency" much as a public utility regulator would use "the public interest": as the fundamental principle justifying their work, and as a pursuit for trained experts. In short, to engineers, streets were public utilities to be regulated in efficiency's name. When downtown business interests turned to engineers to solve traffic problems in the 1920s, engineers' classification of city streets as public utilities became the prevailing construction of city streets. Almost as soon as they began, however, other social groups interested in city streets challenged the public utility model. This chapter introduces the cultural setting that gave engineers their opportunity to treat streets as public utilities. In chapter 5, the application of this approach in city streets will be examined in detail.

The new traffic crisis resembled older problems in the supply of city services. For decades, municipal engineers had been devising ways to manage heavy loads on city service networks of limited capacity. In 1920 the traffic crisis looked to them like an old problem in a new form. City streets were, to them, like water supply, sewers, or gas lines: a public service to be regulated by experts in the public interest.[2]

By the 1910s, expert boards had long been supplementing or displacing elected politicians as the regulators of modern public services. Analogously, the rise of the traffic engineers reflected disenchantment with police traffic regulation. Police attacks on disorder did little to ease traffic congestion, and often worsened it. As city business leaders came to see their problem in terms of congestion, they lost faith in the police approach. They turned to engineers. Police would remain important as foot soldiers, but engineers replaced them as the generals in the fight against traffic congestion.

There was little chance to build facilities to accommodate the traffic, but nevertheless the engineers were confident. They had already achieved an impressive record in problem solving in other city services, and by the time automobile traffic began clog the streets, many people had learned to turn to engineers for the answers not just to technical problems, but to complex new social problems as well. Through their actions, if not in their words, a new generation of engineers defined and set before themselves a new class of sociotechnical problems. In these years, engineers increasingly applied their expertise to the control of problems that, in modern times, no longer could be trusted to solve themselves. Sanitation, public utility regulation and scientific management were three such new fields for the application of engineering expertise. A fourth, and a relatively late arrival, was city traffic. To fight traffic, however, engineers needed the same resources they had won in their efforts to control disease, high utility rates, and shop floor inefficiency. They needed a chance to experiment and a wide scope of action.

The Progressive Era gave engineers their chance. Progressive innovations in law and economics gave engineers room to adapt old remedies and to devise new solutions to treat new problems. To the new problem of dense motor traffic in cities, engineers proposed professional traffic control.

In their earlier work in public health and scientific management, engineers showed what measurement and empirical study could accomplish in social fields once the province of law and custom. The example of public utility regulation was still more specifically suited to street problems. The problem of supplying safe and efficient safe street service was much like problems in the provision of other city services, such as water and gas. Professional traffic control was the application of the principles of public utility regulation to city streets.

## Social Organization

In their efforts to get hold of the slippery subject of Progressivism, a number of historians have found a handle in the Progressives' quest for social control. In *Social Control* (1901), Edward Ross expressed the growing doubt that the good of the whole was safe in the unguided hands of the many.[3] In the face of the countless technological and social disruptions accompanying mature industrialism, Progressives substituted expert control for imperiled traditional or natural restraints.[4]

The experts who formulated the elements of social control were often engineers. In the Progressive Era, the language of engineering was extended to include the ordering of society; no longer was it confined to the design of masonry abutments, iron drive shafts, or electrical windings.[5] Engineers developed their own forms of social control to manage the problems of modernity. "We are trying to accommodate human nature to new physical facts," *Engineering News-Record* editorialized in 1924.[6] Traffic control was one such effort.

Progressive engineers were indeed practitioners of a kind of social control. Traffic engineers soon identified their incipient profession as "traffic control" because they thought of their mission as an effort to impose technical control on a social problem. Later, as the progressive foundations of traffic control declined in favor of a new model, the term "traffic control" declined with it.

Even more than they were social controllers, however, engineers were social organizers. "Social control" can suggest intrusive control, for example in matters of public morality. But with "social organization" many engineers hoped to get efficiency without such strictures.[7] Though they did not always succeed, the difference remains crucial. Scientific management experts, for example, did not merely want control workers' motions, they sought to organize the work environment so that the most efficient motions would follow of their own course, without the constant intrusion of the supervisor. Engineers sought to organize public utility regulation so that optimum service at minimum cost followed naturally. With such organization, there would be no need for intrusive policing. For example, the engineering innovation of the water meter freed water customers from the inspectors who had formerly tracked down water waste. In an ideally organized social arrangement, the supervisor would function like the governor on a steam engine: always present as an agent of efficiency, but never clumsily intrusive.

The engineer's brand of social control was not that of the temperance crusader. The prohibitionists' real analogues in city traffic were the police departments; both exerted social control intrusively and inefficiently. The engineers' method—the method that set them apart from most other progressive reformers—was social organization. Among the social organizers were the traffic engineers. "Scientific organization of traffic," wrote one of them, "could cut traffic congestion at once by half."[8]

Professional traffic engineers were not social planners. Nearly all of them preferred to leave the city's destiny in the hands of the myriad participants in the urban marketplace. Instead, traffic engineers sought to maximize the opportunities for exchanges in this marketplace. Yet in the name of efficiency, engineers often restricted the demands of some street users for the benefit of others.

When city business leaders turned to engineers, engineers' most useful tool was not their building skill but their expertise at regulation. To some, the engineers' reliance on regulatory measures to the exclusion of expensive reconstruction was settling for less than the best way to fight traffic. Many civil engineers doubted that regulation could be an adequate substitute for the building art.[9] Yet to most engineers, regulation opened new frontiers for the application of expertise to social problems, including city traffic. A New York engineer of national repute, with a confidence characteristic of his colleagues, claimed in 1923 that with regulatory measures alone "there is no case of congestion that cannot be bettered from fifty to one hundred per cent."[10] Engineers found regulatory means for a wider scope of social ends, and with their record in scientific management and public utilities they convinced others this modern way would work in new fields.

## Chambers of Commerce

Cities could afford to do little about the traffic crisis. "Everywhere the people are demanding more service and better service," the New York Bureau of Municipal Research reported in 1917, "and in response to that demand cities are performing an increasing number of functions."[11] With war in 1917 came shortages of materials and labor, and inflation; after the war unemployment further burdened city budgets. And just in time, Prohibition eliminated "one of the most productive sources of revenue": liquor taxes.[12] A shortage of fuel from 1914 to 1922 exacerbated the cities' financial problems.[13]

Merchants and businessmen stood to suffer most from traffic congestion, and they stepped in where city governments could not. In so doing, they changed the prevailing construction of the traffic problem. Once chiefly perceived as a threat to order, it was redefined as a threat to commerce. In each city, new business associations united their efforts. Such associations were known by various names, including "Board of Trade" and, most often, "Chamber of Commerce." Rare outside the very largest cities before 1900, new chambers of commerce grew from "booster clubs" or exclusive fraternities into more open and businesslike organizations in the new century's first decade, while old chambers modernized. In nearly all matters of municipal administration the chambers grew in importance, sometimes serving as the initiators, financers, and executors of city functions.[14]

The American City Bureau coordinated local efforts. Its consultants advised member chambers, especially those too small to keep experts in salaried positions. The Bureau promised to help the client chamber turn its city into "an efficient industrial machine."[15] It reorganized chambers nationwide in the early decades of the century, rationalizing chains of authority and increasing revenues and enrollment. In 1913, Lucius Wilson, the energetic manager of the Bureau's campaign department and former head of Detroit's chamber, stepped up the movement's pace. In a typical case, Wilson and the Bureau boosted membership in the Syracuse Chamber of Commerce to record levels while increasing dues from $10 to $25.[16]

Other new institutions, especially the Bureau of Municipal Research and state municipal leagues, helped organize commercial interests in cities early in the century. Their "willingness to abandon independent and wasteful individual efforts" would promote "the better plan of bringing together common experience . . . for the study and solution of common problems."[17] In 1912 the founding of the Chamber of Commerce of the United States gave chambers of commerce a distinct national voice. Yet chambers remained primarily local organizations, and they mattered most in local affairs—including street traffic problems.

When new social groups join an existing attack on a problem, they tend to begin by accepting the prevailing framework. The chambers were no exception. They began by lending support to police departments. In both great metropolises and small towns, local commercial associations helped overburdened police by taking over some traffic functions themselves. They funded some of the ad hoc mechanization of traffic regulation to which police resorted in the 1910s. New York's elegant traffic towers, erected in 1921 by the Fifth Avenue merchants' association, were perhaps the most striking example,[18] but chambers paid for similar improvements in numerous cities. In many cities local chambers of commerce erected traffic signs of their own design, after police departments had failed to do so.[19] Other chambers put up street lights themselves; the chamber in Wichita Falls, Texas, reported in 1921 that it "promoted the idea of a new lighting system, collected the money from property owners . . . and put over the entire project as a property owners' movement." The chamber then turned the completed system over to the city.[20] Elsewhere, local business leaders recruited volunteers to supplement the police efforts as "traffic vigilantes."

Yet during the 1920s many business leaders lost faith in police methods. What was missing, according to such critics, was expertise. Neither the police nor the chambers had it. In its absence, police departments and local business leaders could not agree on the best way to handle traffic. "In matters of traffic," a manufacturer of traffic control devices observed, "each man has a different opinion."[21]

The chambers wanted experts. By the time of the traffic crisis, chambers had experience finding experts who could promote efficiency by targeting waste. "Efficiency" meant different things to different reformers, as Samuel Haber has shown.[22] To chambers of commerce, efficiency meant, above all, securing "the maximum of service at the minimum of cost."[23] The chambers therefore avoided elaborate, expensive plans that stood little chance of implementation. They were pragmatic, not visionary. "We make it a rule never to go after anything unless we expect to get it," remarked the head of an Iowa chamber in 1914. Their credo was "If you want your reform movement to succeed, get the business men back of it."[24] "There has been a real effort to introduce efficiency and economy in the conduct of municipal affairs," reported the New York Bureau of Municipal Research in 1917, and chambers of commerce were at the forefront of this drive.[25]

## Efficient City Government

In the quest for efficiency, local chambers of commerce were often ahead of their own city governments. The chambers tended to share the progressives' distrust of city governments. Yet they were unsympathetic to reforms of the Lincoln Steffens variety, for more democratic government was certainly not the sure road to more efficient government. As Samuel Hays argued, lasting reform in city government originated in "the leading business groups in each city and professional men closely allied with them."[26] Leonard White, a contemporary authority on city government, affirmed that "the opposition to bad government usually comes to a head in the local chamber of commerce."[27]

To the chambers, the problem was waste and inefficiency, the solution was more businesslike government, and the way to get there was expert government—and the experts were often engineers. In practice this meant making city organizational charts more like those in business, and putting engineers in the new administrative posts. These were the motives behind the rapid spread of city manager government, begun at Staunton, Virginia, in 1908, but established as a feasible plan for large cities with the Dayton charter of 1913. For the same reasons government by city commission spread quickly as well.

Advocates of businesslike government saw city manager government as a way to get it. They argued that "the stockholders of a city are its citizens." A Philadelphia mayor compared a city to a corporation of "citizen shareholders."[28] Leonard White ascribed business leaders' support for city manager government to its "resemblance ... to their corporate form of business organization,"[29] while the secretary of the National Municipal League described the plan as "similar to that followed by successful business corporations."[30] In 1922, the president of the Chamber of Commerce in Norfolk, Virginia, praised the city's new form of government as "a business management consisting of five businessmen" (five at-large city councilors) "with a general manager" (the city manager).[31] Unlike businesses, however, city manager cities often were run by engineers. In 1919, fully 48 percent of city managers were engineers.[32]

## Enter the Engineers

The chambers concluded that their best hope lay with engineers, the efficiency experts of the Progressive Era. Henry Gantt and other champions of efficiency agreed that "the man who knows what to do and how to do it is preeminently the engineer."[33] Engineers promised to replace police trial-and-error methods with systematic analysis. According to one of them, with the advent of engineering methods in the public sector "the 'rule of thumb' or the 'hit or miss' method gradually gave way to more scientific planning and methods" in government.[34]

Yet many still regarded engineers as skilled mechanics, not trained to handle larger social problems. Arthur Blanchard, a civil engineer, addressed an assembly of chamber heads in 1915. "The people," Blanchard told the gathering, "... must be shown that engineers are broad-minded, well-educated men, capable of holding with credit the highest administrative office, and do not constitute a tribe of human beings capable only of running a transit, turning a lathe or wiring a house."[35]

By the time of the traffic crisis, engineers were proving they had these talents, as city engineers, city managers, and regulators of public utilities. They were proving that they could solve problems with adding machines and file cabinets as well as with transits and lathes. A contemporary political scientist held that the modern city had several "administrative functions" that required an engineer's skills, including "streets, sewers, water supply and all public utilities."[36] Well before 1900, cities began hiring "city engineers" or "municipal engineers" whose responsibilities united these problems. Smaller towns turned to consulting engineers, who often traveled from town to town.

Engineers showed cities how they could both improve services and reduce costs, and in this promise lay the fundamental difference between police and engineering methods. Police had tended to make traffic rules according to an "adversary model of regulation," to use David Nord's apt term.[37] Engineers proposed instead that efficiency could benefit all. Experts (many of them engineers) in sanitation, public utilities, conservation, and scientific management showed people that with the right regulation they could alleviate many problems—especially city problems—to the benefit of all interested parties. In the new model of regulation, individual persons or businesses would sometimes pay dearly. But as groups, landlords and tenants, service suppliers and service users, loggers and hunters, managers and workers could at least in theory benefit together.[38]

Today we expect economists to be the experts of choice in regulatory matters that involve financial questions. The economists William Hausman and John Neufeld have found, however, that in the pricing of electricity "the dominant rate structures that have long been used by the industry were the products of the minds of engineers, especially those working during the first five decades of the of the industry's existence, roughly from the 1880s to the 1920s." Only since the 1970s, they argue, have economists succeeded in altering these structures.[39] Thomas McCraw found that not until the 1970s did regulatory practice "come to be dominated by an entirely new set of dramatis personae: professional economists."[40] The growth of public utilities gave the engineers this opportunity, and in all such city services, including sewers, water service, and electricity, they were the experts city governments and state public service commissions turned to. Engineers, for the most part, determined the regulations and franchise agreements that would deliver efficiency.[41]

### Scientific Management: A Model of Social Organization

In scientific management, engineers attempted to solve human problems through social organization. Frederick Taylor's methods, and gleanings from them, found wide application in business, where they promised industrial peace through a restoration of the union of interests between capital and labor. Engineers found the technique useful elsewhere too.

Scientific management was a technosocial technique serving technosocial ends. Its practitioners called for the substitution of "system" for "rule-of-thumb methods."[42] Business led the way in the systematization of social processes, and it was business that Taylor particularly addressed. Yet many reformers—Taylor among them—recognized the wider applicability of scientific management. The technique, Taylor maintained, "can be applied with equal force to all social activities," and indeed reformers sought to replace expediency with system throughout much of public life.[43]

For most of its first century, America's legal and political tradition had largely treated each owner and each worker as independent agents. Throughout the nineteenth century, however, the perception of capitalists and workers as classes with distinct interests grew. In 1886 the Supreme Court famously ruled that corporations were in some respects the legal equivalent of persons.[44] As Carnegie and Rockefeller began to extol the social value of entrepreneurs as a class, Powderly and Gompers likewise spoke for the grievances of workers as a class. In both law and politics, mature industrialism seemed to bring with it the superseding of the individual by the class.

Taylorism proposed that both earlier models—the precedence of the individual and of the class—obscured the true unity of interest of all. Many engineers agreed that the achievement, through regulation, of optimum motions and speeds, would benefit both workers and capitalists and thus bring industrial harmony.[45] Efficiency was in the interest of all parties. In this faith lay a reversal of the motivation of much of earlier regulation in the "adversary model," which was an effort to stamp out the sort of abuses that the muckrakers attacked.

Scientific management evolved in the private sector, where employers enjoyed broad authority over their workers' actions. Any who wished to apply scientific management in the public sector had to confront the limitations of American state power. The emerging administrative state, despite its dramatically increased capacities, never overcame them.

To bring efficiency to public life, engineers tried to overcome these obstacles. In their frequent demand for more "businesslike" government, engineers sought a state unimpeded by the machinery that made it representative or by a strict construction of individual natural rights. They sought a state that could, like the Ford Motor Company, apply the advice of efficiency experts. Ends in the polity would remain in the hands of the voter, just as customers held the final say in the fate of Ford. An engineer working for the state of North Carolina in 1921 expressed some envy of "the success of the great modern corporations," and ascribed it to their "excellent engineering organizations."[46] Engineers wanted city and state governments that could put their expert advice to work, and a number of reforms, from commission-manager government to expert accountancy, began in this conviction.[47]

Though they worked in the public sphere, traffic engineers found useful parallels in the Taylorites' principles. The traffic engineer A. G. Straetz found "analogous ideas and methods" in the two systems.[48] The Taylorites were irrepressible quantifiers, universally caricatured as stopwatch-wielding measurement fanatics.[49] In traffic surveys, engineers applied their own versions of time and motion studies to the problem of city traffic, with a meticulousness entirely unknown to their predecessors in city police departments.[50] To engineers, scientific management was proof that they could achieve the reputable results of the scientific laboratory even in the clinically imperfect conditions in which social problems are found. As emulators of science, traffic engineers—unlike police—measured first and recommended afterward. Taylorism contributed to an expanding concept of a common interest to be achieved by positive means with the techniques that industrial development provided.

Yet traffic engineers needed a justification for control that scientific management alone could not provide.[51] City engineers found it in public utilities regulation. The public utility model gave engineers a path to regulatory control that the courts had already cleared. Cases that defended state regulations in areas "affected with a public interest"[52] opened the possibility of extensive traffic control in the name of efficiency. A separate body of case law—that culminating in the judicial approval of zoning codes—strengthened the legal foundation traffic engineers needed.[53] Yet of these two regulatory models, public utility regulation was most applicable to city streets.

## Making Room for Positive Regulation

By the end of the nineteenth century, the American polity was making room for social organization.[54] Increased state capacities and a changing political culture gave organizers of social arrangements much wider latitude. The waves of reform known as the Progressive Era were only the most apparent manifestation of this trend. There was no room for social planning, but in the first quarter of the twentieth century the opportunity for social organization was greater than ever.

In law, in economics, and in public opinion, engineers were finding a newly hospitable atmosphere for their proposals. Principles not found in John Locke or Adam Smith were uncovered in Greek, German, and biblical sources. Trends in law and economics reflected a new doubt that the public interest was safe in a state that acted only (in Locke's metaphor) as an "umpire."[55] Most of the nation's founders had perceived state action as a threat to the people's prosperity and liberty, but now many experts in law and economics said it was necessary to secure either.

As umpire, the state was largely limited to negative regulation: it stopped intrusions by some on the rights of others. The state as umpire acted intermittently. Like a coachman with a whip, the Lockean state would seldom have to intrude; when it did, it was as a punisher of violations. Social organizers preferred to use the reins. They hoped that constant minor checks would guide the state along the right path. Social organizers would still punish violations, but they hoped to use their expertise to prevent them.

The methods of police traffic regulation were largely confined to negative regulation. As social organizers, traffic engineers depended upon positive regulation.[56] Nevertheless, in both of these regulatory styles, the destination was for the passengers to decide. Few engineers were social planners.

## The Law and the Engineers: The Legal Means of Positive Regulation

Over about 40 years, American judges gave engineers the legal tools of positive regulation. The new kind of regulation allowed under the law by 1920 convinced chambers of commerce and traffic engineers that strictly regulatory measures could accomplish at little expense what visionary planners could deliver only for millions. For some city business leaders the goal was a city of stable land values with elegant shopping districts and neighborhoods, which regulation could secure in the form of zoning.[57] Others wanted to relieve the growing traffic crisis. Both saw opportunity in the new regulatory tools.

To exploit their opportunity, progressives revived an old phrase: "the public interest."[58] Chief Justice Morrison Waite had brought the phrase "the public interest" back into notice in *Munn v. Illinois* in 1877, but after 1900 the law resorted to it far more readily.[59] Waite had used this principle to justify positive government action. The principle was readily applicable both inside and outside of matters of law. Others have shown how, in the Progressive Era, the law increasingly recognized "community rights."[60]

Jurists more often invoked the public interest to the detriment of individual interests. The public interest principle was the foundation on which Progressives hoped to build their version of the good society.

American cities functioned under authority state legislatures delegated to them by charter, and such delegated authority included police powers. With them, cities had wide latitude to secure the public health, safety, morals, and welfare.[61] The police power had been the legal basis for negative regulation, but progressives adapted it to the purposes of positive regulation. Progressive historians have described jurisprudence circa 1900 as hostile to the broad application of the police power, but Melvin Urofsky has found that from the 1890s to about 1920 state judges—far from restraining the police power—tended to expand the acceptable limits of its application.[62]

A concurrent trend expanded the realm of the "public interest," thus extending the reach of the police power that protected it. Beginning in 1877 with the U.S. Supreme Court's ruling in *Munn v. Illinois*, the courts gradually and erratically opened new areas of public life to state regulation. To the champions of laissez-faire, the Fourteenth Amendment had come just in time to buttress common-law traditions and defend a newly vigorous "freedom of contract." Inspired by *Munn*, others dusted off common-law precepts justifying public regulation of common carriers and applied them to new areas of commerce "affected with a public interest."[63] In *Munn* the "common carriers" in question were grain elevators, but the case opened legislative doors to other new "public utilities." "The business is one of recent origin," Waite wrote in his majority opinion, and he upheld the legislature's power to extend the law to accommodate "this new development of commercial progress."[64] Such new developments proliferated. The law admitted many of them into the special category of enterprises "affected with a public interest" and therefore regulable. Thus the law gave cities and states the flexibility they needed to manage the burgeoning new city services of the end of the nineteenth century and the beginning of the twentieth: water, gas, "light" (electric power), street railways, and telephones.

Engineers and chambers of commerce found in such positive regulation an affordable means of accomplishing the reforms they sought. Of these, zoning and traffic control are perhaps the leading examples. Charles Ball of the Chicago City Club heralded the spread of zoning as a means of social organization that "substitutes method for chance, symmetry for confusion, progression for patchwork, and order for chaos in city development."[65]

Traffic control promised the same blessings in another area of city life. A number engineers linked traffic control and zoning as related elements of a larger reform. Ernest Goodrich, an engineer and a traffic consultant to the New York Regional Plan Committee, observing that existing streets could not be "easily widened," recommended instead that zoning be used "to preclude the creation of traffic congestion."[66] Above all, however, engineers saw in the positive regulation common to zoning and traffic control a way to make city functions efficient.

The new model of positive regulation, based on the "public interest" and the "rights of the community," was also useful to engineers who found that the short road to efficiency went by way of serving aggregate goods, even at the cost of individual goods. Engineers subordinated those individual demands for street space that in their judgment were at odds with the good of the whole. In 1926 the editor of *The American City* surveyed leading national traffic experts and found them in agreement that "streets are primarily provided for general public use" and that therefore "the rights of the different classes of traffic to unlimited use of the streets" were "subject to the public and civic welfare."[67]

## The New American Economics

Beyond the courts, developments in economics contributed to this widening field of action, and engineers found justification for their work in the economists' writings.[68] The invisible hand seemed unable to restrain railroads and other "center" firms, and their excesses gave the progressives' new political and legal views a receptive audience. The discoveries that monopolies corrupt free enterprise, and that they can be a normal consequence of a largely unregulated advanced economy, contributed to the popularity of progressivism at least as much as "the discovery that business corrupts politics."[69]

These trends were shadowed by a decline in the influence of British classical economics in America, as institutional and marginalist economic ideas spread. American economic practice had never quite lined up uniformly with Adam Smith and David Ricardo. Trade protectionism was a mainstay of nineteenth-century American economic policy, and much of the little that Americans published on economic subjects was devoted to the defense of protective tariffs.[70] Still, before the emergence of the profession in America in the 1880s, the leading authorities in economics in America were the British classical economists, notably Smith, Ricardo, and

John Stuart Mill.[71] But beginning with Mill himself, changes in and challenges to the classical orthodoxy later gave traffic engineers a theoretical backdrop for their methods.

Mill introduced two important amendments to classical economics in *Principles of Political Economy* (1848). One of these was the first clear conceptualization in English of the principle of natural monopoly.[72] Through its application to the regulation of public utilities, this idea would shape positive regulation—and ultimately traffic engineering—early in the twentieth century. Second, Mill held that certain public interests could not be secured except by government intervention in the economy—especially the protection of dependent persons, of working conditions, and of the poor.[73] Mill's formula for determining these interests—utilitarianism—was itself a fundamental principle of twentieth-century positive regulation, including traffic engineering.[74]

New schools of economics further diluted classical economics in America. In the 1880s the most important influence came from Germany, not least because Americans wishing a doctoral degree in economics had to go there to get one. One of them, Richard Ely, was the apostle of the German economics to the Americans. A founder of the American Economic Association (1885) and a writer who could reach a general educated audience as well as his academic colleagues, Ely best represents the new economic thinking in America from the 1880s to the early years of the twentieth century. With German "historical" economics, Ely loosened American economics from its British moorings. Renamed "institutional" economics and colored by the American social gospel, the German school influenced a generation of intellectuals who maintained that the state must intervene to mitigate the worst consequences of advanced industrialism, and that a collective good exists distinct from the sum of individual goods.[75] In their 1885 "Statement of Principles," the founding members of the American Economic Association declared: "We regard the state as an agency whose positive assistance is one of the indispensable conditions of human progress."[76] John Dewey, one of their number, concluded typically that "social and humane ideals" were insecure in the American polity and "demand the utilization of government as the genuine instrumentality of an inclusive and fraternally associated public."[77]

The institutional school of economics was controversial. By 1900, however, even its leading critics no longer wanted to limit the state to the umpire's job. Latter-day American classical economists accepted a fundamental revision of the tenets of Smith and Ricardo.[78] They replaced the classicists' labor theory of value with their own theory of marginal utility.

Neoclassical economists continued to take their cue from Britain, but by the turn of the century their new master was not Mill but Alfred Marshall, described as "perhaps the most eminent and widely influential economist" of his time in "the English-speaking world."[79] To save classical economics, Marshall and his American followers accepted considerable amendments to its principles, sacrificing its neatness to preserve its applicability to an advanced industrial economy.[80] They agreed with the institutional economists that the pure competition upon which Smith's analysis depended could not exist, and that in many exchanges the unassisted invisible hand did not fashion the optimum result. Even the neoclassical economists allowed room for positive regulation.[81]

### Natural Monopolies

Mill's modification of classical economics came on the eve of a great surge in the category of enterprises that did not behave according to the price law Smith had expounded. These exceptions lay in the realm Mill designated "natural monopolies," which included most of the services lawyers and regulators came to term "public utilities."[82] Nineteenth-century cities grew to depend upon natural monopolies for the amenities by which civilization measured itself. Bridges and canals were old examples, but soon they were joined by steam railroads, street railways, grain elevators, water and gas service, telegraphs and telephones, and finally electric power. Once a small corner of economic life, by 1900 natural monopolies composed much of its vital structure. The origins of the efficiency model that traffic engineers applied to city streets in the 1920s lay in these earlier problems in the regulation of public utilities.[83]

Beginning with Mill, nineteenth-century economists showed that certain enterprises tended to become monopolies. If more than one firm entered one of these fields, in time only one firm would survive. Mill called these "natural monopolies," by which he meant "those which are created by circumstances, and not by law."[84] Others developed a theory from Mill's hypothesis.[85] In America, the idea of natural monopoly found fertile soil. The rapid spread of powerful corporations, especially the railroads, stimulated the search for a theory of monopoly. Before the late nineteenth century, however, economists treated monopoly largely as a problem stemming from exclusive government charters. New economic conditions required new examinations of monopoly, including a fuller development of the idea of the natural monopoly.

Most nineteenth-century Americans believed in the necessity of competition to the well-being of consumers.[86] Yet railroading repeatedly proved to be an environment in which direct competition seldom lasted. The evidence of this problem was in well before there was a satisfactory explanation, and critics of the railroads supplied this lack by accusing the rail barons of treachery. Yet when Charles Francis Adams took charge of Massachusetts' railroad commission in 1869, he had already developed a thesis that in railroading competition *cannot* succeed, and ought not to be attempted, since on the rails "competition and the cheapest possible transportation are wholly incompatible."[87]

Popular intellectuals in America, including Henry Demarest Lloyd, Henry George, and Edward Bellamy, offered explanations of and prescriptions for the spread of monopolies. But these widely influential amateurs could not shape opinion in professional circles. Though Lloyd and others criticized public utilities, popular writers failed to confront the problem of the natural monopoly. Yet at least two writers, Thorstein Veblen and Richard Ely, united professional authority and popular influence in economics, and of these it was Ely who brought the idea of natural monopoly to the notice of the American public. Ely's *Problems of To-day*, first published in 1888, soon became a standard textbook in economics.[88]

Ely was less a developer of the natural monopoly theory than a publicist for it, both to professional economists and to a more general educated audience. Already in 1887, the American economist Henry Carter Adams limned the essential attributes of the natural monopoly.[89] Ely enthusiastically brought Adams's findings to the pages of *Harper's*, and "natural monopoly" entered the American lexicon.[90] In 1900 an economist concluded that Ely "had perhaps done more than any other single writer" to disseminate the natural monopoly idea.[91]

In *Monopolies and Trusts* (1900), Ely extended the work of Henry Carter Adams and others[92] to arrive at a complete formulation of theory of natural monopoly. Ely concluded that three conditions contributed to the making of a natural monopoly, of which two are of special importance here. First, in a natural monopoly, the commodity or service "is furnished in connection with the plant itself—railway service, for example, cannot be shipped," for the service in demand is shipment itself.[93] In practice this meant that competition was unfeasible in services depending on fixed conduits of finite capacity under conditions of high demand, such as in telephone service or urban transportation. Since duplication of expensive and spacious conduits was impractical, economists and engineers found that they had to regulate such services. Second, the profits in a natural monopoly follow the "law of increasing returns," or, as later economists would say, great economies of scale lead to diminishing marginal costs.[94] That is, additional service can be provided at smaller cost, usually because the new demand can often be met on existing facilities. Providing the second thousand travelers with streetcar service costs far less than serving the first thousand. In the words of a modern economist, the result is that "one company can supply a market at less cost than can two or more firms," and competition fails.[95] Engineers found that both of these laws applied to city streets, and once given the chance, they soon treated streets as they had other public utilities.

According to the economist William Sharkey, Henry Carter Adams and Richard Ely "managed to establish the most important characteristics of natural monopoly industries."[96] Even seven decades later, the economist and regulator Alfred Kahn did not alter theory as Adams and Ely had left it.[97] Theory of natural monopoly was well established before engineers turned their attention to city streets, and served them as a model for traffic regulation.

Ely was an active propagandist for the regulation of natural monopolies. The historian Benjamin Rader found among Ely's papers reams of letters to "almost every major American city and hundreds of smaller ones." Ely, writes Rader, "converted an inestimable number of municipal reformers to the idea that 'gas and water socialism,' or at least limited franchises, should be a cardinal tenet of city reforms."[98] Ely called for government ownership of public utilities, and indeed most cities owned their water works by 1900. While most other utilities remained in private hands then as now, states and cities began to regulate their municipal services. After 1907, state public service commissions proliferated rapidly.

## From Legal and Economic Theory to Engineering Practice

What is a public utility? Lawyers and economists proposed various fixed definitions. In practice, however, the category was like a political office: socially determined and hotly contested. Traffic engineering began with the claim that busy city streets are public utilities.

To the law, a public utility was simply an enterprise "affected with a public interest"; jurists could agree on little more.[99] Economists offered a neater definition: a public utility was not just an enterprise "of real public importance," but also one in which competition was unfeasible.[100]

In practice, engineers often judged which enterprises were public utilities and which were not. In the infancy of professional economics, engineers supplied most of the applied economic expertise. "Of necessity," Hausman and Neufeld concluded, engineers "became applied economists."[101] For example, when in 1914 Houston found that its electric company was netting 20 percent per annum with no competition, its mayor called in a New York consulting engineer. The engineer worked out a franchise agreement holding the company to an 8 percent return.[102]

To most engineers, regulation was not a substitute for the free market, but a means of restoring its benefits where competition would not last. Public utilities grew to accept regulation of rates and service as an acceptable price for exclusive franchises—that is, for state-enforced monopolies. A lighting executive maintained that public utilities are "most efficiently and economically conducted as monopolies," but he admitted that they "are properly subject to regulation by the state and by the municipality."[103] The hope was that the state, through regulation, would protect private investment in risky but vital city enterprises. Like the flanges on the wheels of a locomotive, ideal regulation would keep an enterprise on its tracks without impeding its progress.

### Streets as Public Utilities

Municipal engineers insisted on a definition of "public utility" broad enough to include streets. With the frequent advent of new utilities, engineers needed conceptual flexibility. The engineer George Roux contended in 1916 that prejudices among non-engineers "erroneously" confined the category of public utility to "[railroad] transportation, water, gas, telephone, telegraph, or electric light and power," and urged instead a broader and more adaptable definition.[104] Referring to major urban streets, one engineer declared that "a highway is a publicly owned utility" for which users should pay rates that discouraged inefficiency.[105] Engineers' views were seconded by some other experts. Lent Upson, a prominent expert in city government, classified streets as utilities that were not "revenue producing."[106] The economists Eliot Jones and Truman Bigham held that the "limited amount of space available in the city streets" justified the regulation of street uses.[107]

Economists' definitions of natural monopolies served engineers who wanted to treat streets as natural monopolies. Like other natural monopolies, streets were subject to extreme economies of scale, they had limited capacity, they were impractical to duplicate, and they offered a service inseparable from street capacity itself. The problems of street traffic were like those of other public utilities. There were severe demand peaks at some times and excess capacity at others, the physical plant was vast and the extremely expensive to expand, and it was difficult to charge users.

Engineers attacked these problems in street traffic as they had attacked them in other public utilities.

Like public utility regulators, engineers invoked "the public interest" to justify traffic control. Components of this justification included social utility, equity, and especially efficiency. These principles occasionally appear explicitly in engineers' reports; they are nearly constant in the unstated assumptions behind their methods and recommendations. In invoking them, engineers claimed to put the interest of the whole over the interest of each. Traffic engineers were sympathetic to a street railway executive's complaint that "most people approach the problem with the idea of 'How can I get from where I am to where I want to go,' instead of 'Where can the greatest number of people get in the shortest space of time?' "[108]

Engineer regulators, including traffic engineers, were utilitarians. They sometimes expressly defended their actions as serving "the greatest good for the greatest number." In 1911 the superintendent of Cleveland's water department justified the extensive water service regulations by claiming that they were for "the best good of all the people."[109] The city engineer George Herrold saw his goal in managing traffic as achieving the "good of the greatest number," even at the expense of the few.[110] The equity principle was closely related. Engineers were not afraid to weigh the importance of various demands for street capacity, and to restrict those uses they considered an undue burden on other street users.[111] Finally, traffic engineers, like Taylorites, pursued efficiency. They did not equate the sum of individual street users' travel choices with the optimum city transportation system. They were willing to go to some lengths to shape demands—just as the scientific management reformers had not trusted workers or foremen, but insisted on prescribing and enforcing "the one best way."

## The Lessons of Water Supply

Engineers who wanted to treat American streets like public utilities faced technical and social obstacles similar to those that confronted earlier regulators. Municipal water supply is a useful example. In the nineteenth century, clean water was difficult to secure, especially in cities. Individualistic, private methods could be dangerously unreliable. Rooftop cisterns ran out; shallow wells went dry or were contaminated. Large-scale, collective water service would correct these problems, but entrepreneurs who would supply it faced daunting barriers. Water mains had to cross private property. Demand had to be reliably high to pay expenses; thus there could be no competing water mains on the same street, and few non-subscribers. Users had to be charged according to their consumption. Water service in cities, in other words, required "collective organization" and an absence of real competition, in direct opposition to "democratic postulates, competitive ideals, and liberal individualistic traditions."[112]

There was no reconciling the contradiction. The consequence, as Maureen Ogle has shown, was delay in the provision of water service in American cities.[113] Joel Tarr found in the case of city sewer systems that inefficient ad hoc rule-of-thumb methods ruled until the 1890s, when a "rational engineering model, based upon empirical considerations of urban need" finally succeeded.[114] Until then, cities that could not ban private, individual water supply (for all its deficiencies) could not attract private, large-scale water supply systems, because even a minority of non-subscribers was sufficient to make a large-scale system uneconomical. According to Ogle, too many "clung to their cisterns . . . even in the face of both superior alternatives and an urban crisis."[115] Yet together the necessity of reliable water supply, better water supply technology, and a better understanding of the propagation of waterborne disease led cities to use their police power to protect water supply entrepreneurs or to build systems themselves. Few objected to the substantial collective, regulatory and anti-competitive measures by which city people have been supplied with water ever since.

Street traffic was closely analogous to water supply. Before the automobile, individualistic means of street travel were inefficient. Walking was slow; a horse was expensive. Omnibuses divided the expense of a horse among several passengers, much as the expense of a well could be shared by several neighbors. Horse-drawn streetcars, like water mains, were far more expensive, but their speed and capacity could repay investors—if they could be assured of enough passengers. Such assurances came in the form of franchises, which protected investors from competition. With the electrification of street railways in the 1890s, franchises extended city wide. In street transportation and in water supply, relatively reliable and inexpensive service came through anti-competitive regulation.

Engineers compared city traffic problems with problems in water supply and sanitation. To make his case for one-way traffic, St. Paul's city planning engineer, George Herrold, compared streets to two public utilities, observing that "we do not attempt to pass water in two directions in a water main, nor to carry off sewerage in two directions through the same pipe."[116] In most cities for most of the 1920s, engineers fought traffic by restricting the private automobile, much as they had restricted private water supplies. They treated the private car less as an intruder upon the rights of street railways than as an abuse of the street as a public service. In so doing, engineers were attempting to repeat the successes they had won by similar methods in other city services.

Similarly, shortly before the traffic crisis, engineers had shown cities that with the application of engineering expertise they could reduce water consumption, reduce rates, and increase water pressure, thus satisfying the utility, the customer, and the city government. New engineering techniques for metering water and cleaning mains exemplified this approach. Metering was rare before 1900. Flat rates for water service were common; customers were typically charged by the tap. The method rewarded waste. Suppliers' returns were low, users' rates were high, and water was chronically scarce. Suppliers' only remedy was to police users with inspectors.[117] The method was analogous to police traffic regulation: it was intermittent, heavy-handed, inefficient, and adversarial.

Water meters changed all this. By 1914 the Neptune Meter Company, a New York manufacturer with national distribution, could boast that metering was "becoming the *universal* method."[118] Neptune endorsed engineering methods in the manner of Taylor's condemnation of the rule of thumb. "Some things should naturally lie outside the realm of guesswork," the company argued, and "selling water" was such a thing.[119] One expert assured cities that "with universal metering and other methods of waste prevention, nearly all the cities in which the cry of 'water famine' was prevalent in the past year would have had an ample supply of water," even "without enormous expenditures for extensions of plant."[120] Most water customers found their bills reduced.[121] Water supply engineers devised other solutions as well. Burt Hodgman, an engineer, found cities ascribing their water problems to "increased demand," which would have required costly capital investment to supply, when "the fact is that the main will not deliver enough water" because deposits had "decreased the efficiency."[122] Meters, especially, showed municipal engineering at its best. They worked constantly but unintrusively. They were inexpensive, and they united the interests of apparent adversaries. Traffic engineers hoped to repeat this success in the busy streets of America's cities.

## Water, Sewers, and Streets

To fight traffic, city engineers drew on their experience in water supply and in other city services. The story was repeated in the careers of thousands of city engineers.[123] Oscar Claussen, St. Paul's chief city engineer, apprenticed in railroads, water supply, sewers, and electric power. His colleague in St. Paul, George Herrold, had worked in railroads and municipal engineering; by the early 1920s his chief responsibility was street traffic. The engineer W. B. Bates began his career in Virginia railroads, telegraph, and telephone lines; he brought this experience to his later work as city engineer in Roanoke, where he was responsible for streets, sewers, sidewalks, and garbage. Later Bates was city manager of Portsmouth, Virginia.[124]

Consulting engineers did similar work for smaller towns. Vincent Clarke, for example, served several Connecticut towns as a civil engineer and as an expert advisor on electric railways, water, and other services. In Cheyenne, Wyoming, the city engineer was expert in public utilities valuation; other Wyoming cities consulted him on diverse city service problems in water supply, sewers, and electric power. Consulting city engineer William Bryant Bennett worked with cities of all sizes and with state commissions on street railways, electric power, gas, and water.[125] C. C. Williams, a civil engineer, applied his expertise in city water supply to street traffic problems. He classified streets as a public utility, and he argued that street users should pay rates proportional to the street capacity they used.[126]

Some engineers who worked in public utilities before the traffic crisis went on to important careers in traffic control. Morris Knowles, a member of the American Society of Civil Engineers, advertised his skills in 1914 as including "efficient and economical operation of water works, valuation and rate studies; drainage and disposal investigations and reports to commercial and civic organizations."[127] By the early 1920s, however, he was chairman of Pittsburgh's City Planning Commission, and in this capacity was concerned largely with traffic control.[128] Traffic control engineers with no personal background in utilities regulation worked in a profession shaped by this framework.

Engineers borrowed their chief tool for street traffic work from a public utility. The traffic survey was introduced as a survey of street railway service. In early surveys of city rail traffic, engineers introduced most of the techniques they later used in the more complex surveys of street traffic.[129] Engineers used the results of these surveys to regulate streets in much the same way as they had regulated public utilities. Their response to increasing demand for street space was much like their response to increasing demands for other services—and entirely unlike the usual response in the free market. Engineers recommended increasing the supply of the service only as a last resort; efficiency achieved through regulation was their first and great commandment. *The American City* editorialized in 1915 that "the most hopeful line of approach for reducing traffic congestion appears to lie in more scientific methods of traffic regulation."[130] Before he abandoned the public utility model, the traffic expert Miller McClintock expressed the traffic control creed in a typical way: "Major changes and the construction of new thoroughfares are costly and should never be undertaken until full use has been made of existing streets."[131] "Instead of widening streets at an expense of millions," the engineer Ernest Goodrich told a panel at the 1923 National Conference on City Planning, "street capacity can be increased effectively by regulating traffic."[132]

## Engineers and the Traffic Crisis

The chambers turned to the engineers because the latter promised to maximize traffic capacity downtown, thereby securing business investments. Rejecting the "adversary model," engineers assured cities they could get better results for everyone, and they had a record to prove it. Police had sacrificed speed for the sake of safety, and both for the sake of order. But engineers claimed they could deliver speed *and* safety. Though he would abandon the public utility model in the late 1920s, Miller McClintock was until then probably the leading traffic control professional. He advised cities that "rapidity of movement, properly regulated, is not incompatible with a large degree of safety."[133] "It is not enough," one traffic engineer argued, "to stand behind the slogan 'order and safety.' We must have both safety and speed."[134]

Beginning in the late 1910s, the engineering field gradually adopted traffic control from the police departments. In 1923, *Engineering News-Record*, the leading civil engineering journal, claimed there was "no more pressing or interesting subject" in engineering than traffic control.[135] In 1926, the trade journal *Roads and Streets* could maintain that "traffic control engineering is perhaps the latest specialty in civil engineering."[136] By the end of the decade, *Roads and Streets* announced, "cities all over the country" were "recognizing that traffic is primarily an engineering problem" and that there was "less dabbling and more sound engineering progress in traffic control."[137]

True to Blanchard's claim that his colleagues were "capable of more than running a transit," engineers applied their building talents to the traffic problem only sparingly. Instead they used their administrative skills. Most engineers in the 1920s did not take seriously proposals to reconstruct the city for the sake of the automobile. City governments and chambers of commerce seeking efficiency did not engage engineers to reconstruct—or even to alter substantially—the physical city. Cities could not afford such projects. In 1915 *The American City*, the leading journal for city officials and chambers of commerce, expressed a widespread doubt about the usefulness of "proposed plans for overhead crossings of streets, for double-deck streets, and the like," in part because of their "excessive cost."[138] Although some engineers were more sanguine about the feasibility of major projects, few among them worked for city governments or chambers of commerce.[139]

The years of social experimentalism called the Progressive Era gave engineers the opportunity to define streets as public utilities and to seek efficiency through positive regulation. Traffic engineers missed the vanguard of progressivism, for street traffic did not constitute a problem worthy of their best efforts before the late 1910s. Still, the traffic engineers were progressives by training and disposition. Progressivism welcomed the talents engineers had to offer, and it trained them in the kind of problem solving they could later apply to city streets. Progressivism taught engineers that the public interest (as they saw it) required positive state action; in return, engineers taught progressives the limitations of the adversary model of regulation. Both scientific management and public utility regulation stood for the universal benefits of efficiency, and engineers sought to bring these benefits to street traffic in the 1920s.

# 5 Traffic Control

Sooner or later the cities are all coming to realize that traffic control is not a policeman's job.

—*Engineering News-Record*, 1925[1]

The so-called "traffic problem," formerly everybody's business, is now graduated to a higher plane of engineering analysis and transportation practice, where it always belonged.

—J. Rowland Bibbins, consulting engineer, 1926[2]

When city people talked about the new traffic problem, they did not all mean the same thing. Pedestrians complained of the automobile's trespass upon their rights. Parents dreaded the new threat to their children's safety. Police struggled to restore disrupted order. Despite their diverse perspectives, however, these groups shared a kind of conservatism that attached them to long-standing constructions of the city street. In the automobile they saw a threat to established customs. Upholding time-honored ways, these groups tended to perceive the automobile's intrusions as threats to justice.

Downtown business leaders also saw threats in the new traffic problems, but they seldom characterized them in terms of justice and injustice. Traffic to them meant business, and traffic jams were a threat to the bottom line. "Congestion," department store executive F. C. Fox complained to traffic experts in 1926, "is now undoubtedly seriously interfering with our business." But business leaders remained optimistic. "We know," Fox said, that traffic congestion "can be solved[,] and I think it must be solved by men who are expert in traffic matters."[3] To Fox and other city leaders, the experts were engineers.

Engineers' success with public utilities spoke well for their skill at managing modern city service problems. For a time, therefore, business leaders let engineers define traffic problems as technical matters for trained experts. Chambers of commerce in the 1920s turned to the experts for answers, paid for their studies, accepted their recommendations, and got their cities to implement them.

The engineers' proposals were practical and inexpensive. True, their regulations would burden some street users more than others, but they claimed conditions would improve for all. Experts in traffic control therefore won wide support for their work from disparate interests. In recent decades, historians have normally portrayed street railways and automotive interests as natural and inevitable antagonists.[4] By the late 1920s, the two were indeed often at odds. Yet until then the engineers won support from disparate groups, all of them interested in easing traffic congestion. The secretary of the Cleveland Automobile Club, for example, far from calling for ambitious projects to accommodate automobiles, echoed the traffic control professionals: "they shouldn't be grandiose remedies that we seek. They should be practical. They should be within our pocket books' limits."[5] The street railways complained of autos clogging city streets, but they agreed with the automotive interests that the answer was expert regulation. According to the president of the American Electric Railway Association, "There is no conflict of opinion between electric railways and the automotive industry as to the seriousness of the traffic situation."[6]

Traffic engineers considered themselves voices of reason and moderation, and their professional pretensions reinforced this tendency.[7] Engineers stood for the logic that efficiency worked to the benefit of all, that efficient arrangements must be worked out by disinterested experts, and that as such experts, engineers were the rightful authority in traffic problems. Unlike police, engineers could claim to command the power of science in problem solving. In 1923, a Los Angeles police official declared that, because traffic regulation had "reached the point of a scientific problem," the field "should be divorced from police duty and given to engineers."[8]

But engineers were not independent. Their patrons influenced engineers' definitions of the traffic problem. The traffic engineer Maxwell Halsey tacitly acknowledged the engineers' dependence on their merchant patrons. Engineers, he said, must avoid the temptation to approach traffic problems simply "from the engineering angle"; instead, they should use "diplomacy" with merchants. Engineers should "solicit the support of the merchants," and "work with them from the beginning."[9] Merchants' chief concern was to ease traffic congestion. Engineers, therefore, paid relatively little attention to the needs of pedestrians or to traffic safety. Nevertheless, many considered themselves, as the engineer John Beeler did, defenders of "the welfare of the community" from "selfish and individual motives."[10] As long as they enjoyed the backing of a wide array of business interests—including street railways, auto clubs, and chambers of commerce—engineers succeeded in defining traffic as a technical matter for experts. In the 1920s, engineers pursued traffic efficiency in the name of the public interest. They called this mission "traffic control."

## Traffic Congestion and Urban Concentration

Traffic congestion was due in part to the density of city populations, and the concentration of business in small central districts. Tall buildings downtown and the urban railways that brought commuters to them made the American city of the early twentieth century denser and more centralized than ever before. In the 1920s, one-fourth of Chicago's 3 million people entered its central business district each weekday.[11] Most of Robert Kelker's fellow traffic engineers would have agreed with him that "the prime factor in the present street congestion is the traffic that flows to and from our high buildings."[12] Thus, in order to attack traffic congestion, traffic engineers might have joined city planners in calling for measures to deconcentrate cities. But they did not.

To traffic engineers' chief patrons—chambers of commerce—deconcentration was exactly the danger they were seeking to avoid. Some traffic engineers had past or continuing affiliations with street railways, which also had a stake in the preeminence of central business districts.[13] To city chambers of commerce, deconcentration threatened downtown commerce and property values. Merchants in Portland, Oregon, feared a "shifting of the retail center, with consequent depreciation and loss."[14] Businesses on New York's Fifth Avenue worried that deconcentration would lead to a "crumbling" of Manhattan's most elegant retail district.[15] The guiding principles of traffic control treated deconcentration accordingly. Most traffic engineers saw their goal as easing congestion without reducing density. Two factors reinforced this position.

First, traffic engineers and their sponsors wanted practical, affordable remedies for the traffic problem, the benefits of which could be realized in short order. The failure of police departments' short-term remedies taught chambers of commerce patience, but they could not wait for the city to be rebuilt, and they could not afford to rebuild it. Academic planners, particularly those influenced by the garden city movement, proposed numerous plans of urban deconcentration. The American City Planning Institute traced a catalogue of urban ills to excessive urban density, and so recommended planned deconcentration. Yet for all of their proposals, city planners contributed little to the deconcentration of cities.[16] In the first half of the twentieth century they did not have the means to achieve planned deconcentration. Zoning was their best new instrument. With it, downtown business interests and professional city planners pursued their ambition to limit central business district densities.[17] In 1922, the *Chicago City Club Bulletin* reported with satisfaction that "zoning will flatten out the human pyramid, which congestion has created."[18] Yet zoning was of slight practical effect in the first decades following its implementation, especially as a means of urban deconcentration.[19]

Second, traffic engineers' ideal of efficiency led them to doubt the value of deconcentration. Some engineering traditions exalted grandeur. Traffic engineers, however, were heirs of the municipal engineering tradition, where efficiency was the ruling principle. Apart from the practical obstacles in the way of deliberate deconcentration, traffic engineers sometimes identified concentration as a positive good. Optimum use of limited space demanded concentration. From this point of view, the problem was not really to solve the nuisance of congestion, but rather to manage it. Most traffic engineers would have agreed with a consulting engineer for the city of Portland, Oregon, that "there seems to be no permanent solution."[20] "We have not attempted to solve the problem," St. Paul's city engineer admitted.[21] To most traffic engineers and chambers of commerce, urban deconcentration was not a desirable way to ease traffic congestion, but a dire symptom of inadequate traffic control. It was, one engineer claimed, a "disintegration" stemming from the "strangulation" of the central business district by extreme traffic congestion.[22]

The leading expert on street traffic, Miller McClintock, called deconcentration the "chief threat" of traffic congestion. McClintock warned the Chicago Association of Commerce of its danger to "established retail business values in the central district."[23] He cited the "economies resulting from business solidarity" downtown. He held that the "economic advantages accruing from large scale merchandising" and from "the grouping of similar establishments" necessitated dense concentration. Conversely, deconcentration would "result in smaller and more scattered retail units and multiplied overheads."[24]

To traffic engineers, deconcentration was not a way to fight traffic. It was unconditional surrender. Left alone, congestion would dictate urban form, deconcentrating cities even with no help from city planners. Traffic engineers proposed instead to fight for the efficiencies of concentration without the burdens of congestion. They invoked the public interest to justify the fight. Their weapon was the power of positive regulation, buttressed by recent trends in law and economics. Engineers intended to use positive regulation to achieve traffic efficiency. In this way they hoped to keep vital traffic moving while preserving the supremacy of city centers.

## Taking Social Organization to the Streets

To fight traffic congestion, traffic engineers proposed to bring social organization to the streets.[25] In Miller McClintock's words, to make traffic efficient, cities would have to make "full use" of "existing streets."[26] Engineers' first task, therefore, was to identify inefficiencies. Many arose, engineers believed, from street users' separate pursuit of individual transportation interests. Individual street users could not see the whole transportation picture. Street railways and motor fleets could plan routes and schedules but had no control over the behavior of other street users. Some engineers, however, hoped to coordinate all vehicular transportation.[27] A Detroit traffic engineer with experience in Connecticut's Public Utilities Commission saw his task as curbing street users' "personal whims or convenience" through engineering.[28] Engineers reasoned that if they could curtail the most inefficient practices they could make traffic better for everyone.

### Identifying the Problem: The Traffic Survey

Traffic engineers substituted a formal approach for the common-sense methods of their non-professional predecessors. To avoid "costly experiments," engineers measured first and recommended remedies later.[29] This innovation took the form of the traffic survey, a younger cousin of public utility surveys and time-and-motion studies. Faulting the earlier ad hoc approach to traffic tangles, John Nolen, a renowned city planner, said "you do not solve a problem by putting a traffic policeman at the point of congestion." Instead, "you find the reason why that congestion occurs."[30]

Even in small cities, but especially in the great ones, traffic engineers counted traffic in all its dimensions. Having begun in the 1910s with simple vehicle counts, surveyors soon recorded speed, turns, pedestrian movements, curb parking, and traffic accidents. They examined travelers' origins and destinations, and they distinguished local from through traffic. Engineers claimed that with survey results "deductions" could "be made in a scientific manner."[31]

Engineers' "deductions" reveal how they constructed the traffic problem. They saw their mission as optimizing traffic capacity. To this end, for example, engineers recommended clearing sidewalks of obstacles to pedestrian traffic flow, both for the sake of foot traffic itself and to prevent pedestrians from resorting to the pavement to avoid impassible sidewalk obstructions. In Baltimore, traffic engineers found that "cellar openings, light wells, doorsteps, waste paper boxes and kiosks" interfered with pedestrian traffic and therefore recommended that the city "remove all obstructions from the sidewalk and utilize the full width from building line to curb for pedestrian traffic."[32] Judges generally approved.[33] One legacy of this success was that sidewalks in 1925 or 1930 were recognized as more nearly exclusive channels of pedestrian travel—and thus roadways, by default, were more exclusive channels for vehicles. This sidewalk reform was therefore a small step toward the automotive city.[34]

The traffic survey was also a political tool, giving those who wielded its results an advantage over others. One engineer urged cities to refrain from issuing traffic regulations "unless the traffic authority is prepared, and has the facilities to make an adequate study before and after to prove beyond a doubt the effect of the change." Otherwise, he warned, "the man with the loudest voice or the one who can get the greater number of persons to the 'hearing' will win his case regardless of the merits of it."[35] An official of the U.S. Chamber of Commerce told department store executives in 1928 that surveys could change minds about such things as curb parking: "My general observation would be that unless the retail merchants have had the benefit of a thorough survey of this question of parking, they are very apt to be completely hostile to any parking restrictions."[36] Backing from local chambers of commerce gave engineers the means to conduct careful, thorough traffic surveys. But engineers were resourceful, and they found ways to get the information at modest expense. In many cities, engineers and commercial associations turned to local Boy Scouts for traffic survey duty.[37] Engineers also used Boy Scouts for safety education and parking enforcement, and in at least one city (Newark) they used Girl Scouts for such work.[38]

### Engineering Efficiency: Coordination of Traffic Lights

To fight congestion, engineers abandoned police departments' "adversary model" of traffic regulation, working instead to reconcile speed and safety. A leading traffic expert noted that increasing the speed of traffic on a street increased the street's capacity.[39] Engineers believed that expeditious traffic was safe traffic.[40] They modified police techniques of imposing order on traffic, reinventing them as tools of congestion relief. Police, manufacturers of traffic signals, and downtown businesses devised the first traffic light systems, not engineers. Police often controlled such signals manually; when they were coordinated, they changed simultaneously. To anyone with a bird's-eye view of several city blocks, simultaneous signal changes yielded breathtaking order. To an engineer, however, the system was cloddish and inefficient. It achieved order by worsening congestion. It did not take a traffic expert to see the failure of simultaneous signal changes—but it took an engineer to coordinate signals optimally. "More efficient signal control must be worked out," said one engineer in 1926—"and by engineers rather than by policemen."[41]

The problem was technically complex. Engineers needed to know the volume of traffic on each block of every street involved. From these data they had to derive optimum signal timings, determining intervals that delayed traffic least. And then they would have to find a way to implement the plan.

Nowhere in the world was motor traffic more congested than in Chicago's Loop district. The city's Yellow Cab Company, alarmed by pedestrian casualties that had begun to "break down public good will," implemented the city's first extensive system of lights in 1923. While the lights apparently did protect pedestrians, they also slowed motor traffic enough to annoy motorists.[42] The Chicago Association of Commerce turned to the engineers in 1925. To find a better way, it organized a Street Traffic Committee composed of association members representing a broad coalition of transportation interests.[43] The committee put engineers of the Chicago Surface Lines to work on the problem. They linked signals at 49 intersections to nearby timers, allowing local control. They also wired the signals to a central electromechanical brain in the basement of City Hall. There, clocks, motors and relays gave operators central control of the whole system. At 8 A.M. on Sunday, February 7, 1926, Mayor William Dever threw a knife switch, starting the system.[44]

To many Chicagoans, the coordinated signals were a magnificent fulfillment of engineering's promise. According to one report, "approbation is heard everywhere." Newspapers were "unanimous in their approval of the system," as were "police officials in the traffic department, taxicab drivers, individual car operators, motormen and others."[45] A city traffic expert reportedly was "amazed." Police Chief Morgan Collins said that the lights were "working out far beyond our expectation." All vehicular traffic benefitted. Streetcars could maintain schedules and carry more total riders. Taxis and private automobiles went much faster too.[46]

On the system's first Monday, reporters toured the Loop by car. "From the motorist's standpoint," they concluded, the system was "an instant and unqualified success." When they found that they could drive through four successive intersections without stopping, they compared the experience to a "fairy tale." At midday a cornerman at State and Madison told them that motorists "slip through here like oil. You don't hear the horn tooting and fuss we usually get at noon. Look at the drivers. They've all got smiles on their faces. I found myself smiling a little while ago."[47]

By delivering efficiency, engineers had once again united apparently disparate interests.[48] Other cities followed Chicago's lead, and engineers at General Electric's National Lamp Works (maker of bulbs for traffic signals) simplified the task of calculating optimum signal timings for coordinated systems by producing a special slide rule for the task.[49]

But what was "optimum" signal timing? To most traffic engineers, well-timed signals maximized a street's vehicular capacity. Pedestrians were left out of their equations. Many city people faulted coordinated signals for

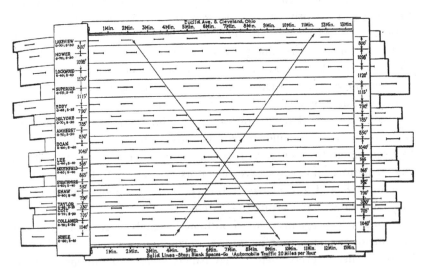

**Figure 5.1**
With slide rules like this one (used in Cleveland in 1927), engineers could coordinate signals to keep traffic moving at a given speed (in this case, 20 miles per hour). Source: "Slide Rule Chart Determines the Timing of Traffic Signals," *Engineering News-Record* 98, no. 6 (February 10, 1927), p. 231.

**Figure 5.2**
A Chicago street in 1929, before coordinated signals and no-parking. Courtesy Schenectady Museum and Suits-Bueche Planetarium.

**Figure 5.3**
Another 1929 photo, showing the same Chicago street as in the preceding photo after coordinated signals and no-parking. Courtesy Schenectady Museum and Suits-Bueche Planetarium.

making it more difficult than ever to cross streets. Cars sped along faster, giving pedestrians fewer safe opportunities to cross at mid block or at red lights. In the Chicago system's first week, the *Tribune* reported that "the walker found life one succession of heart thrills, dodges, and jumps." Signals were timed for vehicular traffic, so sometimes green and amber intervals were too short for those on foot. "At almost any moment in the afternoon or evening," reported the *Tribune*, "a score or more persons could be seen stranded in the middle of these intersections between two steady lines of vehicles."[50] Before, when the cornerman's attention had been absorbed by vehicular traffic, pedestrians had enjoyed relative freedom. Now police were "busy regulating pedestrians."[51] Chicago traffic officials found pedestrians would not conform to the system.[52] Because signal timings in coordinated systems were based on vehicle speeds, they helped to redefine streets as motor thoroughfares where pedestrians did not belong.

Around 1920, with the backing of local chambers of commerce, traffic engineers began to pursue such social organization of city streets. Engineers formalized customary rules (vehicles should keep to the right of the center of the street; pedestrians should not congregate in the middle of streets), urged better enforcement, and developed new rules. Engineers segregated traffic into lanes and imposed one-way streets, loading zones, through-traffic streets, and pedestrian sidewalks. They narrowed the functions of both the sidewalks and the roadway. This formalization of traffic forever changed the character of city streets. Within a decade, streets and sidewalks—venues of myriad public activities as late as 1920—had largely been redefined as exclusive transportation ways, subject to regulation in the name of efficiency. Cincinnati's city manager—a trained engineer—put the new principle this way: "as traffic demands grow more acute the use of streets for other purposes must be more and more restricted."[53]

### Traffic Engineers and Automobiles

By correcting inefficient vehicle movements, optimally coordinated signals eased traffic for all vehicles. Yet many engineers identified another source of inefficiency—this one inherent in certain vehicles themselves. As the engineer Robert Kelker recommended, when a traveler's actions "delay the movements of hundreds or thousands of people he then should be compelled" to forgo individual preferences for the sake of the "public welfare."[54] On these grounds, many engineers found that the worst contributors to congestion were motorists.

Engineers ranked the automobile and urban concentration as the two leading causes of the problem. In traffic congestion, Miller McClintock contended, "the skyscraper and the motor car play the principle rôles."[55] Engineers sometimes cited the automobile as the sole necessitator of professional traffic control.[56] Like his fellow engineers, Clarence Sherrill perceived a "necessity of giving precedence to vehicles in the streets in the order of their capacity for moving the greatest number of riders in the least possible time and using the least possible space in doing so." Sherrill, Cincinnati's city manager, put "private motor cars" at the bottom of his ranking of vehicle efficiency.[57] Passenger cars, engineers said, were less efficient than other modes, and some added that the errands they served were less important. One engineer argued that "the commercial vehicle is of far more value to the economic life of the community than the pleasure car."[58]

Engineers seldom looked at traffic from the auto driver's point of view. For example, most engineers were indifferent to the driver's need to park. One journalist found that New York traffic "experts" were "concentrating on the problem of how [curb] parking can be reduced to a minimum, or banned completely."[59]

Above all, engineers faulted automobiles for their prodigal use of space. They "occupy, either while in motion or while parked, space altogether out of proportion to their transportation efficiency," one engineer wrote.[60] A St. Paul engineer found that an occupant of an automobile required 10.7 times as much street space as a streetcar rider. Other engineering estimates were even less flattering to motorists.[61] Automobiles crowded streets everywhere in the 1920s, but engineers found that they were carrying only a small fraction of city travelers.[62] As defenders of urban concentration, traffic engineers could not forgive motorists for taking up so much room.

## The Parking Evil

To the engineer Clarence Sherrill and to many of his colleagues, the "crux" of the traffic problem was "the parking evil."[63] They sought strict control of curb parking, or outright bans (generally called "no-parking"). St. Paul's traffic engineer, for example, called for "the most drastic regulations on parking."[64] A journalist concluded that parking restrictions were the engineers' first and favorite resort.[65]

To hear that curb parking was an "evil" must have come as something of a shock to most non-engineers. Custom decreed that the centers of streets were to be kept clear for travelers, but the verges were places for almost any kind of public use. Though the curb seemed to mark a distinct boundary between sidewalk and roadway,[66] in practice the border between them was far less distinct. Boys hawked newspapers, shined shoes, or played marbles. Merchants stored crates, acquaintances chatted, horses drank, pushcart vendors sold fruit, and builders stacked bricks. Few of them had any formal grant of privilege. And towns often put telephone poles, fire hydrants, and watering troughs in the roadway near the curb.[67]

Laypersons defended curb parking as every citizen's prerogative. Observing engineers' pursuit of "absolute prohibition of parking in the congested center of the city," some were reminded of the Eighteenth Amendment.[68] A letter to the editor of a Detroit newspaper compared those who abused "parking privileges" to "bootleggers." Even engineers made the comparison.[69] Defenders of the automobile resisted the new prohibition. Auto club managers, such as Cleveland's F. H. Caley, were not alone in describing curb parking as "a fundamental right."[70] Where cities restricted curb parking, motorists "promptly rushed into print with charges of discrimination."[71]

The clash reflected fundamental disagreements about what a street is. Most city people, including motorists and pedestrians, clung to custom. To them, the verges of streets were for diverse uses. Efficiency-minded engineers reconceived the street, from curb to curb, as public infrastructure for transportation. The entire roadway, they said, was intended exclusively for moving vehicles. Engineers repeatedly claimed that they were merely returning the street "to the purpose for which it was intended."[72] They faulted towns for allowing "misplaced traffic lights and numerous other obstacles that reduce actual roadway capacity to a small fraction of what it [sic] was intended."[73] If, as McClintock argued, streets were for "public travel," then "any use, including parking[,] which interferes with this primary use must be considered . . . an abatable nuisance."[74] Engineers maintained that "parking is not a 'right,' it is a privilege," subject to the discretion of the experts.[75] Judges usually agreed.[76]

Drawing on the public utility model, engineers based their attack on curb parking on their definition of the public interest, citing three principles: social utility, equity, and efficiency. Curb parking, they said, violated all three.

First, engineers claimed, curb parking served no public purpose—indeed it was an appropriation of public space for private use. Traffic engineers therefore reminded their clients that "streets are public property."[77] St. Louis's Director of Streets and Sewers described curb parking as "the use of public property dedicated to transportation needs, for storing [private] property," and traffic engineers expressed countless variations of this view in the 1920s.[78] Engineers defined public use as transportation and deemed other uses private. Engineers working on Baltimore's traffic problems in 1925 and 1926 complained of "push carts and temporary huckster and vendor stands" interfering with transportation.[79] The first step in securing the public interest in city streets, engineers held, was to discriminate between public and private uses and to subordinate and restrict the latter for the sake of the former. The parked car, McClintock argued, "must follow the peanut wagon, and the sidewalk showcase into oblivion."[80]

Second, curb parking was inequitable: it served the convenience of a small minority at considerable cost to all. In congested districts, engineers said, curb parking led to "the interests of thousands" being "sacrificed" to "the convenience, real or imaginary, of the comparatively few."[81] They found that curb parkers represented a tiny fraction of street users; toleration of curb parking therefore amounted to "preferential treatment." "It seems unreasonable," a Chicago engineer complained, "that a comparatively few people can utilize the most valuable street space in our cities, practically at will, for their own pleasure and convenience and to the serious inconvenience of thousands of their fellow citizens."[82]

Few motorists could find space at the curb. Rochester, for example, had downtown space for 3 percent of its autos.[83] Parking meters did not appear until 1935. Even time limits were scarce, so motorists everywhere filled up curb spaces early in the morning. The lucky few often left their cars parked all day. With dozens of autos for each available space, only a small and shrewd minority could benefit from the free space. Engineers condemned the result as "unfair" and branded the few who found spaces "a privileged class."[84]

Through surveys, engineers got numbers with which to fight such abuses. In Detroit in 1929 they found that the 22 percent of curb parkers who parked for more than an hour hoarded 60 percent of the daily capacity of the curbs.[85] In Chicago, six autos entered the central business district for each curb space, and in more motorized St. Louis 22 autos went downtown for each space there.[86] Thus, even among motorists, engineers said, curb parkers were "a specially privileged class to be allowed to occupy so large a percentage of the street area with such a very small percentage of the motor cars."[87] According to McClintock, engineers could right such inequities by regulating street curbs "in such manner that all who desire to use them may be given an equal opportunity."[88]

Finally, curb parking was inefficient. The space occupied by the curb-parking minority could be put to better use. Washington (second only to Los Angeles as the most motorized large city in the United States) calculated in 1928 that 29 percent of the surface area of its streets was occupied by parked motor vehicles.[89] Engineers found such allocation "a highly wasteful use of expensive streets."[90] Clarence Sherrill estimated that in the downtowns of large cities the 100 square feet of space that a curb-parked auto used was typically worth about $3,000.[91] That motorists used such valuable space gratis was all the more galling to engineers when they found that three-fourths of drivers reported they would be unwilling to pay any money to park.[92] Another engineer estimated that the cost per vehicle of providing street space for parked cars was about 14 times the cost per vehicle of providing space for moving cars.[93] "The joke here," an engineer concluded, "is surely on the taxpayers who foot the bill, not on the 'wise guy' who gets there first."[94]

Curb parking, engineers said, aggravated the problem they sought most to relieve: congestion. A single car at the curb made a whole lane almost useless.[95] Unrestricted curb parking attracted autos downtown and encouraged motorists to cruise the streets looking for spaces. Washington motorists arriving downtown on weekday afternoons added nearly a mile to their trips searching for free spaces.[96]

"The streets are for moving traffic," said George Herrold of St. Paul. Other uses were subject, he claimed, to strict curtailment in the interest of this essential purpose.[97] A Detroit traffic engineer called for "the return of the streets to the purpose for which they were originally intended—that of moving vehicles instead of using them for storage purposes."[98] In this effort to redefine streets, engineers directed their first attack against curb parking.

### Regulating Curb Parking

As engineers inherited traffic regulation from police, they transformed police techniques to fit their own technological frame. Before the introduction of the parking meter in 1935, curb spaces were free. To ration the scarce spaces, police quickly introduced time limits. Detroit was already restricting parking in this way in 1915.[99] By 1920, Boston had a city-wide 20-minute limit and was "tagging" (ticketing) violators.[100] Where local police could enforce limits well, time limits could triple a city's curbside parking capacity.[101]

Most cities did not give their police departments what they needed to enforce parking time limits. South Bend, Indiana, a city of 71,000 people, assigned only two policemen to parking enforcement in the early 1920s.[102]

Though manpower was scarce, enforcement was laborious. To simplify the task, police early hit upon the idea of marking tires with chalk. Before the development of special legal procedures for traffic violations, however, police had to issue a court summons in person to each violator.[103] Boston's police commissioner admitted that it was "difficult for the policeman to locate the person actually responsible."[104] Compounding these problems, ticket fixing was rife. A Kansas City reformer claimed that his city was "fixing tickets at the rate of 25,000 or 30,000 a year" and that the accumulated unpaid tickets numbered 43,000.[105]

Time limits for curb parking were often "absolutely unworkable."[106] "In no city," McClintock concluded in 1925, "has it been found possible strictly to enforce limited parking."[107] In 1929, New York traffic surveyors discovered that "most of the machines" parked in two-hour zones stayed "three hours or more."[108] Three-fifths of the cars at Rochester curbs in 1928 were illegally parked, and 38 percent of the cars parked in 10-minute zones remained there more than an hour.[109]

Traffic engineers put little effort into improving the enforcement of parking time limits. They did even less to compel motorists to pay for curb space. They wanted to clear the curbs entirely. "There has been much energy wasted on the question of 'parking time' by police, civic and commerce associations, city councils and merchants," George Herrold complained. Even with limits, a curbside lane with parked cars "can not be used for moving vehicles."[110] A Chicago engineer warned that "limiting the time of parking to a half hour or to an hour does not do away with the nuisance."[111] A New York businessman agreed, arguing that a time limit "does absolutely nothing toward clearing parked cars from the streets; it merely shuffles them around" in a "chalk mark game."[112]

Traffic engineers saw proposals to widen important thoroughfares as wasteful. Since "in most cities the building of new streets is not economically possible," engineers argued, it was "necessary, when the traffic flow becomes too heavy, to take back the privilege of parking and utilize that space for travel."[113] Engineers shared a special distaste for reconstruction to accommodate curb parkers. Leon Brown condemned it as extravagant.[114] Another engineer, Hawley Simpson, complained of "instances without number" in which cities spent enormous sums to widen streets for "needed traffic relief" by "buying private property and condemning valuable buildings, only to have the added space pre-empted by a line of parked cars."[115] One consulting engineer recommended that "as soon as a street becomes so congested that either the street must be widened or parking on the street eliminated, the latter course should be pursued."[116]

Engineers therefore recommended that curb parking be banned in congested districts. "The ultimate and only satisfactory solution of this parking problem," one engineer contended in 1922, was to "prohibit all parking on streets in business sections" and hope that lots and garages would take up the slack.[117] Engineers admitted that they would be quite pleased if "no-parking" discouraged one mode of transportation to the benefit of others. Miller McClintock was unconcerned about motorists barred from parking at the curb, since they could "resort to some other means of transportation." Because of "the adequacy of rail and bus transportation in most cities," motorists unable to find curbside spaces merited little sympathy. Curb parkers, McClintock reasoned, put little value themselves on a practice for which they paid nothing; banning parking might therefore ease congestion "by weeding out those vehicles which have no real business in the area."[118]

### Merchants and Parking

Traffic engineers' hostility to curb parking reflected the interests of their clients. The chambers of commerce for which most traffic engineers worked represented big downtown businesses, especially department stores. Downtown department stores risked losing business to merchants in less congested districts. Even without the help of engineers, some stores saw an answer in parking bans. When a large Brooklyn department store found that fewer than 1 percent of its customers arrived by automobile, it joined the cause.[119] Large department stores found their curbs lined with automobiles belonging to customers of nearby small specialty shops. "Everything that does not move must go," the New York merchant W. W. Arnheim, chairman of the traffic committee of an association of Manhattan retailers, demanded. (Arnheim explained this paradoxical demand by compounding the paradox: "The parked car must go."[120])

But the parking question divided merchants. Big merchants' fear of deconcentration led them to back parking bans. "Unless something is done," F. C. Fox worried, "the downtown shopping district will soon be a thing of the past." "We have made a huge investment in real estate and in building up good will," he added. With deconcentration, "the money we have spent for the good will and invested in our plant, would be seriously jeopardized."[121] But many small merchants fought "no-parking." They thought of crowded curbs as a sign of thriving retail sales. "No parking means no business," said one city business leader.[122] Colonel A. B. Barber of the U.S. Chamber of Commerce told department store executives in 1928 that merchants resisted parking restrictions because they were "apt to think that all parked cars are cars of their customers."[123] Low-volume, specialty, and upscale merchants dreaded parking bans. Their critics blamed "the opposition of some newspapers to sensible parking regulations" on "merchants, usually small storekeepers."[124] Others blamed the "considerable protest" against Chicago's 1928 ban on "a few small merchants and several owners of second-class office buildings" and other "very minor interests."[125]

Upscale merchants with small shops had reasons to fear "no parking." They saw the "carriage trade," now getting around in automobiles, as their best customers. Engineers offered these shopkeepers supporting evidence. Surveys showed that in Washington curb-parked motorists paid 52 percent more per purchase than shoppers who had not arrived by automobile.[126] Shopkeepers generally supported time limits on parking but fiercely opposed "no parking."[127]

Shopkeepers disagreed with traffic engineers' construction of traffic problems, especially parking. They agreed that congestion was choking downtown commerce, but they explained congestion differently. In 1925 and 1926, the Commerce Department's Bureau of Foreign and Domestic Commerce surveyed 1,500 merchants around the country. The merchants complained of business losses from traffic congestion ranging from 1 percent to 20 percent. Yet merchants did not blame automobiles for these problems. To them, the greatest contributors to congestion were "faulty traffic regulations." They also blamed not parkers but the lack of parking space, and not traffic lanes relegated to auto storage but narrow streets. They declared streetcars worse traffic cloggers than automobiles.[128]

### "No Parking" in Practice

Yet engineers' numbers persuaded the leading downtown businesses, which dominated the chambers of commerce. In the early years of traffic control, engineers held together enough support from business associations to get some parking bans implemented.

In banning parking, Chicago's congested Loop district was again an example to the nation's cities. While Chicago Surface Lines engineers were working on their plan to coordinate the Loop's traffic signals, the Association of Commerce called in Miller McClintock to work on parking and other traffic problems. In January 1926, McClintock's staff began a comprehensive traffic survey for the city.[129] Until 1926, Loop district motorists could even park perpendicular to the curb, but police required parallel parking soon after the coordinated signal system went into operation. Many motorists ignored the 30-minute parking limit, even after the *Tribune* began publishing violators' names. Police stepped up enforcement but could not clear the curbs.[130]

With the resources of the Association of Commerce behind him, McClintock directed what was then the most extensive traffic survey in history.[131] McClintock's team spent a year and $50,000 gathering information and assembling a 300-page report. The Association of Commerce, not the city, paid the entire cost.[132]

Armed with the survey results, McClintock returned to the Association of Commerce with his advice: Ban daytime curb parking. Surveyers observed how 96,000 Chicago shoppers arrived downtown and found that only 1.57 percent of them reached stores from curb-parked autos. Although McClintock's case persuaded the Association of Commerce, many Chicago merchants attacked the proposed ban. They feared losing the carriage trade. Amid controversy, the city decided to try McClintock's plan for 90 days. On January 10, 1928, the city banned daytime parking in the Loop district, eliminating 1,743 curb spaces. The results convinced many skeptics, and the city made the ban permanent.[133]

McClintock found that "the success of the 'no parking' rule in the Chicago 'Loop' district has led to a more open-minded attitude in regard to prohibited parking by business men and motorists alike."[134] Similar efforts elsewhere often succeeded, especially when engineers began with surveys. Results from a Pittsburgh parking survey, for example, "aided materially" in overcoming merchants' resistance.[135] Improved traffic flow and sales receipts in towns with restricted parking persuaded other towns to give parking restriction a try. In small cities, too, many business owners found that parking restriction improved their trade.[136]

Traffic engineers found rush-hour parking bans far more effective in fighting congestion than time limits. Such limited bans grew common in the mid 1920s. In 1926, after New York's police found the enforcement of parking regulations "impossible," the city banned rush-hour parking on major north-south avenues below Fifty-Ninth Street.[137] "No parking" simplified enforcement and cleared lanes for moving traffic.[138] Even merchants who prized their curb-parking patrons were often pleased. They found that morning rush-hour bans prevented some all-day parking by commuters, leaving the curbs free for their customers.[139]

### A Fragile Consensus

For most of the 1920s, traffic engineers found their recommendations respected and adopted, if not always admired. Engineers earned the

approval of most of the business associations that hired them, and sometimes they united streets railways and auto clubs behind traffic control.[140]
Traffic engineers were confident. Engineers had solved the water waste
problem. They had stopped typhoid fever.[141] The ultimate success of the
engineering model was "not to be doubted."[142] Their patrons often agreed.
F. C. Fox, a Brooklyn department store executive, was sure that congestion
could be solved by "men who are expert in traffic matters, and not by
political committees."[143]

Yet this traffic control consensus was fragile. In isolated places, and in a
disorganized way, many merchants dissented from the local chambers of
commerce that had accepted restrictive measures, especially "no parking."
Traffic congestion and measures taken to relieve it divided merchants. As
street railways organized to call for stricter control of traffic, auto clubs
courted motorists who found traffic control a nuisance. Others in motordom worried that restrictive measures in cities would cap the demand for
automobiles among the fast-growing urban population.

The "political committees" that F. C. Fox criticized were never far removed
from the deliberations over the traffic problem. Engineers claimed special
authority as disinterested experts, but they depended entirely on their
patrons in business. When members of these coalitions found elements of
traffic control threatening, they withdrew to form new bodies claiming
expert authority. As chapters 6 and 7 will show, the role of such "political
committees" in traffic questions grew in the 1920s, and often their members
were less coy than engineers about representing interested parties.

In the mid 1920s, some of the critics of traffic control proposed another
new construction of the problem of city traffic. Forgoing the "public interest" language of the engineers, they drew from the lexicon of natural
rights, especially in the form of classical political and economic liberalism.
The new rhetoric helped them to challenge the engineers' claim to special
authority in traffic matters, legitimized the voice of other interested parties
and laypersons in traffic questions, offered a new diagnosis of congestion,
and promoted a new tolerance for the traffic demands engineers had tried
to restrict. The dissenters' culturally potent rhetoric challenged the engineers' public utility model of street traffic.

# 6 Traffic Efficiency versus Motor Freedom

We . . . want the operation of all motor transportation more thoughtfully and rigidly regulated and controlled.
—B. C. Cobb, American Electric Railway Association, 1926[1]

When the cities and their traffic commissions begin to find that the traffic problem cannot be solved by putting drivers in jail, they will turn their attention to the streets.
—Alvan Macauley, president, Packard Motor Car Company, 1925

Through the 1920s, professional traffic control measures joined and partly displaced police traffic regulation. Wherever congestion squeezed commerce, surveys, "no parking," coordinated signals, and other professional measures soon followed. Traffic lights replaced cornermen at big-city intersections and even spread to small-town crossroads. The standard, professionally timed traffic light slowly displaced silent policemen in all their variety. By 1938, the sociologist Louis Wirth could name "the clock and the traffic signal" as the two symbols "of the basis of our social order in the urban world."[2] As tokens of urban progressivism, traffic lights so enamored small-city boosters that engineers had to warn local authorities to use them only at intersections that actually needed them.[3]

Traffic control principles appeared to have prevailed. Even today, in the dense centers of cities, away from the limited-access urban highways, the traffic signal remains the icon it was for Wirth in 1938. Yet even before Wirth made his choice of symbols, signals and the other regulatory methods of traffic control were sometimes identified not as emblems of transportation progress and urban progressivism but as relics of backward or even un-American thinking.

Even before professional traffic control's day, some academic city planners and public intellectuals had suggested more glamorous alternatives to bland, regulatory traffic management techniques. Yet nearly all such proposals were strictly visionary.[4] In the mid to late 1920s, however, a new model of city traffic management emerged. It lacked the impressive drawings of the visionaries, but its radically new perspective on urban roads and streets inspired enthusiasm from social and business groups—local, regional, and national—that saw in traffic control a threat to their future. Their conceptual reconstruction of traffic congestion as a shortage of street capacity would lead to a physical reconstruction of American cities for the motor age.

### Diverging Interests: Regulated and Unregulated Transportation

Fast growth in the demand for scarce street space continued apace in the 1920s. Under the pressure of ever-increasing motor traffic, the engineers could not repeat their stunning accomplishment in water supply, when they had met the needs of all interested parties.[5] Despite their best efforts, city traffic remained too congested to keep all concerned satisfied.

But engineers' greatest obstacle was growing resistance to traffic control in principle. Such resistance stemmed not from engineers' failure to ease congestion and to satisfy travelers' demands efficiently, but from the very kind of success it promised. When engineers urged the restriction of inefficient street users, above all they meant motorists. Those with a stake in automotive transportation, therefore, soon withdrew the trust they had placed in engineers. Such resistance prevented the full implementation of traffic control.

Under these stresses, the loosely woven traffic control consensus of the mid 1920s unraveled. Automotive interests and street railways had joined with business associations to take the traffic problem away from police and give it to engineers. By the mid 1920s, however, motordom was faulting the engineers' methods. As street railways continued to back traffic control, contention between automotive interests and railways grew bitter.

It was not a simple case of modal rivalry—a "struggle between road and rail."[6] The distinction between the modes was not so clear. Street railways enthusiastically adopted buses when reliable models were available, and considered them useful supplements, not competitors, to streetcars. Buses improved street railway service by feeding commuters at the urban fringes to streetcar stops in the city. Bus manufacturers considered electric railways their chief customers, and assured them that the bus was an adjunct to streetcars.[7] Street railways operated most city buses; their industry association estimated that "practically all major city bus operations in the country" were run by the railways and their subsidiaries.[8] Railroads bought trucks from manufacturers who marketed them as specialized supplements to freight trains. "Bulk and distance haulage is exclusively a steam railroad function," said a truck manufacturer in 1924. "We would much rather have the railroad for a customer than a competitor."[9]

The front line of battle was not between road and rail; rather, it divided regulated and unregulated modes. The most regulated vehicle in cities was the streetcar; for a time its greatest rival was the unregulated jitney.[10] Their fight was an early sign of growing rivalry between those modes of transportation regulated as public utilities (whether on rails or not) and those not so regulated. Under the stress of increasing traffic congestion, regulated modes tried to expand the sphere of such regulation to other modes, while other modes fought against the application of public utility principles in urban transportation.

Beneath this clash of interests lay a wide divergence of perspectives on streets and their legitimate use. Few articulated either perspective explicitly, but both sides' positions on more particular matters nevertheless reveal them, just as a mosaic of photographs reveals the position of a stationary photographer. Street railways saw urban transportation as a city service, like other services (such as water and gas). From this perspective any use of the street network, as a publicly owned component of all urban transportation systems, was fundamentally subject to regulation. In time operators of unregulated vehicles developed a sharply different perspective. To them, streets were a marketplace, a setting in which the fittest modes would survive provided political influence, in the form of regulations, did not protect the unfit. These perspectives grew clearer as stubborn congestion raised the stakes for both sides.

This is to say that street railways had no fundamental objection to regulation. This claim may come as a surprise to those acquainted with explanations of the decline of American railroads, including street railways. Historians have shown that burdensome and sometimes inequitable regulation discouraged investment and drove many companies into bankruptcy.[11] Street railways objected to many particular regulations. They denounced obligations to maintain pavements between rails, to clear snow along their routes, and to carry city officials free of charge.[12] Railroads complained about regulations—but they believed in regulation. Like other natural monopolies, railways depended upon the public utility principle—and the regulation that went with it—to protect their enormous investment in infrastructure and to attract new investors. To the railways, the danger was not regulation, but "the toleration of unregulated competition in utility service."[13]

Street railways constructed traffic congestion as a problem caused by inefficient and inequitable modes operating in the absence of proper regulation. Buses operated by small, independent companies were generally unregulated, and often targeted the most profitable portions of the street railways' routes at the most profitable times, setting fares without the meddling of state utility commissions. The street railways, meanwhile, were forced to charge the same fare at peak hours and at lulls, and to run unprofitable, low-demand routes. Accepting such regulations, the American Electric Railway Association attacked unregulated competitors.[14] Instead of demanding simply a reduction in their own obligations, in the 1920s street railways called for the regulation on public utility principles of other vehicles.

Bus and trucking companies, as common carriers, were obvious targets for such efforts. A railroad executive told manufacturers of auto parts that "trucks and buses which hold themselves out to be common carriers for hire should be treated as other common carriers," such as railroads, "and placed under the jurisdiction of public regulatory bodies."[15] A civil engineer with public utilities experience condemned as "iniquitous" bus and truck companies' "free use of public highways in competition with [regulated] railroads and street railways."[16]

Like the traffic control engineers of the early and mid 1920s, however, street railways believed the public utility principle should reach further. They sought to extend such regulation to the streets themselves, and thus to all vehicles on them. George Baker Anderson of the Los Angeles Railway suggested that "the scientific use of the streets secured by modern schedules on the electric railways might be extended to free-wheel vehicles."[17] B. C. Cobb of the American Electric Railway Association said street railways wanted "the operation of all motor transportation more thoughtfully and rigidly regulated and controlled."[18] The Chicago Surface Lines called for the extension of regulation to "other traffic groups" on the city's streets, in efficiency's name.[19] C. A. Copper of the Los Angeles Railway argued that because of the limited capacity of city streets, the expense of widening them, and the consequent public interest in the efficiency of the traffic they carried, "there should be extended to all urban traffic the principles of control and restriction that apply to public utilities generally."[20] As such calls grew from a murmur to a chorus, unregulated modes replied by staking out a new position of their own.

## Saturation: The "Floor Space" Problem

In 1920, automotive interest groups were still diffuse, and their position on city traffic problems was indistinct. Local auto clubs and dealers backed traffic control as an improvement upon police regulation. Soon, however, many of them began to see in the traffic control principle a Trojan horse. Traffic control threatened to lure the unwary among them with its promise of congestion relief, while within it hid the forces that would destroy them.

The change reflected a new uncertainty in the industry's future. In 1916 there were more than 30 Americans for each passenger automobile. Industrial mobilization for war, materials shortages and a postwar recession interfered with consumer demand for the next few years. By 1922, however, with these problems behind them, manufacturers set new sales records. Americans bought more 3.6 million automobiles in 1923, most of them in the first half of the year. There were then only 7.4 Americans for each car.[21] Industry leaders were optimistic. When some spoke of a day when Americans' thirst for cars would be slaked, and the market "saturated," manufacturers publicly scoffed.[22] Yet before the end of the year, despite a robust and growing economy, sales flagged. In 1924 dealers sold 12 percent fewer vehicles, unsold cars crowded lots, and dozens of overextended manufacturers failed or took buyouts. Survivors in the industry anxiously looked for an explanation. A growing market for used cars was part of the problem—but it could not explain everything. Had manufacturers begun to saturate the market? Would Americans soon own all the cars they wanted? Coinciding with the sudden sales slump, trade journals in the automobile industry raised the bogey of market "saturation." Lest they undermine confidence, the writers denied the threat—but the frequency of their denials confirmed the industry's fear.[23]

Until the crisis, manufacturers had little to say about city traffic. Saturation fears ended their silence. Quite suddenly, beginning in 1923, the industry sought a new and direct say in traffic matters, and its role grew swiftly through the rest of the decade. It began to listen when engineers predicted that "saturation" was coming. The St. Paul engineer George Herrold estimated that year that the ratio of persons to vehicles would probably not fall below five, and others concurred.[24] Auto industry people, a trade journal now admitted, "feared the time when the power of the public to buy motor cars" would become "almost infinitesimal."[25]

Some in the industry blamed the messengers. A trade journalist faulted the misleading forecasts of the "would-be prophets."[26] Some engineers deepened the rift by welcoming saturation with unconcealed eagerness. Saturation, they said, would keep traffic congestion from getting worse, and it would bring with it the day when traffic accident deaths per capita would "cease to increase and possibly decline."[27]

Engineers usually understood the saturation point as a simple function of the satiation of market demand for a consumer durable. At least in public, manufacturers were unwilling to concede that such an inflexible, natural barrier to their market existed, and such forecasts annoyed them. In a manner typical of his colleagues in the industry, Fred Fisher, president of the Fisher Body Company, attacked the saturation point idea. "There ain't none," Fisher said, because Americans "won't walk."[28] The manufacturer Edward Jordan agreed. "There is absolutely no end" to the demand for automobiles, he claimed; "they will keep on building automobiles until everybody has a good one and none ever wears out." He advised his colleagues: "When you hear anybody talking about next year's volume falling way off . . . just remember that every single man and woman [and] every family that possibly can dig, scrape, borrow, beg or steal enough money is going to have an automobile."[29] Others in the industry joined the attack on the "fallacy of buying-power saturation."[30]

But how to explain the flattening sales? To the auto industry, the first clue was the urban character of the sales slump. The growth rate in auto registrations was markedly lower in cities than in the country as a whole. For example, between 1915 and 1925, while Baltimore's population growth rate was about twice that of the nation as a whole, its motor vehicle registrations grew at only about half the national rate.[31] Clearly much of the difference must be credited to the greater utility of motor vehicles in rural areas, and to a large population of recent immigrants in Baltimore. But the size of the difference suggested, at least to manufacturers, that more was at work. And even those residents of Baltimore who owned automobiles began to avoid driving them into the central business district. Of vehicles registered in the city, engineers found a smaller proportion going downtown each year from 1922 to 1925.[32]

To explain exceptionally slow sales in cities, motordom reconstructed "saturation." There was no "buying-power saturation," it said. The real bridle on the demand for automobiles was not the consumer's wallet, but street capacity. Traffic congestion deterred the would-be urban car buyer, and congestion was saturation of *streets*. In the spring of 1923 an industry journalist, still expecting motor vehicle sales to "double," nevertheless warned that this doubling "will not take place in the city districts!"[33] A year later, *Automotive Industries* warned industry leaders that traffic congestion "tends to choke off the sales of automotive vehicles today."[34] *Motor Age* warned of the "danger" in "a condition of highway saturation, especially in and near the great cities."[35] Paul Hoffman, a Studebaker executive, later named this proposition the "'physics, rather than economics' theory," explaining that "the so-called saturation point of automobile ownership in any large city is controlled" by street capacity, not wallet size.[36]

Another auto executive described saturation as a matter of "floor space," a metaphor others in the industry soon grew fond of. "The problem," wrote Edward Jordan, president of the Jordan Motor Car Company, "is not one of temporarily policing, arrangements of streets, and traffic signals and everything of that sort, but the fundamental problem is one of floor space."[37] The manufacturers' trade association, the National Automobile Chamber of Commerce, soon took up the cause. In the summer of 1923, the group's general manager, Alfred Reeves, urged his industry to act to protect its market. "This city congestion is proving a great obstacle to the use of cars," he said; "nothing could be more helpful to the industry than to see the situation cleared up."[38]

Under traffic control principles, saturation of street space would justify further restriction of auto traffic, and even limited bans. An alarmed auto industry asked itself "Will passenger cars be barred from city streets?" Its answer was that as long as traffic engineers were in charge, auto bans were "not improbable" on "many streets" of major cities.[39] This was no exaggeration. For "extreme instances" of congestion, a Philadelphia engineer recommended "the prohibition of all privately driven vehicles from the central area during business hours."[40] For such positions, engineers had the support of the street railways. The president of the American Electric Railway Association called for a "reduction of the volume of unnecessary traffic"—automobiles—in city streets.[41] This threat spurred the industry to find a new model of traffic, one which would allow more, not fewer, automobiles on streets. The new model was based on a new construction of city traffic and the streets it ran upon. Alvan Macauley, president of Packard, was one its promulgators.

Macauley was through with the traffic engineers' regulatory habits. "Apparently the prevailing theory is that as traffic congestion increases, all that is needed is simply more rules, more one-way streets and more signal towers—and bigger traffic courts." The "obvious lesson from the burdensome tangle" of traffic regulations was that, so long as the traffic control model prevailed, "regulation and restriction increase" in response to congestion. Such regulation might help department stores get more customers to their doors, but this was not Macauley's affair. He wanted to sell cars to city people. And under the traffic control model, "As automobile use is handicapped, so inevitably must automobile selling and automobile production be handicapped."[42] Macauley attacked the "engineers of some cities" who "feel that this fundamental difficulty" of traffic congestion "is due to the automobile." Their attitude was one of "hopeless pessimism." The problem was quite the opposite, he wrote; "the difficulty is due to the street." Traffic control "is only a temporary palliative, not a cure."[43] By redirecting responsibility for congestion from automobiles to the streets they ran upon, Macauley was proposing a radical reconstruction of the city traffic problem.

The saturation crisis coincided with a worsening public relations problem for the industry, as the number of dead in motor vehicle accidents first exceeded 20,000 in 1924.[44] The problem was worst in the cities, compounding the threat from the "floor space" problem. Thus, in 1923 and 1924 the industry faced two new threats that quickly raised city traffic to one of its leading problems. In the summer of 1923, the general manager of the National Automobile Chamber of Commerce spoke to industry colleagues of the slackening pace of sales that had already begun. He put safety and congestion at the top of his list of the challenges that manufacturers would have to overcome to improve sales.[45]

The industry discussed plans to fight for its future. In doing so it acquired a new, more cohesive agenda. It hoped that cheaper cars could reach those who had been unable to afford them. It sought to increase exports in a world where, outside of the United States, autos were still scarce. The industry hoped to encourage two-car homes, it introduced annual model changes, and it worked to overcome the "used car evil" by restricting sales of such cars.[46] Such efforts, industry leaders reasoned, could overcome the danger of saturated buying power. And it fought to take control of the safety issue.[47]

Yet none of these measures could ease the "floor space" shortage. Solving it would require a new paradigm of urban transportation planning, one that would fight congestion by increasing the supply of "floor space" in response to (and even in anticipation of) demand for it. Macauley called for wider streets and for "wholesale replanning and rebuilding."[48] Engineers warned that the construction of new traffic facilities would not in itself end congestion, since new capacity would invite new demand. But automakers saw no threat in this. Referring to a New York City engineer's warning that new roadways "would be filled immediately by traffic which is now repressed because of congestion," *Automotive Industries* calmly observed that this was "an interesting thought from a sales standpoint."[49]

## Mobilizing to Fight Traffic

Fighting traffic control and its efficiency model would not be easy. Traffic engineers had powerful backing from downtown business associations and street railways. And supplying more "floor space" would take money cities did not have.

### Street Railways

As a source of support, chambers of commerce were unsteady. Engineers could not win chambers' consent for their more drastic proposals, such as auto bans, even in the most crowded streets. They were more successful in keeping some curbs free of parked cars, at least during rush hours. Where they existed, however, parking bans and restrictions were often poorly enforced.[50] Street railways were usually represented in the business associations that sponsored traffic control, and they were its most reliable source of support. When business associations wavered in their support of traffic control, street railways sometimes took matters into their own hands, taking up the cause of traffic control themselves. They did so by building a more reliable base of engineers committed to traffic control, and by selling the traffic control model directly to city people.

Some railways used their own engineers and the prestige of their national organization to position themselves as legitimate traffic experts with a direct role to play in fighting traffic. Frank Coates, president of the American Electric Railway Association, urged street railway executives "to regard themselves as transportation experts rather than as strictly electric railway experts."[51] "Our industry," wrote George Baker Anderson of the Los Angeles Railway, "has the most extensive traffic experience of all users of crowded highways and its codified wisdom should be the guide in working out the problems."[52] The street railways had at their disposal thousands of trained engineers, and on this basis the American Electric Railway Association made a claim for authority in city traffic problems. In the 1920s this claim was far easier for railway engineers than for others, since before 1930 "transportation engineer" was a term synonymous with "railway engineer." Meanwhile, electric railways, like the auto industry, learned to speak with one voice in transportation matters. By 1926, the president of Baltimore's electric railways could declare: "We are now articulate as an industry, and daily becoming more so."[53]

While the auto industry confronted the saturation crisis, street railways faced serious threats of their own. First, automobiles interfered with streetcar movement, slowing travel and disrupting schedules. Railways nationwide tried and failed to secure laws forbidding motorists from driving on streetcar tracks. They doubted that engineers could achieve such relief in law, and about 1923 they began to resort to direct public relations campaigns. In such publicity efforts, street railways sought to give engineers' construction of traffic congestion as an efficiency problem a wide basis of popular support. For two weeks in December 1923, the Los Angeles Railway displayed large posters on the exteriors of its streetcars, addressed to motorists. One read: "Mr. Autoist: Give the Street-Car Riders Fair Play. Please Don't Block Traffic. Thank You." The railway assigned a "young lady trained in traffic" to ride its streetcars and chide any motorist who ignored the poster and blocked a streetcar's progress, or who was "in any way interfering with a street-car rider's rights." The woman "called the driver's attention to the appeal" on the poster and "gave him a few hints on respect for the time and property of others."[54]

The Twin City Lines of Minneapolis-St. Paul launched a similar publicity campaign in 1925. Its slogan was "Give the street cars the right of way." Through newspaper advertisements directed at those who "cannot be effectively reached in more dignified copy," it introduced "Motorman Bill," an avuncular, pipe-smoking character who hid his irritation at motorists behind a broad smile. In his homespun way, Bill told Minnesotans what traffic engineers had known for years: Streetcars use "mighty little street space per person," and straphangers "deserve some consideration in the use of that part of the street where the track is." Other company advertisements asked readers to "give the street cars elbow room."[55]

Through public relations, street railways also had to fight a growing popular perception that streetcars were an old and, at best, quaint technology, outclassed by modern automobiles. Reflecting this perception, a daily cartoon syndicated in 250 newspapers depicted the "Toonerville Trolley" as a rickety electric wagon piloted by an aged curmudgeon. The pair also had a career in vaudeville and in film, and as popular toy.[56] Railways complained of "all the talk one hears about the passing of the trolley."[57]

In response, the street railways converted traffic engineers' dry survey data into vivid, intelligible, and occasionally misleading images, and took them to the public. Most depicted the superior spatial efficiency of streetcars. The Kansas City Railways used pictures to show newspaper readers that "one street car has the carrying capacity of many automobiles and occupies very little of the pavement space" and urged them to "remedy" congestion by riding street cars and buses.[58] At cinemas in Birmingham, Alabama, audiences saw a short movie distributed by the local street railway, "Mrs. Birmingham Goes Shopping." By making fun of the frustration of motorists fighting traffic and looking for a place to park, it cast the streetcar as the smart transportation mode for modern shoppers.[59]

Like the railways, streetcar manufacturers began to spread the traffic control word beyond Chambers of Commerce, conference rooms, and city engineering offices. Westinghouse bought full-page advertisements in national magazines, including the *Saturday Evening Post, Forbes*, and *Nation's Business*. "Street cars relieve street congestion," announced one, adding that "the larger the proportion of people using street cars, the faster the traffic moves." The ad cited traffic engineers' findings, giving them a wider audience. General Electric conducted similar campaigns to "reach the individual" with the traffic control message.[60] By the mid 1920s, streetcar interests were getting the word out.

## Motordom

In the face of the saturation and safety crises, the auto industry organized. The sales slump hastened a flood of mergers, acquisitions, and bankruptcies, increasing the possibilities for cooperation within the industry.[61] Surviving manufacturers exchanged information, looking to the benefit of the industry as a whole. Alvan Macauley, president of Packard, noted in 1926 "an unusual freedom in the exchange of ideas among the automobile companies." It was "not uncommon for one company to solve a manufacturing problem in the factory or laboratory of another company." Because of a new spirit of cooperation, "Factory doors are open and the old-time manufacturing secrets, with the automobile industry at least, are things of the past."[62] Auto clubs, manufacturers, and dealers wanted motorists to be able to get downtown. When they saw such access constrained, they began to publicize their own positions on the congestion problem, challenging the traffic engineers.

Los Angeles set the trend. By 1920 the automobile already had a mass constituency there. Before 1923 there was already one automobile for every three Angelenos—more than twice the national rate.[63] The Automobile Club of Southern California sponsored traffic studies in 1920, taking a role that elsewhere belonged to chambers of commerce.[64] The club was influential not just because of the high rate of automobile ownership, but also because the city's electric railways undermined their own position. The railways, organized in the Board of Public Utilities and influential in the city council, helped pass an ill-planned daytime parking ban in 1920. In

**Figure 6.1**

Four advertisements for street railways. Sources, clockwise from upper left: *Chicago Tribune*, May 12, 1924; *Chicago Tribune*, November 12, 1925; *AERA* 16 (August 1926), p. 26; *Chicago Tribune*, October 27, 1925.

**Figure 6.2**
A Westinghouse ad promoting street railways as relieving street congestion. Source: *Nation's Business* 14 (July 1926), p. 75. Courtesy George Westinghouse Museum.

a city of motorists, the ban touched off angry resistance. Traffic engineers would probably have recommended only a rush-hour parking ban, but the railways wanted more. The ban was in effect just one week when the city council backed down and allowed 45-minute parking.[65]

In the wake of the botched parking ban, the auto club, like chambers of commerce elsewhere, hired consulting engineers to conduct traffic studies, and even organized its own engineering department.[66] From 1920 to 1922, the club conducted perhaps the most impressive traffic study of the early 1920s. It submitted the results to the city in a report titled The Los Angeles Traffic Problem. Concurrently, the auto club, the electric railways, and other organizations interested in traffic organized and funded a new Los Angeles Traffic Commission. The commission endorsed the auto club's traffic plan as "one of the greatest civic contributions ever made to the City of Los Angeles" and supplemented it with a second study of its own.[67] When neither plan prompted city action, the Traffic Commission resorted to "the employment of nationally prominent experts" in city planning to survey the city's streets and propose a plan that could not be ignored. The city planners would be responsible to a "Major Highways Committee" of the Traffic Commission, and would confine their attention to "the city's main thoroughfare needs, rather than matters of particular detail." Drafted under the direction of Frederick Law Olmsted Jr., Harland Bartholomew, and Charles Cheney, the new plan incorporated the two earlier studies, adding more recommendations for extensive street widening and major new thoroughfares. It was submitted to the city as the Major Traffic Street Plan in 1924 and, endorsed by voters in a referendum, the plan was adopted by the city.[68]

The Major Traffic Street Plan was a departure from the conventions of traffic control. Drafted at the height of the traffic control consensus, after the animosity of the 1920 parking ban had subsided, the plan was not a promotional tract for highways. The street railways were represented on the Traffic Commission, while the commission's president, the owner of an auto dealership, endorsed traffic control methods downtown. In the manner of traffic control, the planners called first for securing "maximum use of existing street space" through "traffic regulations." Yet they broke with the traffic controllers in identifying the city's "greatest immediate need in solving its street congestion problem" as street widening and the opening of new thoroughfares. Only after a major building program, the planners argued, should the city consider applying the traffic control principle of discriminating between the efficiency and value of particular modes and restricting the inefficient street users. This important revision of traffic control meant that efforts to accommodate automobiles would precede efforts to favor streetcars, despite the planners' admission that streetcars offered "economy of space and low cost of operation per passenger." The city planners accepted the Los Angeles Railway's findings that at rush hour streetcars averaged 77.7 passengers each while autos carried 1.67, but would endorse discrimination of modes only after the construction of more and wider streets.[69] The plan would subordinate traffic control to a new effort to make room for automobiles in the city's streets.

## Studebaker and McClintock

With the help of the leading figure in traffic control, Los Angeles's distinct approach to traffic influenced the nation. In 1923, the year the national auto industry's saturation scare began, the Los Angeles Traffic Commission chose a local Studebaker dealer as its new president. Paul Hoffman, just 32 years old, had been selling Studebakers in Los Angeles since he was 19. He was extraordinarily successful. Hoffman admitted that he was alert to opportunities for "making some quick and easy money."[70] At 27 he had opened his own dealership, and in six years he had multiplied the assets of his business 25 times.[71] Hoffman took a deep interest in his city's traffic problem, seeing in it a threat to the future of his business. He wanted to secure the place of the automobile in the city, but he feared that his proposals lacked stature and credibility. He needed a spokesman outside the auto industry with credentials who could lend some prestige to the cause.

In the first summer of his presidency of the Traffic Commission, Hoffman met a Harvard graduate student who had come to the city to study its traffic. Miller McClintock, age 29, was working on a doctoral thesis in municipal government. His subject was city traffic, and his dissertation was becoming a traffic control treatise. Its first principles were equity and efficiency. McClintock had come to Los Angeles as part of a research tour and had won an invitation to speak before the Traffic Commission. There Hoffman and McClintock met.[72]

The two young men were quite different. Hoffman, three years older and many times richer, was a salesman by temperament, outgoing and forthright. McClintock, a former Chaucer scholar, was serious, studious, and reserved. McClintock's early work on traffic was thoroughly in the traffic control vein, and indeed he would publish his dissertation as a book titled *Street Traffic Control* in 1925.[73] Though McClintock was not an engineer, his book was in its time the leading textbook in traffic engineering, and based solidly on an efficiency model. Hoffman agreed that traffic control could buy cities urgently needed time by letting them make better use of their streets, and saw in McClintock someone who could help the Traffic Commission do this for Los Angeles. But Hoffman was developing bigger ideas for the city of the future, a new city for the motor age. In time, McClintock would join Hoffman's cause.

McClintock completed his doctorate in June 1924,[74] and in July the Traffic Commission hired him as a consultant. To complement its new

Major Traffic Street Plan, the commission wanted a new traffic code for the city, based on extensive research in the streets and applying the traffic control principles McClintock had expounded in his dissertation. The commission at first tried working through the City Council, offering McClintock's free services. Some councilmen were suspicious of the offer, however, and it was refused.[75] The commission proceeded on its own. McClintock's job was to work out efficient traffic flow downtown by regulatory means, since the planned new thoroughfares the commission proposed would not reach the central business district. As he promoted large road projects throughout Los Angeles County, Hoffman endorsed McClintock's mission downtown of "securing maximum use of existing street space through better regulation of traffic."[76]

With the commission's ample funding, McClintock was able to conduct a thorough survey of downtown traffic. To ease congestion without new construction, he scraped away the accumulated layers of police measures and replaced them with professional regulations based on careful surveys. McClintock, for example, replaced what Hoffman called "the thoroughly ineffective and annoying school stop ordinance" by confining children within new painted crosswalks, thus freeing motorists to keep moving elsewhere. McClintock similarly regulated pedestrians in town, and restricted the manner in which motorists could pass streetcars. McClintock's code went into effect in January 1925.[77]

McClintock's task gave him a chance to implement traffic control's more modest tenets. The new code introduced an evening rush-hour parking ban downtown, regularized turns at intersections, and required the city to post signs informing street users of rules. Yet McClintock was also willing to accommodate the preferences of his employers. He did not include any of the more stringent measures he was contemplating in his dissertation. The finished code included no provisions for modal discrimination, and recommended no general parking bans. It did more to restrict other street users. It confined pedestrians downtown to new crosswalks and required them to obey traffic signals, and it excluded horse-drawn vehicles from the central business district during the evening rush hour.

McClintock said that a new age justified new ways. "The old common law rule that every person, whether on foot or driving, has equal rights in all parts of the roadway must give way before the requirements of modern transportation," he told the press.[78] His ordinance included strict pedestrian control measures, with fines for jaywalkers. In downtown streets pedestrians would have to keep within crosswalks. Where there were no signals, they would have to raise a hand to halt oncoming motorists.[79]

The code won backing from the street railways and from local motordom, and the city quickly adopted it.[80] To both parties, willful pedestrians were obstructions and thus a threat to the efficiency of their mode. Traffic fatalities soon declined, and both automobiles and streetcars traveled more quickly through the central business district.[81] The rules thereby helped redefine downtown streets as vehicular thoroughfares. They became the basis of a model code McClintock prepared for ten Southern California cities.[82] By easing traffic congestion in America's most motorized city, McClintock was earning a name as a traffic expert.

Hoffman was going places too. He was Studebaker's most successful salesman, and in April of 1925 the corporation chose him for its new vice president of sales. Hoffman moved to corporate headquarters in South Bend, Indiana. He took to his new post a keen conviction that the auto industry's future was at stake in the city traffic problem, and that McClintock, as an expert one step removed from the industry and thus ostensibly impartial, might help him make his case. Hoffman soon summoned McClintock to South Bend.[83]

McClintock later recalled that at Studebaker headquarters Hoffman told him that he had "demonstrated that your principles can bring relief" to city traffic. "The solution," he added, "is the ultimate question of the future of the automobile. Tell me just what I can do to help you." McClintock, with characteristic faith in expertise and investigation, answered that an organization for conducting research and training engineers in urban transportation would be the best beginning to fighting city traffic nationally. Hoffman took McClintock's answer to Studebaker's president, Albert Russel Erskine, and by the summer of 1925 he had won a corporate commitment to establish a new traffic foundation, the Albert Russel Erskine Bureau for Street Traffic Research.[84]

Studebaker financed the Erskine Bureau, at first with two annual engineering fellowships of $1,000 each, but soon with annual grants of $10,000. In the meantime McClintock had taken a teaching job at the University of California's Southern Branch (later UCLA), and for his convenience Studebaker made this the Erskine Bureau's first home. In 1926, however, Studebaker moved the Bureau and its director to Harvard University. There it presented itself as a Harvard University research institute, seldom volunteering that its financial lifeline stretched to South Bend. McClintock's traffic control principles soon evolved; his definition of efficiency changed and he began to attack the "floor space" problem. Each year, Bureau graduates went to work in cities throughout the country, presenting themselves as experts unaffiliated with industry. In time McClintock's institution was recognized as "the No. 1 U.S. authority on traffic control," a vanguard in the auto industry's fight for more room in cities for its product.[85]

McClintock's transition from traffic control principles to the "floor space" school was not instantaneous, but by 1927 it was complete. Financially insecure, he certainly needed the job Hoffman gave him. The young traffic expert had just finished graduate school, and in six years he had been through a string of jobs. His master's thesis was a rarefied analysis of Chaucer's *Troilus and Cressida*.[86] He had been a newspaper reporter and an instructor in English and "Financial Publicity" at four universities. With his new doctorate, McClintock taught municipal government at the future UCLA. With such flux in his early career, and with a new baby besides, it seems likely that McClintock was eager to accept the opportunity for security that soon came his way.[87] Within two years, the traffic control expert was openly fighting for the future of the car in the city.

## The New McClintock

McClintock's conversion was part of a larger trend. McClintock, Macauley, and others who led it gradually developed a more sophisticated model in which Jordan's crude "floor space" metaphor was only a component. They proposed that traffic congestion was a problem of supply and demand. They argued that experts should not manage supply problems by trying to control demand. Instead, the supply of traffic facilities should expand with rising demand, like the supplies of other commodities in a free market.

While motordom remained divided throughout the period on numerous particular matters, this new conception of traffic problems brought its members closer together. Sporadic attacks on traffic control in 1923 and 1924 soon evolved into a positive case for the rights of cars in cities. Motordom began to promote its cause in the terms of classical political and economic liberalism. Its rhetoric opposed freedom to regulation, individual rights to collective goods, and abundance to efficiency. While the engineers held the rhetorical advantage of claiming to stand for science and progressivism, they were up against a culturally potent rhetoric of free enterprise and individual liberty applied to twentieth-century social problems.

The change in traffic engineering was not the simple result of motordom's rise. Local auto interests either ignored or backed the traffic control engineers until 1923, and the industry's growth alone cannot account for their change of position. When in 1919 Baltimore banned rush-hour parking downtown, for example, the auto club and the street railway both backed the plan.[88] Five years later auto interests feared such restrictive measures. This transformation in the industry's conception of the city traffic problem, and of city streets themselves, stemmed from its fear that traffic control alone would limit the automobile's usefulness in cities.

The change is embodied in McClintock. After he accepted Studebaker's offer, traffic control remained a necessity in congested downtowns, and chambers of commerce remained important sources of funds. The largest project of McClintock's early career in traffic engineering, the $50,000 Chicago traffic survey of 1926, was paid for entirely by the Chicago Association of Commerce. The association showed the predilection of other chambers of commerce for the efficient, modest, and inexpensive methods of traffic control. McClintock's work in this survey shows no departure from the principles he set forth in *Street Traffic Control*. His mission in Chicago, he explained, was to find "the manner in which existing street facilities can be used to their fullest extent."[89] In a 1927 traffic survey for San Francisco, McClintock again sought "to gain more efficient use of existing street facilities."[90]

Yet in the Chicago survey's introduction, removed from the particular recommendations he addressed to the Association of Commerce, McClintock expressed views quite different from those in his textbook on traffic control. He accepted the "floor space" idea. "The most basic solution for street and highway congestion," he contended, "lies in the provision of greater street area."[91] He repeated this claim in his work for San Francisco.[92] But in practice he retained traffic control principles. In the spring of 1927, however, he gave up the balancing act, and a new McClintock went public. For the first time he began to seek publicity and tried to appeal to a wider audience—especially motorists. "We must drop our prejudices and be willing to readjust ourselves to conditions," he told a Kansas City audience in April.[93] Two years earlier, in *Street Traffic Control*, the old McClintock had maintained that widening streets would merely attract more vehicles to them, leaving traffic as congested as before. The automobile, he wrote, was a waster of space compared to the streetcar, noting that "the greater economy of the latter is marked." "It seems desirable," McClintock wrote, "to give trolley cars the right of way under general conditions, and to place restriction on motor vehicles in their relations with street cars." He described the automobile as a "menace to human life" and "the greatest public destroyer of human life."[94]

Two years later, all had changed. McClintock wrote of "the inevitable necessity to provide more room" in the streets. He called for "new streets" and "wider streets."[95] In 1925 he referred to grade separation of street intersections as an "ultimate" answer which was, except in a few cases, too expensive for consideration in the short run.[96] In 1927, McClintock declared that "grade separations at important highway intersections can and should be put into practice immediately."[97] In 1925 McClintock virtually ruled out elevated streets as expensive and impractical; two years later he urged that they be considered.[98]

As director of the Erskine Bureau, McClintock began to see his task as helping cities to "adjust their physical layout . . . to the requirements of an automobile age." He began to address motorists directly: "When these adjustments take place the motor car owner will profit greatly in increased safety and convenience."[99] He adopted what was for him an entirely new use of the term "efficiency" when he noted the "growing demand that changes be made to make it possible for the automobile to be used to its greatest efficiency."[100]

The new McClintock also found new words to promote this automobile age city. The automobile was not merely a mode of transportation; it was an expression of American ideals. "This Country," he told Society of Automotive Engineers in 1928, "was founded on the principle of freedom. . . . Now the automobile has brought something which is an integral part of the American spirit—freedom of movement." He ridiculed traffic control engineers for resisting the new truths of the motor age. Some of them "still act as though they were living with whip-sockets and dashboards and hitching-posts."

A final striking change in McClintock's rhetoric is found in a new conviction that the traffic problem could be not just managed, but solved. *Street Traffic Control* is a judicious weighing of costs and benefits, and in it McClintock acknowledged the benefits of costly measures but saved his recommendations for inexpensive adjustments. Two years later, he argued that "the millions of dollars which have gone into planning projects have never failed to come back in the form of improved property values." With such measures, McClintock now promised, traffic problems "will be solved to the satisfaction of the public within the next quarter of a century."[101]

In Miller McClintock, the auto industry had, by 1927, an articulate and credible spokesman, the first traffic expert with a doctorate in his field. He was insulated from any obvious affiliation with industry by a Harvard byline. And at the industry's expense, he was turning out more such experts each year.

## New Experts

Studebaker did not act alone. In the face of the saturation crisis, the auto industry instilled new life in its old organizations, and formed new ones. The National Automobile Chamber of Commerce, the trade association of auto manufacturers, formed a Safety and Traffic Department late in 1923.[102] Unlike the Erskine Bureau, the new body consisted entirely of industry executives, who unapologetically publicized their positions on traffic questions with no pretense of impartiality. Soon the automobile manufacturer Edward Jordan was using his post in this group to promote grade separations at urban intersections as a superior alternative to traffic lights.[103]

The National Automobile Chamber of Commerce soon sponsored its own investigations of the traffic problem. In April 1927 it hosted a traffic conference in Chicago, bringing engineers and manufacturers together. Miller McClintock addressed the conference.[104] The industry journal *Motor Age* reported with satisfaction that "there was a noticeable departure from the suggestions of restrictive control of motor vehicles such as used to feature any gathering of this kind." Participants deemed "many of the regulations of today experimental and temporary." Instead, the "able minds" there turned to "the matter of facilitating the movement of passenger cars and trucks for the greater convenience and safety of the public."[105] The day after the Chicago conference adjourned, a new automotive interest group, the National Highway Traffic Association, held a similar meeting in New York.[106]

In its attack on the problem of "floor space," motordom helped to foster a more prominent role for city planning. Through zoning, the city planners' greatest accomplishment in the first quarter of the twentieth century, planners constructed traffic congestion as the consequence of uncontrolled building densities. Since it largely confined its effects to new construction, however, zoning influenced city form only at a glacial pace, and left city planners little positive role in the shaping of cities. Beginning in Los Angeles, local motordom began to lend new support to city planners. Los Angeles's Major Traffic Street Plan confined its attention to thoroughfares, but it gave prominent academic city planners a chance to see some of their plans get past the drawing boards. Such opportunities had been rare since the vogue for new civic centers and fairgrounds had waned a decade earlier.

The American City Planning Institute, however, refused admission to non-professionals, and shunned promotional "propaganda and education work." Such professional exclusiveness left no room for interested amateurs, such as auto industry people.[107] When in 1928 the Russell Sage Foundation funded a less exclusive alternative, the Planning Foundation of America, such men saw it as an opportunity to get city planning credentials. The foundation straddled the border between a professional association and a congress of interest groups. Although funded by Sage, it had to raise matching funds from corporate donors. Prominent professional city planners served on the foundation's advisory council, including two of the three drafters of Los Angeles's Major Traffic Street Plan. Among their amateur colleagues on the council were J. C. Nichols, a pioneer of the automotive suburb, and Studebaker's vice president for sales, Paul Hoffman.[108] Unlike the City Planning Institute, the Planning Foundation embraced "propaganda and education work." Eighty percent of its budget went toward promotion, publicity, and education. The foundation called for the reconstruction of cities. It urged "New Cities for the New Age." It characterized traffic control regulations as "only emergency measures." "Local regulation will not solve the traffic problem," the foundation announced; it called instead for incorporating city streets into great highway systems.[109]

Automobile clubs joined in. The jumble of local, state, and national clubs in 1920 quickly evolved into an orderly federation under the American Automobile Association. In 1924 the AAA became the sole national organization of motorists and, under its dynamic president, Thomas Henry, was on the way to fulfilling its mission of securing "one national association, one national policy and one national service."[110] In 1933 the AAA hired a prominent traffic expert and the founding president of the Institute of Traffic Engineers, Burton Marsh, to serve as director of the Association's Traffic Engineering and Safety Department. Marsh brought prestige in traffic matters to the association. He was by most accounts the first full-time traffic control engineer, having gone to work on Pittsburgh's traffic problems in 1924. Marsh retained his position as the institute's president while beginning his 30-year career with the AAA, which noted with pride the prestige the president of the Institute of Traffic Engineers brought it.[111]

Increasing traffic congestion drove street railways to demand more regulation, and motordom to reconstruct congestion as a shortage of street capacity. Under the resulting strain the traffic control consensus collapsed.

For the rest of the decade, engineers continued to work for chambers of commerce to ease traffic congestion through a quest for efficiency. Beginning about 1924, however, motordom and street railways dispensed with expert mediation in an effort to influence the city traffic problem directly. Both groups formed organizations of their own. These soon began to supplant the traffic engineers, whose authority, prestige and autonomy ebbed. As will be shown in chapter 7, transportation interests' scope of action quickly grew further as a new approach to government invited trade associations to pursue their various interests in new national forums.

# III  Freedom

Americans are a race of independent people, even though they submit at times to good deal of regulation and officialdom. Their ancestors came to this country for the sake of freedom and adventure. The automobile satisfies these instincts.
—Roy Chapin, "The Motor's Part in Transportation," *Annals* 116 (November 1924), 1–8 (4–5)

We shall no doubt see automobile traffic . . . more and more collected in great arterial ways, from which the pedestrian will be excluded, being overpassed or underpassed, and being kept from playing or wandering in the roadway.
—William Cox, paper presented at conference of American Society of Civil Engineers, New York, May 1927; reprinted as "Population Density as a Factor in Traffic-Accident Rates," *American City* 37 (August 1927), 207–209 (209)

In the typical quest story of folklore, the hero begins as a naive youth, ignorant of his destiny and sometimes even of his own name. Often through a threat or a disaster, he learns of his higher purpose. With this painful discovery he first sees his true mission, finds his courage, and sets out to do battle.

Automotive interest groups discovered their destiny in the 1920s. Lacking self-awareness in 1920, they joined with chambers of commerce and local safety councils to fight accidents and congestion on the terms of Safety First and efficiency. In 1923 and 1924, looming threats to the automobile's future in the city shocked them into self-discovery. Finding their identity, they named themselves "motordom." For the first time they saw their enemy: prevailing constructions of the problems of safety and congestion.

In the mid and late 1920s, motordom began its quest to reconstruct city traffic problems for the motor age. After a period of defensive and piece-meal criticism of safety reformers and traffic engineers, motordom found more positive programs. Traffic control's efficiency model of congestion could never serve motordom well, because automobiles could not hope to compete with other modes on spatial efficiency. Neither could motordom secure its future in the city as long as the terms of the safety problem were those of the local safety councils.

Instead, motordom found its own perspective. Chapters 7 and 8 trace the development of this new model. The Los Angeles case, introduced in chapter 6, is continued—from the angles of congestion (chapter 7) and safety (chapter 8). As chapter 7 shows, motordom learned to construct congestion not as an excess of cars but as a scarcity of street space, to be remedied by a supply-and-demand model of street capacity. The new model required civil engineers to supplement traffic engineers.

In the traffic safety institutions that motordom founded, new experts promoted new ways to fight accidents. Chapter 8 recounts how motordom worked to reconstruct street casualties so they would no longer be the sole responsibility of motorists. Instead, accidents could be a failure of pedestrians to adapt to a new age, or a failure of the streets to adapt to technological progress. Motordom organized to spread the responsibility for safety so that pedestrians, including children, would no longer be presumed innocent. Beginning in Los Angeles, local auto interests worked to redefine streets as places where pedestrians did not belong. Through safety education it found ways to prevent child casualties without impeding motor traffic. To exonerate the average motorist, it redirected hostility to a newly identified minority of reckless drivers. With new sources of public funds, it worked to accommodate cities to automobiles, and it used persistent traffic casualties to argue for more such accommodation. Chapter 9 follows the application of the new motor age model of city traffic into the 1930s and beyond.

To unite its perspectives on congestion and accidents, motordom constructed a new application of a culturally resonant rhetoric of freedom. And to justify fundamental departures from time-honored customs of street use, motordom characterized the 1920s as the dawn of a new era to which old ways had to adapt. It was the dawn of the motor age in the American city.

# 7 The Commodification of Streets

In the final analysis street traffic congestion is a problem of unbalanced supply and demand.
—Miller McClintock, 1930[1]

In 1925 the leading textbook in professional traffic control advised engineers to favor efficient and restrict inefficient modes and to consider expensive transportation facilities only as a last resort. In 1941 the advice in the leading traffic engineering textbook was different: "If people prefer to drive downtown and can afford it, then facilities must be built for them up to their ability to pay. The choice of mode of travel is their own; they cannot be forced to change on the strength of arguments of efficiency or economy."[2] The position set forth in the second book was already old news; it had prevailed throughout the 1930s. The rapid reversal of traffic management principles in the mid and late 1920s marked the introduction of new ways to fight traffic. Regulatory traffic control was giving way to traffic engineering for the motor age.

The traffic control consensus dissolved, traffic engineers found new patrons, and the regulatory model was relegated to a subordinate role. By 1930 the most prominent traffic engineers no longer treated city streets as conduits of practically fixed dimensions. They showed a new reluctance to single out inefficient users for restriction. Traffic engineers had treated streets as public property under state control, but just a few years later streets were more often treated as a half-public, half-private territory where the pursuit of efficiency could not justify state intrusion. In 1930 engineers no longer presented themselves as the shapers of traffic demands, encouraging some and restricting others; instead they their job became to supply street capacity as demanded. Engineers, furthermore, redefined "demand." In the early 1920s, they counted persons and weighed the importance of their trips; a few years later they counted vehicles and considered the fees street users incurred.

In short the engineers of 1930 no longer conceived of the street as a public utility, regulated by the state in the name of street users collectively. By then streets and roads had been redefined as commodities bought privately by users to be supplied not according to the judgments of engineers, but in automatic response to demand.

## The Decline of the Public Utility Model

When the "floor space" problem was prodding the automobile industry to join in the debates over city traffic, circumstances were making its entry easier. The prestige of the public utility model they challenged was declining. Regulators, including traffic control engineers, had presented themselves as disinterested, apolitical experts. They depended, of course, on interests prepared to pay for their surveys and recommendations, but their prestige and hence their influence also required a credible claim of scientific impartiality. Expert, ostensibly apolitical regulation of public utilities "in the public interest" was the Progressive Era's answer to the problem of the natural monopoly. The popular reaction then to dissatisfaction with utilities was the resort to regulation. Before World War I, the prevailing popular view was that state power could (as Mayor Carter Harrison of Chicago said in 1914) "relegate the utilities to their proper position of servants instead of masters of the people."[3] Those suspicious of the street railways' power could still hope in 1913 that regulators could be "consulting civic statesmen."[4] And public utilities usually went along. For example, in the virtual absence of inflation from the beginnings of electrification until 1915, electric railways could live easily with fixed fares (almost always five cents).

In the 1920s, however, room opened for a kind of representative expertise that openly argued its sponsors' case. Business spokesmen with no plausible claim to impartiality could, without embarrassment, argue the case of the interest they represented. The trade association movement of the mid 1920s gave interest groups new opportunities to participate directly in the management of public problems, including city traffic.

Confidence in expert regulation fell in the late 1910s and the early 1920s. Before this transitional period, regulation was often accepted as the cure for flaws in public utility service. Afterward, regulation was more often blamed as the cause of poor service. Regulation protected companies from the competition that would spur them to offer good service at a good price.

In 1917 and 1918, preparedness and war brought with them labor shortages and higher costs for power. Railways requested fare increases, which often came only long after increases in operating costs. Older regulatory failures persisted—for example, railways had to charge the same for long and short hauls and for peak and off-peak service. But with the new pressures, the railways could no longer afford such inefficiencies. Bankruptcies crested in 1919, when 51 street railways were turned over to receivers.[5]

These stresses cast a shadow of popular suspicion over regulation itself. Street railways took refuge in their franchise protections, bandied them at upstarts such as jitney operators and independent bus companies, and demanded fare increases, while the quality of service declined. State public service commissions generally granted fare increases. Riders smelled a rat. Defenders of state regulation complained of a "public attitude toward utility commissions" that "has been critical rather than constructive, reflecting a feeling of uneasiness and distrust." They noticed with frustration that the wartime fare increases "were looked upon with suspicion, even though the cost of other commodities and services was rising." There was "a widespread assumption that the commissions were, after all, too friendly to the utilities," leading to "a general reaction . . . of complaint against the commissions." They despaired of the "widespread discontent" with the public service commissions.[6] "That which was the protection of the public a few years ago," a utilities lawyer observed in 1921, "is now considered by the public to be its menace."[7]

The public utility model suffered an even greater loss of prestige than other forms of state regulation. The street railways exemplify this change. Street railways were a regular source of irritation to riders, investors, and regulators. Yet criticism of street railways changed fundamentally in the years immediately after World War I. They were generally doing well before the war—too well, said many, since they were legally protected monopolies. The Progressive Era response was a call for more regulation to limit profits and to improve service. But the economic stresses of 1917–1920 forced city railways to seek fare increases to survive.[8] Government added to the burden with wartime controls, which persisted long after the armistice.[9] By 1920 the reputations of the railways and of the state public service commissions that regulated and protected them had hit bottom. This time, however, critics ascribed the railways' problems not to a lack of regulation, but to regulation itself. The public utility model was falling into disrepute.

Beginning in the mid 1920s, new participants in urban problem solving tended to avoid the tarnished model of regulation by ostensibly independent experts. They tended not to look to engineers to choose social ports of call or to chart paths to them (as engineers had done for chambers of commerce); for them, engineers were navigators who kept their vessels true to courses charted by others. *Public Works* editorialized in 1924 that engineers' task was to identify demands and then "provide these wants to the fullest extent possible." It was not the place of engineers to shape demands, favoring some and discouraging others.[10] Though such a role was more modest, the chances for prestige were in some ways greater. Engineers affiliated with major corporations had less freedom of maneuver, but they could join in more ambitious national projects.

### Hoover: "A New Conception of Government"

Herbert Hoover, U.S. Secretary of Commerce for all but the decade's margins, exemplified the new, limited role for engineers. Hoover was, as the engineer Morris Cooke described him, "the engineering method personified."[11] A trained engineer, Hoover treated his Commerce post as an engineering assignment, relentlessly tinkering with commercial problems to work out more efficient arrangements. At the same time, he shunned direct governmental intrusion in the private sector. He hoped to foster direct cooperation within industrial sectors, and thus to obviate government intervention. Hoover's conception of the engineer-administrator, as Edwin Layton has shown, was limited: he "imagined the engineer functioning as a sort of social catalyst, producing action by others," and he "did not view the engineer as an independent force in national affairs."[12]

The late progressive ideal of expert direction—an ideal exemplified by city manager government, by reforms in public budgeting, and by in the regulation of public utilities—was in decline. In its place a newly confident interest group pluralism arose. Its advocates made no apologies for the interestedness of those participating in it.[13] With the fault of interest came the virtue of action. When interest groups agreed to a plan, they could see it through to completion.

Defenders of a direct role in social problem solving for interest groups argued that they, like legislators, were representatives, and that gatherings of interest groups were their representative bodies. Progressive engineers had depended upon interest group support too—for example, from chambers of commerce. But to them there was a professional obligation to keep above the fray. "Selfish and individual motives must not be allowed to interfere with the best interests of the community," one of them declared.[14] By the mid 1920s many disagreed. They argued that interested parties were best able to secure larger community interests.

One advocate described this new approach as "the method of the 'miniature industrial legislature.' "[15] Paul Hoffman of Studebaker celebrated the Los Angeles Traffic Commission as "representative of all groups vitally affected by traffic." "Every effort is made," Hoffman declared, to keep the commission "so representative in character that a true cross section of opinion can be obtained." The "deep interest of its members" would ensure both the legitimacy and the viability of the commission's plans.[16] Alvan Macauley of Packard recommended similar organizations of interested groups, noting that "such plans should produce results."[17]

Thus the repulsion from public utility model of social organization was accompanied by an attraction to a new model, known to historians as "associationism."[18] By mobilizing private interests, associationism would help government meet the enormous demands of the twentieth-century administrative state while remaining merely "an umpire and not a player in the economic game."[19] This model gave private interest groups—notably trade associations—a direct role in solving social problems, sometimes to the exclusion of expert mediation. At other times it gave the expert only the task of hired advocate, like a lawyer trying a case for a client.

The sources of this model were several. In 1925 the Supreme Court signaled a new toleration of exchanges of information between competing enterprises. Shortly thereafter, the new chief of the Department of Justice's Antitrust Division announced a "decided change in the attitude" of the department "towards trade associations and their activities"—a change from suspicion to outright support. The Federal Trade Commission, with new Coolidge appointees, joined the trend. In just a few months in 1924 and 1925, interest groups were granted far more freedom to work together on problems of mutual interest. Robert Himmelberg has described the mid and late 1920s as a time of "radical cooperative plans" that "constituted a more thoroughgoing relaxation of antitrust than associationists had dared to hope or ask for earlier in the decade."[20]

In the new representative democracy of organized interests, automotive groups were more followers than leaders. The real initiative lay in two national organizations in Washington: the Chamber of Commerce of the United States and Hoover's Commerce Department. The National Chamber (as it was then often called) was formed in 1912 as a national organization of local chambers of commerce. In 1920 both it and the Commerce Department offered diverse services to local and national organizations, but they soon became potent agencies in their own rights, taking the lead in areas where earlier they had followed their clients.[21]

The National Chamber and the Commerce Department worked together closely to promote a new approach to the problem of big business in a democratic society. Although they often spoke for "individualism" and against "collectivism," they rejected both nineteenth-century laissez-faire and extensive state regulation and offered instead a middle way, which Hoover called "associational." Both organizations pursued these principles by bringing trade associations, business representatives, and other interest groups together in large national gatherings to work out agreements. To Hoover the great advantage of such arrangements was that business itself would enforce what its representatives had agreed among themselves; state power would not be needed. Hoover therefore spent much of his Commerce Department career arranging "an enormous number of tiresome conferences with the officers of hundreds of business associations and groups."[22]

Just blocks away from Hoover's Commerce Department, the National Chamber promoted "home rule" and "self government" for business. The National Chamber understood that "if business doesn't keep its affairs in order, Government will step in and arbitrarily regulate business." If business, the Chamber argued, could regulate itself, without government interference, the interests, both of business itself and of the public generally, would be served.[23]

Hoover needed the National Chamber's help to plan his national conferences. In 1924, Hoover and the Chamber cooperated to arrange a National Conference on Street and Highway Safety. The conference received considerable publicity and became known, despite the great number of other conferences Hoover arranged, as "the Hoover Conference."[24] Its full name is misleading. It was more a lasting organization than a conference. Its participants were called "members," and its three meetings (in 1924, 1926, and 1930) were only high points in more than six years of activity.[25] Neither was safety its exclusive or even its primary subject. Participants paid about equal attention to traffic congestion and considered other related problems as well. Opening the second meeting in 1926, Hoover admitted: "We started out to solve the problem of traffic accidents." But, he added, "We have passed through the door of the problem of urban transportation."[26]

The National Chamber and the Commerce Department worked together on the conference from the start. One conference member, an auto executive, described the division of labor: The conference "originated with Secretary Hoover," and "under his direction" it depended upon "the machinery of the United States Chamber of Commerce."[27] The Commerce Department issued some of its press releases expressly for the National Chamber's journal, *Nation's Business,* and the meetings were held at the National Chamber's Washington headquarters. The conference was officially a Commerce Department affair, with the secretary himself serving as chairman. Hoover, however, asked the Chamber's top transportation authority, Colonel Alvin Barber, to serve as director of the conference; Hoover's own role as chairman was largely ceremonial. The 1924 conference had eight traffic committees, and National Chamber men served as secretaries on four of them. The Chamber's president, Elliot Goodwin, chaired an additional committee that was responsible for the conference's finances.

The conference reflected Hoover's associational tenets and the National Chamber's related principle of business self-government. On December 15, 1924, Hoover opened the first conference by telling delegates that he hoped the membership included "representatives . . . from every interested element" and that success depended on "the American spirit of cooperation."[28] The next day, as the conference closed, Hoover cited it as an example of "a new conception of government": a government that would refrain from the exercise of "central authority" and would instead accomplish its end through the "intelligent cooperation of the entire community."[29]

The Hoover traffic conference was not a forum for independent experts, or even for experts affiliated with interest groups. Engineers were notably absent. At most, 7 percent of the participants were engineers, and many of these were affiliated with state highway commissions, which were, as Hoover acknowledged, responsible for "our rural traffic."[30] Only one of the 607 participants was a city "traffic engineer"; about twenty more were professional engineers.[31]

## New Expertise, New Authority

The Hoover Conference's very existence was predicated on a new construction of expertise, one not based on technical training or specialized knowledge. Hoover argued that spokesmen for modern industry did not need mediation through engineers or other recognized experts, because they were experts themselves, members of "a new profession, business administration," who bore the essential trait of other professionals, "a responsibility to the community and insistence upon a high sense of service."[32] Conference members proposed that they were transportation experts not because of special training or professional affiliations but because of their experience and rank within a transportation business. The rhetoric of the conference equated "experience" with expertise. For example, President Coolidge, addressing members at its opening, described the conference as an effort "to mobilize the best experience in each part of the country."[33] The secretary of the Cleveland Automobile Club proposed that "men of traffic experience" were better able to address traffic problems than experts.[34] Evidently because conference members were indeed persons of experience, even if not of special training or professional independence, Commerce Department press releases repeatedly described them as "experts."[35] The press obligingly referred to the gathering—however implausibly—as a conference of "technical experts" seeking "technical findings."[36] Operators of street railways considered themselves "transportation authorities"; the "railway company," they argued, was "as competent as any one to judge the suitableness of a proposed measure."[37] The American Automobile Association described the traffic conference as a "great body of traffic experts."[38]

The new experts did not depend on traffic surveys or other investigations of street conditions. A Statistics Committee offered other conference committees an empirical foundation for their work. Overwhelmingly, however, the committees drafted traffic recommendations from a consensus of their members' opinions. Engineers could apply their expertise later, to execute schemes the new class of experts devised themselves. Members took these positions as representatives of the interests of their sponsoring organization. No engineer served as an expert mediator. Conference members had learned to distrust allegedly apolitical experts. The members were representatives of the businesses that sent them and that covered conference expenses. Hoover achieved his goal of "not representing any special group or interest" not through seeking an ideally apolitical body of experts but by opening the doors to "the representatives of all of the industries that bear upon this problem."[39]

## Interest Group Democracy

Interest groups claimed authority as representatives not merely of trade associations but also of mass constituencies, and for authority they appealed not only to business experience but also to democratic principles. Most people in cities traveled on streetcars, so beginning in the mid 1920s the street railways tirelessly claimed authority as the representatives of the majority. "We must regard ourselves as representing the interests of the masses of the people," the president of the American Electric Railway Association told the National Chamber in May 1926.[40] In December, a New York rail executive claimed priority for streetcar patrons by arguing that "the rights and well-being of the greatest number of people affected by traffic strangulation ought to receive first consideration."[41] A Chicago Surface Lines engineer reasoned that street railway management had "an unmistakable obligation to represent successfully the interests of the people whom it serves" and that these people constituted "the major portion of the citizens" in cities. Therefore, "each railway management" should "assume leadership in working out the traffic problem of the city in which it operates."[42] Railways publicly promoted this claim of majority representation. For example, the Philadelphia Rapid Transit Company campaigned for congestion relief under the slogan "Give the 80 per cent a Square Deal."[43]

Motordom generally did not challenge the street railways' claim to represent the majority. Like the railways, however, it did claim to represent motorists and others interested in automotive transportation, and it argued that motorists deserved special consideration in traffic matters. It based this case in part on motorists' financial contribution to roads and streets through registration fees, taxes on gasoline, and automotive excise taxes. Just as important, however, motordom claimed to stand for the protection of the rights of a minority threatened by legislative tyranny.

Auto clubs claimed to represent these motorists. Club spokesmen did make occasional (and unconvincing) claims to represent a majority of travelers. The secretary of Cincinnati's auto club prematurely claimed in 1926 that "the automobile club really represents the public, for most of the public own motor cars."[44] An auto dealer in Washington made a similar claim: " 'What is good for the automobile owner is good for the public,' because, practically speaking, the opinion of the automobile owner can be said to represent public opinion on matters relating to motoring generally."[45] Yet such claims were rare.

Far more often, auto clubs claimed only to be fighting for the rights of the motoring minority. They countered the street railways' majority-rule rhetoric by appealing to minority rights. As early as 1915 auto clubs were instructing motorists in "their 'inalienable rights' of owning and driving their cars without the harassing complications" of regulation.[46] They

## HOG-TYING

## *the* AUTOMOBILE

The Latest Proposed Law Would Do This With a Vengeance

**Figure 7.1**
Especially after the Cincinnati speed governor war, motordom characterized restrictions on automobiles as tyrannical. This cartoon appeared in *Ohio Motorist* 16 (October 1924), p. 8. Courtesy Ohio Conference of AAA Clubs.

championed motorists as a law-abiding population whose rights were constantly imperiled by two forms of tyranny. One was the tyranny of the unaccountable expert—the professional traffic engineer. The secretary of Cleveland's auto club complained of such "busybodies and half-baked self-styled experts," arguing instead for the authority of "men of traffic experience, and preferably for those who are deeply concerned from the viewpoint of their business pocketbooks" and who were thus accountable for their decisions.[47] Auto clubs also characterized their fight as a resistance to majority tyranny. Beginning in 1923, clubs claimed motorists were a persecuted minority, suffering under restrictions that deprived them of the advantages of car ownership. Clubs complained of a "tendency to freak legislation" and of "the fifty-seven different varieties of fool traffic regulations that are being put into effect over night in some communities."[48] Clubs attacked new traffic control measures as "traffic reform waves" that "seem to sweep over our larger cities every time a man with a wild idea and a John the Baptist manner come to lead us out of this wilderness," and ascribed them to "popular hysteria."[49] The clubs represented themselves as twentieth-century Minutemen, ready to fight for their members' rights.

### The Fewer Laws Club

Associationism, like the Committees of Correspondence of the early 1770s, helped dissatisfied auto club officials organize to fight for motorists' rights. Herbert Hoover's traffic conference, like the first Continental Congress, served them and others as a convenient national forum. The conference shows the truth of Edwin Layton's observation that, in applying his administrative principles, Hoover "abandoned the dream of a society led, if not dominated, by engineers."[50] Hoover saw engineers like himself as facilitators and coordinators of private cooperation, not as philosopher kings whose expertise made them fit to rule.

Hoover's belief in a dynamic but limited role for engineers suited a growing mood at the U.S. Chamber of Commerce in favor of business self-regulation and against regulation imposed—even at the behest of business—by experts. Hoover and the National Chamber shared a distrust of state regulation, suggesting that "natural law" worked best unencumbered by needless legislation. Hoover liked to declare that "it is not possible for you to catch an economic force with a policeman."[51]

The National Chamber showed its anti-regulatory mood by forming an informal "Fewer Laws Club," arguing, like Hoover, that "natural law" is "as wholly outside of legislative control as are the seasons and the weather."[52] It is nothing new to hear of the 1920s as a decade of business and a time of retrenchment from the state intervention of the Progressive Era.[53] Yet engineer-regulators continued to pursue an expert-determined public interest through state regulation into the middle of the decade. It was only then that the saturation and safety crises led the auto industry—the nation's largest—to doubt the expediency of leaving traffic matters in the hands of engineers who were only too willing to use the state to restrict motor vehicles.

## Uniformity

Motordom positioned regulatory approaches such as traffic control as violations of basic American principles. "There are already too many laws," complained H. W. Slauson of the Kelly-Springfield Tire Company in 1923. Slauson warned of the danger of "legislation which will prevent the efficient and effective use" of automobiles in cities. "The automobile is too effective a tool to have its efficiency curtailed in the slightest by half-baked ideas such as have already been proposed, and which have as their object a reduction in the number of cars which shall use the streets and a restriction of the areas in which they may be kept."[54] Alvan Macauley described traffic control as a "burdensome tangle of restrictive legislation" that "surrounds" motorists.[55] Traffic control's critics developed a broad anti-regulatory critique, directed loosely at "legislation"—that is, public regulation in general. Though Congress was the most obvious target, state and local rule makers were yet more irritating because their regulations varied from jurisdiction to jurisdiction.

The critics demanded uniform rules. As World War I ended, the Council of National Defense was working to promote uniform traffic regulations. In 1920, a congress of interests consisting of representatives of National Automobile Chamber of Commerce, the American Automobile Association, the Highway Industries Association, and the American Association of State Highway Officials proposed a uniform vehicle code for adoption by the states.[56] Traffic control engineers and street railways agreed with motordom that the inconsistent regulations were a needless encumbrance on traffic. Under the police, traffic regulations had often been contradictory, even within cities. Traffic engineers had done much to correct this problem. Street railways also showed early interest in uniformity.[57]

Yet the broad consensus in favor of uniformity concealed important matters of contention. Which rules would become the uniform rules? Which would prevail everywhere? Among engineers, discussions of uniformity took the form familiar to them: the establishment of standards. Engineers tended to approach the problem as a straightforward and tedious search for the most efficient arrangements. At national conferences, engineers debated national standards on practical details ranging from the colors of signal lenses to the rules for left turns, and they reported their decisions to traffic signal manufacturers and state legislatures.

Motordom, however, made its fight for uniformity an attack on the *quantity* of traffic regulations as well as their multiplicity. The secretary of the Chicago Motor Club complained that there was "no uniformity" and that traffic laws were "contradictory and conflicting." Yet he also complained of states drafting "many new laws" and of "every city, village, town and hamlet enacting laws regulating the motorist." For the motorist, the auto club should be "the Moses to lead him out of this confusion of trouble."[58]

This interpretation of the uniformity issue was at first confined to the meetings and journals of auto clubs and trade associations. Yet the U.S. Chamber of Commerce soon recognized a mutual interest. In 1924 and 1925, as the legal barriers to cooperation by trade associations fell, the Chamber launched a national publicity campaign against state regulation. In advertisements run simultaneously in several major newspapers, it attacked those who were in the habit of saying "There ought to be a law" whenever they identified a public problem.[59] In this campaign the Chamber often took a special interest in traffic regulation. The editor of its journal, *Nation's Business*, lamented that the spread of the automobile was followed by "a cry of 'More Laws! More Laws!' "[60] In a cartoon, the journal depicted Uncle Sam tied like Gulliver by bonds labeled "ordinances," "regulations," "statutes," and "laws" while a lilliputian legislator approached with another restraint labeled "more laws." A traffic policeman admonished the immobilized captive by pointing to a sign marked "No Parking."[61] The journal complained of the "muddle of our motor legislation," and the danger that legislatures would soon "pile up" more regulations on the "harassed traveler by car."[62] The Chamber's campaign was widely echoed elsewhere; one Rochester editorial cartoon compared lawmakers to stray cats in their proclivity to quantity production of (legal) offspring. Another cartoon of 1925 depicted a shackled citizen and the tablet of the Ten Commandments, both dwarfed by a lofty mountain of legislation; it won a Pulitzer Prize.[63]

The cause of uniform rules justified a rewriting of principles. As a well-represented host of the Hoover traffic conference, the National Chamber was influential in its deliberations. The Chamber ensured that uniformity achieved practical preeminence among conference objectives. At the opening of the second meeting, in 1926, Herbert Hoover told members that the "outstanding feature" in the conference reports for the previous two years was "the lack of uniformity in our traffic law and regulations."[64] The quest for uniform regulation ended in the demise of positive regulation in city streets.[65]

## Rewriting the Principles of Traffic Control

The Hoover traffic conference overturned the prevailing principles of traffic control. In the name of uniformity, the conference submitted these principles to the scrutiny of committees of transportation interests. But the committees pressed on beyond uniformity, devising new traffic principles of their own.

Representatives of motordom held most of the leading positions on the committees that worked out the new principles. Yet street railways and other interests were represented too, and at first these committees worked largely within the broad bounds of traffic control. By 1927, however, conference members had drafted a new model of city traffic management, published by the Government Printing Office and represented to states and cities as the Commerce Department's official plan for uniformity in traffic rules.

The director of Hoover's transportation conference, Colonel Alvin Barber, was also the head of the Transportation Department of the U.S. Chamber of Commerce. After the first meeting of the traffic conference adjourned in December 1924, Hoover, Barber and other conference organizers assembled a Committee on Uniformity of Laws and Regulations. It was chaired by a lawyer, who was assisted by a representative of the American Automobile Association. They agreed to draft a model state vehicle code (including a model driver-licensing act) to replace the disparate existing laws, and to develop guidelines for uniform municipal traffic ordinances. They would then submit the model state code to states and cities for enactment, with the official endorsement of the U.S. Commerce Department.

To draft the proposed model state code, the committee turned to a staff lawyer of the Automobile Club of Southern California. Committee members presented their "Uniform Vehicle Code" to the second meeting of the Hoover traffic conference in March 1926. Local and state auto clubs organized to get the code through their state legislatures.[66]

### Winning Back the Trust of Motordom

Though the model vehicle code promised uniformity in state vehicle laws, it did not inspire motordom. The committee that drafted the code was diverse. Of 33 members, only five represented auto interests and only two were from street railways. The committee draft permitted motorists the unusual right to pass streetcars unloading passengers where there was no safety zone for them, but this provision did not get into the final draft.[67] The final version won the railways' support.[68]

Many in motordom were disappointed. They had seen a chance to get real change with the model state code, and the final compromise document had not delivered. Many states had already devised their own statewide uniform codes, and auto clubs had sometimes drafted these codes themselves. They were reluctant to discard them entirely for the code developed by the Hoover traffic conference.[69] Within a few months, conference director Barber found that "automobile dealers and automobile club secretaries in a number of states have taken a rather unfriendly attitude toward the Code." Such discontent jeopardized its chances "in a number of states," Barber told Hoover.[70]

But the dissatisfaction was far worse than a threat to the code. It imperiled the associative tenets the conference was founded upon. Hoover's great hope was that associationism could preclude positive regulation. If interest groups cooperated, the state could remain an umpire only, even in a twentieth-century economy. Hoover staked his career on a bet that associationism could reconcile individual liberty and modern integration, preserving both. If interest groups walked out of his conferences, he could hope for real results only with more intrusive state action. The traffic conference would succeed only, as Hoover later told members, by finding "courses of action upon which all can agree."[71]

To make the conference succeed, Barber had to keep motordom at the table. The uncertain future of the code taught Barber a lesson about the power of "organized motordom," as automobile clubs described their members collectively.[72] By the mid 1920s, automotive interests were far better organized than they had been just a few years earlier, and they could take a stand and fight for it. "Every automobile owner," a trade journalist reported in 1925, "whether he rides in a Ford or a Rolls Royce, is a brother under the skin when questions affecting his welfare or pleasure are brought forth." Dealers too were vigilant: "Legislatures have learned to their cost that automobile owner and dealer associations are a potent factor when it comes to defeating proposed legislation inimical to good of the industry."[73]

Auto clubs could get the attention of state and local officials, in part because of the organized mass constituency behind them and in part because of the valuable information the clubs could deliver. The secretary of the Cincinnati Automobile Club pointed out that "many automobile clubs, by the collection of statistics and valuable data, make themselves so valuable that they are looked to and consulted by public officials." Another auto club official claimed that the legal staff of the Cleveland Automobile Club made the group "recognized as the natural leader in the working out of traffic problems."[74] With such information at the clubs' disposal, legislators paid attention. A Pennsylvania club secretary could credibly claim in 1925 that "every Road Bill, with the exception of one, which was recently passed by the Legislature of Pennsylvania, was drawn up in the office of our Club."[75] "The American Automobile Association alone is in a position to bring about the desired uniformity," the AAA claimed in 1927.[76]

For their part, the street railways claimed to represent the great majority of city travelers. Yet the railways had no organized mass constituency. Owners of automobiles often identified themselves as motorists, but the streetcar passengers who vastly outnumbered them in large cities made no equivalent identification. A Vancouver street railway official warned his American colleagues that "the car rider is unorganized and inarticulate and the street railway company must supply the want."[77] A Brooklyn City Railroad executive was likewise frustrated at streetcar passengers' disinterest in the fate of their railways, concluding that "only until they are awakened to the situation can we expect its real solution."[78] A Chicago street railway engineer also could not understand this indifference. "I have never

been able to understand why the street car patrons do not support prohibi-
tive parking measures," he mused.[79] The railways were afraid to face the
answer: streetcar patrons did not identify themselves with their mode.
Many surely thought of themselves as future motorists.[80]

Barber lost no time applying the lessons he learned from the reaction to
the Uniform Vehicle Code. If the conference could develop a supplementary code and put motordom in charge of it, he believed, he would be able
to retrieve the lost loyalties of irritated groups and win their support for
the Uniform Vehicle Code. In the spring of 1927, Barber met with leaders
of the auto manufacturers' trade association, the National Automobile
Chamber of Commerce. They assured him that the right new committee
would not only win their support but would also be "an important means
of for strengthening the support for the Uniform Vehicle Code." Since the
code was threatened, the promise was attractive. Hoover was away in California, leaving Barber in charge of the planning. Barber began at once to
plan a committee to draft a Model Municipal Traffic Ordinance.

State traffic conferences had already pursued model traffic ordinances
for cities. The Los Angeles Traffic Code of 1925 (the origins of which were
described in the previous chapter) became the basis of a model ordinance
for the cities of Southern California. Championed by the Automobile Club
of Southern California and the California State Automobile Association,
the ordinance had been adopted in 40 cities by the end of 1927. By then
a model ordinance drafted by the Detroit Automobile Club had been
adopted by more than 200 Michigan cities and towns.[81] None of the proposals for a national model for cities had gone far, however. Barber thought
he could beat the odds by putting a powerful coalition behind a new model
ordinance for all states.

In May, Barber wrote to Herbert Hoover. To secure motordom's backing,
Barber said, the new Model Municipal Traffic Ordinance committee could
not be headed by a lawyer, as the Uniform Vehicle Code committee had
been. Instead he advised Hoover to put William Metzger in charge. Metzger
was the director of the Detroit Automobile Club and served on both the
executive committee and the board of directors of the American Automobile Association. He was one of the most recognized champions of the
automobile in America. He was credited with opening the first auto dealership in the world (he had sold Cadillacs in Detroit since 1898) and was a
wizard at selling cars. He launched the annual Detroit auto show. He was
also on the Board of Directors of the National Automobile Chamber of
Commerce, and later he became chairman of the executive committee of
the Hoover traffic conference.[82]

Barber saw immediate practical benefits in letting motordom dominate the membership of Metzger's committee. Interest groups represented on the committee were unwilling to pay for Commerce Department personnel, but the National Automobile Chamber of Commerce, seeing possibilities in Barber's friendly overtures, offered the services of its own staff. Motordom, in return, would be well represented. Besides Metzger, Barber recommended representatives of two automobile clubs and one man from the Mack truck company. He also recommended Miller McClintock of the new Erskine Bureau and Harland Bartholomew, a city planning engineer who had helped draft the Major Traffic Street Plan for Los Angeles. Barber needed other interests on the committee to give it a representative character, so he found two officials of street railways to serve on it. Barber sent the proposed committee list to Hoover for approval; Hoover wired back to Barber to go ahead.[83]

The list of committee members continued to grow, however. The final composition of the new Committee on Municipal Traffic Ordinances and Regulations included Alvan Macauley, Packard's zealous fighter of traffic control and champion of urban reconstruction for automobiles. Ten representatives of street railways also participated, and so Barber could declare that the committee included "all interests affected." They were outnumbered by 24 representatives of motordom, including nine delegates from auto clubs and eight from manufacturers. Auto parts, tire, and rubber companies, the National Automobile Chamber of Commerce, the American Automobile Association, the Automobile Club of Southern California, and the National Automobile Dealers Association all had a say in the drafting of the ordinance.[84]

The committee considered new ways to fight traffic. In a draft of the model ordinance, they recommended that highway engineers—not traffic control engineers—attack city traffic problems. Cities, for example, should prevent conflicting traffic flow by building "overpasses whereby a main highway will travel above or underneath an existing road," as a step "toward solving the problem of city congestion."[85] Yet such proposals were the purview of a different conference committee. Metzger's committee was to draft a list of traffic rules for cities. Larger matters of principle were therefore dropped from the final draft. The result was "modeled largely along the lines of the California Municipal Ordinance," adopted in many cities in the Far West and "containing basic principles as outlined in the Los Angeles Traffic ordinance" of 1925.[86]

The ordinance codified pedestrians' confinement to sidewalks and crossings, leaving to individual cities the choice of how far to go. At a minimum, cities adopting the ordinance would require pedestrians to yield the pavement to motorists everywhere except in a crosswalk. At their discretion, cities could require pedestrians to cross only at crosswalks, even in the absence of motor traffic. Either way, the rules overturned pedestrians' ancient legal supremacy in the street, at least in principle.[87] The Model Ordinance signaled an abandonment of the more stringent traffic control principles. The committee expressly rejected positive regulation, in the name of efficiency, equity, or anything else.

An early draft of the committee report reveals its members' opinions. They decried not just "an astounding lack of uniformity" in traffic rules, but a "super-abundance of regulations." Street users faced "restriction upon restriction," and "restrictive regulation has in many instances impeded the free movement of traffic." Traffic congestion was no longer to be understood as an efficiency problem. Instead the "floor space" principle, devised by auto manufacturers in the wake of the 1923–24 sales slump, became the basis of the Commerce Department's official position on the way to fight traffic. According to one committee member, "modern mass movement of vehicular traffic demands the utmost in street area, which is altogether too inadequate in many municipalities."[88]

The minority of street railway representatives on the committee tried and failed to preserve traffic control principles. E. J. McIlraith, an engineer for the Chicago Surface Lines, was one of them. Though a street railway man, McIlraith had recently won the admiration of motordom for his pioneering coordinated traffic signal system in Chicago's congested Loop district. With it, McIlraith had united diverse street users (such as street railways and taxi companies) in a common cause. Sixteen months later, McIlraith wanted the model ordinance to cement the alliance. He asked that the text accompanying the ordinance advise cities to seek "the greatest good for the greatest number" in their traffic regulations.[89] In its various forms this utilitarian justification for regulation was a favorite of the street railways; they carried far more people in city centers than automobiles. It was their old appeal for majority rule united to traffic engineers' faith in efficiency.

But the model ordinance committee rejected McIlraith's proposal. Fellow committee members said that "greatest good for the greatest number" principles would be used to justify regulations that favored some modes over others. In motordom's appeal to minority rights, such discrimination violated motorists' rights to the street. One committee member, Massachusetts' commissioner of motor vehicles, objected to all attempts to distinguish modes in any way that tended to give precedence to one mode over another. He argued that "there is no reason why a street car should have any more rights than a motor vehicle."[90] Committee member Charles Hayes, president of the Chicago Motor Club, likewise objected to a proposed provision for police emergency regulatory powers on the grounds that police might use them "to discriminate between citizens conducting themselves in an orderly fashion."[91] Though the completed report allowed some emergency rule-making power for police, it made no distinctions between modes.

The model ordinance recommended no power for city traffic authorities to favor or restrict modes on the basis of their traffic efficiency. Traffic engineers had often called for such discriminations in the early and mid 1920s, but by the time the Committee on Municipal Traffic Ordinances submitted its work to the Hoover traffic conference in 1928 such calls had vanished from traffic engineer's reports, not to return for 40 years. In one place, however, the committee did call for "seeking the greatest benefit to the largest number": committee members wanted curbs to accommodate the greatest number of *automobiles*. The committee advised cities that "regulations be as simple and few in number as possible," and warned them that "unreasonably or unduly restrictive measures arouse resentment."[92]

The Commerce Department issued the completed model city ordinance in 1928, and it was widely circulated.[93] The National Automobile Chamber of Commerce promoted it nationally, calling the rules "a forward step in motor transportation."[94] By the end of 1928, more than 100 cities had adopted the ordinance.[95] By 1930, state legislatures in New Jersey and Wisconsin adopted most of it for all their cities. Motordom had promised Barber it would support the Uniform Vehicle Code in return, and it was true to its word. By 1930, 23 states had adopted all or part of it.[96] In the name of uniformity, a congress of interests had overturned prevailing principles of professional traffic control. The engineers, with their easy resort to rule making, gave way to the members of the Fewer Laws Club.

### The Commodification of a Public Utility

The guiding spirits of the 1924 Hoover traffic conference were the National Chamber and the Commerce Department. They worked together closely and they shared some first principles. Among these were the conviction that business prosperity in the twentieth century could not come through competition alone. It required the elimination of waste through the exchange of information, the reduction of duplication, and the establishment of standards. Hoover described such business collegiality as "associational activities." Efficiency would follow from such cooperation, and allow market expansion in a post-frontier age when natural resources no longer seemed infinite. Together Hoover, the National Chamber, and other trade associations worked to establish the cooperation needed to pursue these ends.

### Hoover: Ether Roads and City Streets

Hoover took particular interest in applying his associational principles of business self-government to two new problems of twentieth-century life: radio and street traffic. In the early and mid 1920s, signals in the "ether" (the atmosphere) were a tangled mess—much like the crowded streets of America's cities. At about the same time, both in the ether and in city streets, a sudden surge in traffic within a limited space led to stultifying interference between users. In each case, the congested medium had been a largely unregulated space open to first comers indiscriminately. In each case, expanding the limited space was a technically difficult or expensive prospect. Both fields were conducive to the rise natural monopolies among users. They depended on publicly owned media of limited capacity. They were difficult to expand, and it was difficult to charge those who used them. Unregulated competition therefore led to interference and, with time, tended to give way to monopoly. A comparison of his approach to radio traffic and street traffic will shed light on the origins of the new model of street traffic.

In his first year as Secretary of Commerce, Hoover organized a radio conference to avert hopeless interference in the ether; by the end of 1925 he had hosted four more. In radio Hoover acted as a progressive engineer, working actively to secure his conception of the public interest, because left alone the ether was chaotic. In this field his principles were utilitarian. He would countenance strict regulation if was needed to secure efficiency and equity. Hoover saw a positive role for the state, which through "legislation and regulation" would make radio "of the greatest possible good to the greatest possible number of Americans."[97] "Even if we use all the ingenuity possible," Hoover told broadcasters at the first conference, "I do not believe there are enough permutations to allow unlimited numbers of sending stations."[98] To protect a limited spectrum, Hoover resisted broadcasters' claims to a property right in the frequencies they used, arguing that ownership of one of the limited number of usable frequencies would constitute a "monopoly."[99]

Hoover agreed with most engineers that "public utilities," including "communications," could operate effectively only as state-regulated monopolies.[100] He compared private ownership of a radio frequency to "private ownership of a water navigation channel," suggesting that the ether was no place to apply free-market competitive principles. Hoover also recognized that such principles were all the more difficult to apply because of the difficulty of charging listeners—a problem he called "the weak link in the whole radio development."[101] He claimed instead that "the ether is a public medium," and therefore "its use must be for the public benefit." On these grounds Hoover won jurisdiction of the ether for the Commerce Department, holding that "the celestial system—at least the ether part of it—comes within the province of the policeman." The business-minded Republican therefore freely passed judgment on the "ether hogs" who were abusing public property. Radio broadcasting was, in sum, "a problem of regulation."[102]

Hoover proposed stringent regulation. He assigned to radio "one definite field": the broadcasting of "pre-determined material of public interest," which "must be limited to news, to education, to entertainment, and the communication of such commercial matters as are of importance to large groups of the community at the same time." The state must see that such worthy programming was not "drowned in a sea of advertising chatter."[103]

Many constructed the problem in city streets in much the same terms. George Baker Anderson of the Los Angeles Railway had argued that in city streets "some uses are more important than others," and regulation was needed to sort them out.[104] Hoover might have used the very same words of radio. He never made the comparison directly, but his rhetoric shows that he appreciated the similarity. He described the congested radio frequencies as "ether roads." The solution in radio, Hoover said, was "traffic control."[105]

Despite the similarities between radio and street traffic, however, Hoover approached the two problems entirely differently. Hoover perceived radio as an extraordinary new realm of economic and social life, a realm which defied normal market mechanisms and which therefore required innovative approaches. He compared the ether to a public monopoly (like a "navigation channel") the use of which must be subject to strict public control. Finally, Hoover perceived regulation as the unanimous demand of both of the main interested parties, broadcasters and listeners. "This is one of the few instances where the country is unanimous in its desire for more regulation," Hoover observed.[106] Hoover therefore pursued public ownership of the ether and strict public control of broadcasters. Though he failed to keep the airwaves free of "advertising chatter," the 1927 Radio Act and the 1934 Communication Act recognized ultimate public control of the ether and the necessity of its regulation "in the public interest."[107]

Reluctant to choose sides in more controversial matters, Hoover regarded streets differently. In his statements on street traffic he recognized the appalling toll of traffic accidents. He appreciated the economic waste of congested streets. He criticized the confusion that accompanied the diversity and complexity of traffic regulations. But Hoover never joined the traffic engineers in classifying streets as public property regulable in the public interest. He was quite confident that the interested parties would work out practical solutions, and would do so better than the state could. He put special trust in the automobile industry. Later, as president of the United States, when he needed a new secretary of commerce, Hoover chose Roy Chapin, president of the Hudson Motor Car Company and a member of the National Automobile Chamber of Commerce's traffic department. Hoover did not compare the streets, as traffic engineers did, to a natural monopoly or a public utility. He did not share the indignation of some of them at the inefficient individual street users who cost the mass of users so dearly. Though Hoover himself compared city streets and "ether roads," he did not press the point.

Hoover held back because the interested parties in street traffic did not agree on the necessity of more traffic regulation. Both broadcasters and listeners wanted ether regulation, but by the mid 1920s street users were squaring off to compete for entirely different ways of fighting traffic. By December 1924, when members of the Hoover traffic conference first convened in Washington, the saturation crisis and the maturing of national automotive interest groups were leading to growing criticism of professional traffic control. Other interests—especially the electric railways—blamed the automobile as the real clogger of city streets and called for more regulation. In his traffic conference Hoover scrupulously avoided taking sides in controversial questions. Because he would "never deliberately inject" himself "into a row," Hoover sought "courses of action upon which all can agree."[108]

Timing may also have contributed to Hoover's greater reluctance to intrude in city traffic than in radio. Hoover had been at the Commerce Department less than a year when he convened the first radio conference. He depended on government regulation in his early years at Commerce because then the possibilities of associationism still seemed slight. The dramatic developments of 1924 and 1925 that allowed ambitious associational programs were still years away. When the first traffic conference opened, at the end of 1924, the possibilities of associationism looked dramatically brighter, and they improved still more in the conference's succeeding years. Hoover declared that the traffic conference marked "a new conception of government," a departure even from his own early work at Commerce.[109]

### The Gasoline Tax

For practical reasons, streets have been a public responsibility. Users could not be charged for each use, and the street was the original home of the free-rider problem. Although merchants' associations sometimes improved and even built streets, the main builders were public agencies funded through property assessments, often supplemented by bond issues.

In the 1920s a new kind of charge gave some a reason to hope that users might pay in proportion to their use. Outside city limits, beginning in 1919, gasoline excise taxes allowed roads to function more like other consumer commodities.[110] The tax, in theory, reduced the state's role to that of a broker between road purchasers (motorists) and road builders. In fact road location and building remained notoriously political. Nevertheless the gasoline tax promised to subject road supply to economic laws. Politicians would not have to decide how much to spend on roads; users themselves would decide, as an automatic function of their own demand. The more people drove, the more revenue would accrue, and the more roads would be built. At last the market would rule.

Much of the new gasoline tax revenue was spent on county and state highways. These routes were almost exclusively rural in 1920, but as the new highway funds poured in, counties and states began to extend them into and through cities. Beginning in the late 1920s and at an accelerating pace thereafter, counties and states turned to highway engineers to solve city traffic problems. Highway engineers brought highways into the cities, reducing the role of city traffic control engineers in the congestion problem. The new urban thoroughfares were largely bought and paid for by motorists with gasoline tax money.

State and local governments were quick to recognize revenue possibilities in the growing number of motorists. License and registration fees were universal by 1913.[111] Motorists and their auto clubs resisted. Through the early 1920s, most auto interests fought for low fees, for the use of general revenues supplemented by federal aid in state highway projects, and for the use of bond issues and special assessments of property holders to pay for city streets and county roads. Auto interests rallied to defeat proposals for a federal gasoline tax in 1914 and 1915.[112] They accepted the generally held view that roads and streets were public property, available to all, and on this ground they held that that the burden should be shared by taxpayers in general.

From this perspective, gasoline taxes were merely another way to soak motorists. The tax was proposed first by the backers of good rural roads in the West, including auto clubs representing rural motorists, who saw in it a way to get them. In 1919 Oregon, Colorado, and New Mexico introduced the first state gasoline taxes, and in 1921 the Supreme Court upheld their constitutionality. But except in rural areas automotive interests fought the trend. In 1920 Alfred Reeves, general manager of the National Automobile Chamber of Commerce, told a friendly audience: "As to the tax on gasoline, gentlemen, there isn't a man in this room, not even the able representative of the oil industry, who is present, but will admit that the present price of motor fuel is high enough now without adding anything more."[113] When the state of Indiana imposed a two-cent gasoline tax in 1923, Studebaker protested and filed its share of the tax with a public declaration that it did so "involuntarily, under protest and by compulsion of law and to prevent exaction of the penalties threatened."[114] Motorists, auto interests maintained, did not bear a special responsibility for the financing of roads and streets. Until 1923 gasoline taxes were rare, small, and little noticed. At the beginning of that year, most states still imposed none at all. In the next three years, however, everything changed.

Motordom was beginning to attack traffic control as an intrusion upon the free market. It wanted to recast the problem of traffic into the terms of supply and demand, and for this it spotted an advantage in a gasoline tax. Soon motordom was backing gasoline taxes as way of "buying" roads and streets. With the new support, they spread fast. By 1925 all but four states were taxing gasoline, and a year later almost half of state revenues from motor vehicles were collected as gasoline taxes.[115]

A historian once found "no serious opposition" to gasoline taxes in the 1920s.[116] In fact the petroleum industry angrily denounced them throughout the decade, but it soon lost its allies in motordom. Oilmen wanted to form a broad front of opposition with other auto interests, but were shocked by the mid 1920s to find them happily supporting the tax. Frustrated, *National Petroleum News* resorted to attacking them as "the dumbest flock of geese in the world" because they would "goose step" for the tax collector.[117]

Auto clubs' agreement to back gasoline taxes came gradually and with conditions. Clubs helped get the first gasoline taxes passed in rural western states in the name of good rural roads.[118] In more densely settled states, however, auto clubs remained ambivalent for some years, fearing that most of the money would not end up as pavement. "We favor the gas tax" declared a spokesman for Washington's state auto club, yet he hesitated to "step pretty hard on some of you Eastern States" with this endorsement.[119] When Pennsylvania considered a gasoline tax in 1921, auto clubs opposed it. Yet there, as in the West, legislators learned to link gasoline tax measures to road bills, and thus overcome the resistance of organized motorists. Although the secretary of the Pennsylvania Motor Federation described this coupling of road and tax measures as a ploy of "wily tax-gathering experts," he grudgingly admitted that "we do have the consolation of getting fine roads."[120] By 1923 the president of the Pennsylvania Motor Federation backed gasoline taxes linked to road building as the "fairest method" of taxing motor vehicles.[121] A similar temptation persuaded local auto interests to accept municipal gasoline taxes. St. Louis imposed a tax in 1919. The city's street engineer reported that at first "the various automobile interests and the gasoline companies were inclined to test the constitutionality of this ordinance," but "after considerable discussion" they "agreed to pay the tax with the proviso that the money so derived should be set aside and used for the repairs of important traffic ways." The city cemented local motordom's cooperation by agreeing to spend the revenues "under the general supervision of an informal committee composed of city officials and representatives of the various automobile and gasoline interests."[122]

The American Automobile Association attacked automotive excises in general as a "scramble for the money of the motorist," but with its member clubs either ambivalent or entirely favorable in their opinions, it never condemned gasoline taxes specifically.[123] In 1927 the AAA resolved to "most strenuously oppose any use of the funds from taxation upon gasoline for other than road construction and maintenance."[124]

Except for the petroleum industry, automotive interests by the mid 1920s had grown to accept state and local gasoline taxes, and even endorse them, in return for the commitment of revenues to street and road spending.[125] Alfred Reeves of the National Automobile Chamber of Commerce was among the first converts. In 1923 he told his colleagues: "Sometimes we look upon taxes expended for highways as a burden. The fact is that they are an investment comparable to any expenditure for capital purposes."[126] The new funding mechanism also distanced road expenditures from political deliberations. In 1920, when all road bonds were funded by general revenues, the amounts appropriated were public matters for legislative deliberation. Ten years later, most funds were raised through "user fees" in the form of gasoline taxes, and spending levels were more automatic and less subject to public debate.[127] By 1929, all 48 states and the District of Columbia were taxing gasoline.[128]

Gasoline taxes advanced a new understanding of roads. For the first time some road users were paying for roads in proportion to their use. Because they paid for the roads, they could reasonably claim some rights of ownership in them. Roads were no longer simply public property to be regulated in the interest of the general public. The change was not of practical importance to rural roads, because by the mid 1920s, when gasoline taxes funded about half of state roads, motor traffic already accounted for the great majority of traffic on them. And gasoline taxes contributed little to the funding of ordinary city streets. But by the mid 1920s some major city streets joined sate and county routes, and some new state and county routes entered cities. The trend was promoted as a way to ease traffic congestion. Gasoline taxes funded this new way to fight traffic, and motordom began to use this fact to defend motorists' place in the city and to attack the traffic control model.

The traffic engineers of the early and mid 1920s saw in congested streets inefficient and inequitable uses of a public utility. They ascribed these problems to street users who pursued their individual demands "without regard to the rights of others or stopping to view the whole situation."[129] To engineers, uncoordinated individual uses threatened the collective public interest in the streets, and as a publicly owned utility, streets could be regulated in the name of that interest.

In backing the gasoline tax, auto clubs and other automotive interest groups advanced a new proposition: Roads were a commodity purchased by individual users, and therefore road capacity was not a question for public debate but a matter of supplying a commodity to those who paid for it. Road supply, they said, was a problem neither for legislators nor for experts. Nor would city planners deliver the motor age city.[130] Instead, road capacity should be an automatic result of consumer demand. In principle the formula left little room for engineers, except as the executors of plans drafted by the transportation marketplace itself.[131] Gasoline taxes, in effect, gave road capacity a resemblance to commodities supplied in a free market.

## Parking Meters
Parking meters later did the same for curb parking space. Free curb parking caused shortages, and shortages justified rationing (time limits) and out-

right bans. In 1935 Oklahoma City introduced parking meters. The idea originated in the city's Chamber of Commerce, which wanted to make the best use of limited space.[132] Other cities followed this lead. Auto clubs feared that motorists' nickels, like gasoline taxes, might end up in general revenues.[133] Most state and local auto clubs therefore resisted meters, and by 1936 the American Automobile Association coordinated this defensive "war."[134] Meters spread slowly.[135]

But some auto clubs favored meters from the start. The Texas State Automobile Association regretted "another form of taxation on the motorist," but accepted meters because they "make the 'street hog' put his car in the garage."[136] If meter revenues could be committed to street and parking expenses, they promised to make curb space an undisputed zone for motorists, just as gasoline taxes—once they were committed to streets and highways—helped give drivers proprietors' rights in the travel lanes. Meters increased parking supply by making time limits easier to enforce, by discouraging motorists who did not value the space they parked in, and by rewarding cities that returned no-parking curbs to parkers. By 1940 parking meters were becoming an accepted feature of city streets, and after World War II they proliferated.[137]

As soon as motorists were paying a substantial share of the bill for roads and curb space, automotive interest groups began to make a proprietary claim to them. Already in 1924 *Public Works*, a trade journal addressed to road builders (among others), argued that "automobilists and other users of the highways are the ones to be satisfied," since they "largely provide the funds whereby the highways are constructed and maintained." The traffic expert had no business determining and securing the public interest in transportation, rather "it is the duty of the engineer to determine what these and the other taxpayers want and to provide these wants to the fullest extent possible." The journal asked readers to consider "how large a percentage of the taxes are paid by automobile owners and, therefore, to what a very high degree the wishes of such owners should be consulted in the expenditure of taxes for constructing, reconstructing and maintaining streets and highways."[138] Traffic engineers such as Detroit's Harold Gould considered motorists' unregulated pursuit of their "personal whims or convenience" as the source of traffic problem, and proposed instead that "economics"—that is, expert-determined standards of efficiency—"should govern."[139] Motordom fully agreed that "economics" should guide urban traffic management—yet they understood economics to include the individual demands (or "personal whims") that traffic engineers deplored.

## The Free-Market Model of City Traffic

After the late 1920s, the terms "traffic control" and "traffic engineering" grew scarcer more limited in application. After 1930 traffic engineering was a minor urban adjunct of the growing field of highway engineering. "Control" of traffic remained an important element of traffic engineering, but by 1930 it was no longer its first principle. Engineers retreated to the role of technicians, working out the details of plans drafted in congresses of interests. Some were driven out of traffic control by necessity, others were led out by the new opportunities in city traffic management or highway planning.[140] They relinquished positive regulation, the basic tool of professional traffic control, and learned to treat traffic demands not as raw material to be shaped, but as orders for finished products to be supplied. The rise of associationism and the motordom's mobilization with the saturation crisis gave this new conception of traffic a chance to emerge. It was a radical reconstruction of the problem of urban traffic congestion.

### Highway Engineers Come to the City

"The county highways can do much to relieve city streets," Alvan Macauley argued in 1929.[141] Why not, Miller McClintock asked in 1930, "out of public funds made available by the generous contributions of street users themselves, provide adequate, safe, efficient, and modern traffic facilities so that automobile users will provide their own transportation of a high character at their own operating costs?"[142]

Macauley and others in motordom agreed that congresses of interests should work out traffic engineering policy, leaving only its execution to engineers. But they did not want this to be done by engineers of the sort that had predominated in traffic control, most of whom had earned their experience in public utilities. Macauley divided traffic engineers into two groups: those who ascribed congestion "to the automobile," and those who held that it is "due to the street." The first group included municipal engineers, those who regulated to achieve efficiency. In the second were civil engineers, the builders.[143] If, as Macauley maintained, relieving congestion meant "adapting the City to the Automobile" through "wholesale replanning and rebuilding," cities needed civil engineers.[144]

Civil engineers would soon be ready to help. Editors of their leading journal, *Engineering News-Record*, had long argued for a civil engineering solution to the city traffic problem. Already in 1922 they called for "motor boulevards, second-story streets, under or over crossings for pedestrians" in "the near future." The next year it advocated "structural improvements to relieve traffic congestion," calling them "obvious necessities"; traffic control could supply only "temporary alleviation." In 1924 it called for a "new view of traffic control," one that would remove the problem from chambers of commerce and, instead of regulating traffic, "change the physical factor."[145]

*Engineering News-Record* was ahead of the profession it represented. At a 1925 meeting of the American Society of Civil Engineers, Harvey Corbett presented an ambitious building program intended to allow New York City to accommodate 20 times more traffic. He found the audience skeptical. "Everybody blames the automobile," he complained.[146] Without business backing, Corbett's project never got past the drawing board.

Within civil engineering, however, a new specialization was growing quickly, and it was ready to build to accommodate automobiles. Highway engineers had normally planned, built, and maintained only rural roads for counties and states, leaving city streets to city engineers. But in the mid and late 1920s, highway engineers began to make their influence felt within cities. In 1923 an engineer for Idaho's Bureau of Highways suggested city work to highway engineers as a source of income in the slow winter months.[147] A daring early project in Chicago inspired hope for bigger things. In 1924 work began on Wacker Drive, a two-level downtown motor highway.[148] When it opened in 1926, H. P. Gillette, editor of the trade journal *Roads and Streets*, called for urban traffic to become "a new branch of highway engineering."[149] In the ensuing years, civil engineers answered Gillette's appeal. By the 1930s the American Automobile Association could report with satisfaction that, with gasoline tax revenues, "states are taking more responsibility for secondary highways, and are assisting municipalities in paying for construction and improvement costs of bypasses and main arteries" through cities.[150]

Through the 1920s, the U.S. Department of Agriculture's Bureau of Public Roads nurtured the highway engineering profession. Federal highway spending in the 1920s never exceeded $100 million in any year. Yet Thomas MacDonald, the tireless Chief of the Bureau of Public Roads, made the most of limited resources by winning the support of interest groups. He participated in congresses of interests, including the Hoover traffic conference. MacDonald had been a highway engineer for the state of Iowa, and at the Bureau of Public Roads he enjoyed unmatched prestige in his field. He was committed to road building, to the accommodation of automobiles, and to the extension of highway engineers' purview into city limits. In 1926 MacDonald declared that "the greatest necessity is to take care of traffic congestion in and around the congested centers of population."[151]

In October 1930, prominent traffic engineers gathered in Pittsburgh to form their first national professional association. Highway engineers trained in civil engineering were well represented in the new Institute of Traffic Engineers. The group named the civil engineer Ernest Goodrich its first president and Miller McClintock of Harvard's Erskine Bureau its vice president.[152] Goodrich was a member of the American Society of Civil Engineers and of the American Institute of Construction Engineers. As early as 1924, Goodrich had called for the removal of streetcars from the streets New York City, despite overwhelming support for them among traffic control engineers.[153]

The founding of the Institute of Traffic Engineers marked the completion of a transformation in city traffic management. The change had been under way since 1924 or 1925, when the first National Conference on Street and Highway Safety gave interest groups a direct role in city traffic and when Studebaker had founded the first national traffic planning organization. No sudden transformation of cities ensued, but a cluster of projects in the late 1920s and the early 1930s gave motorists the first urban thoroughfares designed to meet their needs. Motorists bought and paid for these roads through gasoline taxes, and most were reserved for their exclusive use.

In greater New York City a major motor highway system was begun in 1928, and so from this year *American City* dated the beginning of a period of "extraordinary development" of "modern express highways."[154] New Jersey began to clear a highway connecting the Holland Tunnel to Trenton the next year, and by 1930 it was weaving a "fabric of super-highways" west of the Hudson.[155] By then, much of Detroit's extensive network of "super-highways" was already in service.[156] Miller McClintock consulted with Chicago on a major new system of regional urban motor highways beginning in 1929.[157] A 1933 report described these new "express highways" as an application of "new principles of highway planning" in cities, especially the elimination of intersections at grade. Such highways, though "still a novelty in 1928," soon "passed the experimental stage" with projects such as New York's.[158] No longer were engineers to single out automobiles for restriction because of their spatial inefficiency. The new roads were "limited ways" for the exclusive use of motor traffic.[159]

### Motor Age Freedom

The new projects marked an end to traffic control engineers' efforts to use positive regulation to secure efficiency and equity in city traffic. Despite the general trend in the twentieth century of a growth in state power, by 1930 exercises of state power in city streets, such as the discrimination of modes on the basis of their spatial efficiency, had lost their bid for legitimacy. In 1926 George Baker Anderson of the Los Angeles Railway could still claim that those traveling in crowded streets in "vehicles used for the convenience or pleasure of private individuals, or minor special interests" were committing an "abuse of the privilege of the streets." Anderson and others tried to maintain that "the greatest good for the greatest number" was an "American doctrine," but in the competition for the rhetoric of Americanism the railways were at a distinct disadvantage.[160]

Expert control and strict regulation were contrary to more obviously American traditions, especially freedom. A traffic engineer noted the tension in 1930: "As one horn of the dilemma we have so-called human rights, at the other modern efficiency."[161] The intrusive regulations of traffic control were opposed by a new application of natural rights rhetoric. By the 1930s, Miller McClintock, once the apostle of regulatory traffic control, was speaking against restrictions that would interfere with the "freedom . . . of automotive transportation."[162] No organized participant in the traffic debates of the 1920s lived up to such rhetoric. Automotive groups joined Herbert Hoover's associative movement eagerly. They took advantage of the new toleration of business cooperation to reduce "wasteful" competition and to advance mutual interests. Yet in championing American traditions of individualism and freedom, motordom had advantages. The car was the most individual of vehicles, and gave its occupants more freedom than other modes.

Evangelism for motor age freedom was born in the 1923–24 saturation crisis, and the National Automobile Chamber of Commerce codified and led the cause. According to Roy Chapin, president of the Hudson Motor Car Company and a vice president of the National Automobile Chamber of Commerce, "American instincts" were at stake in the transportation question. Chapin argued that the kind of regulation that traffic control depended upon had no place in America. Although Americans "submit at times to a good deal of regulation and officialdom," they "are a race of independent people," whose "ancestors came to this country for the sake of freedom and adventure," and "the automobile satisfies these instincts." Regulation was more than just bad policy, it was contrary to American principles. "The automobile supplies a feeling of escape from this suppression of the individual," Chapin wrote. "That is why the American public has seized upon motor travel so rapidly and with such intensity." With the "American craving for freedom of the individual," the automobile was Americans' natural choice of mode.[163] By the time Chapin joined

President Hoover's cabinet as Secretary of Commerce, he had carried his point.

In repeated addresses, Alfred Reeves, general manager of the National Automobile Chamber of Commerce, also tied the freedom of automobile transportation to broader American ideals. He made the connection as early as 1923, when the saturation crisis was beginning to keep Detroit executives up nights. "America is converted to rubber-tire transportation," he explained, because "the flexible independent transportation afforded by the motor vehicle appeals inherently to an independent people."[164] In Americanist rhetoric, to shrink from the expense of accommodating the automobile now would be unpatriotic. "No physical feat need seem staggering," Reeves assured the timid, "to a nation that can build the Panama Canal, can blast Hell Gate and can construct the Croton Reservoir."[165]

Helped by a new rhetorical appeal to American ideals, motordom and the associative state reconstructed city thoroughfares as commodities in a free market, to be supplied as demanded and paid for by street users. Gasoline taxes helped motordom make this case. Paul Hoffman of Studebaker claimed that business recognized that there was "a considerable area of our economy" in which natural monopolies prevailed, and he conceded that "these natural monopolies must be under governmental regulation."[166] But by 1930 traffic engineers' effort to include city streets in this category had failed. Under the new free-market model of city traffic, traffic regulation was to be minimal and was to serve only as an expedient until highway engineers could bring supply up to the level of demand. Traffic engineers were not to manipulate demands in the name of the public interest, as engineers in other public utilities did. Their job was to identify demands and supply them.

# 8  Traffic Safety for the Motor Age

Pedestrians must be educated to know that automobiles have rights.
—George Graham, auto manufacturer and chairman of the safety committee, National Automobile Chamber of Commerce, 1924[1]

We are living in a motor age, and we must have not only motor age education, but a motor age sense of responsibility.
—John Hertz, president, Yellow Cab Company, Chicago, 1926[2]

The day of the emotional sob sister campaign has passed.
—Charles Hayes, president, Chicago Motor Club, 1926[3]

If an automobile injured or killed a pedestrian, the motorist was responsible. This was the presumptive conclusion under traditional perceptions of the city street as a public space. This perspective was shared to varying degrees by the essentially conservative social groups of pedestrians, parents, police, and judges. It was also a hindrance to the full development of the automobile as an urban transportation mode. By 1920, with common-sense appeals to "safety first," pedestrians were given some responsibility for their own safety as a practical matter. But because the allocation of responsibility in principle was little changed, some found Safety First doctrines unjust. Motorists still shouldered most of the blame, and many looked for long-term solutions that would let pedestrians retain easy access to the street.

Proposals to require speed governors in automobiles were the ultimate expression of this construction of the safety problem. There was no need to ban cars outright if governors could make cars no more dangerous than other vehicles. A governor would end the automobile's unique dangers by depriving it of its unique speed. From motorists' point of view, however, governors would also deprive cars of their chief advantage over other vehicles.

The Cincinnati speed governor petition drive of 1923 taught motordom that its problem lay not in details but more fundamentally in how the safety problem was constructed. As long as motorists bore presumptive guilt, there was no escape from the threat of restrictions so dire that the advantages of car ownership would be lost. Motordom would have to reconstruct the safety problem so that pedestrians—including children—would bear substantial responsibility for their own safety. To do this, it promoted a new construction of the city street itself, as a place where pedestrians do not normally belong. Justifying such a radical reconstruction of the city street, overturning long-standing customs of street use, would not be easy. To do it, motordom claimed that a new age had dawned, making old customs obsolete. Motordom called for a new construction of traffic safety for the new motor age.

## Toward a New Model of Traffic Safety

After its victory in the Cincinnati speed governor war, motordom was determined to prevent similar threats to the car's urban future. It set out to rewrite the terms of the traffic safety problem. The day after the referendum, the National Automobile Chamber of Commerce organized a Safety Committee to seek "ways and means of making it easier for vehicles to operate on city streets and to make the streets safer for pedestrians."[4] The auto manufacturer George Graham served as chairman. The committee intended to take control of traffic safety.

### Transforming Newspaper Coverage

In December 1923, the safety expert Charles Price laid a challenge before motordom: "The whole problem of the accident situation is still in the formative stage, awaiting the leadership of some group of interests—such as the automotive industries."[5] Graham accepted. In the short term, the National Automobile Chamber of Commerce could not hope to reduce total casualties while auto registrations were rising, but Graham's committee did believe it could reallocate responsibility for accidents. The NACC could then clean up the automobile's bloody reputation and legitimize motorists' claim to city streets. Somehow, Graham told his colleagues, "pedestrians must be educated to know that automobiles have rights."[6] He began with the newspapers.

Graham observed that newspapers sometimes reported accident fatalities as " 'Automobile Deaths' . . . whether the driver be responsible or not." It was time to challenge such practices. "To my mind," Graham said, "it is a fair question if the driver is actually responsible for more than half the cases." He objected to the depiction of pedestrians as innocent victims, claiming that "in many cases the driver is a much-to-be-pitied victim."[7] In 1924 the determination of fault depended on many variables, most of which were in flux. Was a pedestrian crossing at mid block to blame—or the speeding motorist? By default, newspapers were the most influential judges of such questions, and their judgment was shaped by old customs of street use that favored pedestrians. Graham saw such newspaper coverage as a threat. In newspaper stories, pedestrians were usually motorists' innocent victims. But what could the NACC do about it?

In some cities, local motordom had long known how to get newspapers to change their coverage. Roy Chapin, a founder and chairman of the Hudson Motor Car Company, remembered that around 1910 "the *Chicago Tribune* would not mention the name of any motor car in its columns." "The dealers in Chicago," he continued, "simultaneously withdrew their advertising from the *Chicago Tribune*. In a mighty short space of time that paper woke up and promised to do almost anything if they could get the advertising, and since that time they have been very decent in their attitude."[8] Other papers with big Sunday automobile sections became newsprint cheerleaders for their sponsors.[9] The *Washington Star* was on very friendly terms with the American Automobile Association, which often used the paper as a publicity agent. The *Star* frequently reprinted AAA press releases as news articles, without comment, under the names of *Star* reporters.[10]

The Chicago Motor Club had recently tried a plan more closely suited to the NACC's needs. The club was alarmed by a crusading city coroner given to recurring "wars on speed," blaming "vampire drivers" for every pedestrian casualty.[11] But the same interpretive flexibility that gave the coroner this power could be used to redirect blame. In most accidents, blame depended on countless factors that varied according to the perspective of the investigator. The club's president, Charles Hayes, saw this as an opportunity to counter the coroner's crusader rhetoric with a scientific posture. He appreciated the mysterious power of statistics. Hayes reasoned that the club could use the flexibility of accident explanations to support definitive-looking statistical claims exonerating motorists and condemning pedestrians. Other voices in motordom were also trying out more or less random, undocumentable statistical claims. During the speed governor war in Cincinnati, the executive secretary of Pennsylvania's Keystone Automobile Club, asserted that "by statistics, we find that the majority of accidents occur through the carelessness of the pedestrian."[12] That same week the Chicago Motor Club began buying space in the *Chicago Tribune* for periodic "Traffic Talks," where it publicized findings purporting to show that the "reckless pedestrian" caused "almost 90%" of the collisions between automobiles and people. The solution? "Don't jay walk."[13] It soon organized its own "accident prevention department" to collect and interpret accident statistics for itself. In newspapers and on the radio the club used its findings for "berating the careless pedestrian."[14]

To reshape coverage from coast to coast, Graham proposed a variation on this technique. Instead of purchasing space in dozens or hundreds of newspapers, the NACC launched a central accident news service. To "make sure that the reporter gets and records the essential facts," newspapers reported accidents on blank forms supplied by the NACC.[15] The NACC assembled the completed forms, drawing its own conclusions about blame. It then reported statistics back to the papers, together with proposals for accident prevention.

Graham explained that the NACC would thus "make the newspaper a clearing house" for the industry's "safety suggestions." The NACC hoped that participating newspapers would "be influenced . . . to give greater publicity to the *real* causes of traffic accidents."[16] If it worked, the plan would put the NACC ahead of the National Safety Council as the most influential national authority on traffic accidents. Even before the first filled-out accident forms came back to the NACC, Graham knew what they would prove: "in a majority of automobile accidents the fault is with the pedestrian rather than with the automobile driver."[17]

Those who carefully followed newspaper coverage of accident statistics perceived a change in tone within months. "It is now the fashion to ascribe from 70 to 90 per cent of all accidents to jaywalking," wrote Bruce Cobb, magistrate of New York City's Traffic Court, in November 1924. Cobb suspected that the auto industry's influence was behind the trend. "I am not sure," he wrote, "but that much of the blame heaped upon so-called 'jaywalkers' is but a smoke screen, to hide motordom's own shortcomings as well as to abridge the now existing legal rights of the foot travelers on our streets." Cobb was not easily misled—accident figures, he knew, were "a matter of the viewpoint of the statistician"—but to others statistics bore almost scriptural authority.[18]

The NACC's "clearing house" was soon joined by another new national authority on accident statistics. The Commerce Department's National Conference on Street and Highway Safety had a Committee on Statistics. Its members were representatives of motordom, and it carried the authority and prestige of the Commerce Department. It frequently reported its find-ings and recommendations to the press.[19] The new committee's relation-ship with Graham's NACC committee is not clear; the new group may have eclipsed the old one, or they may have collaborated closely. In any case, the Hoover Conference gave Graham new ways to shape safety problem. Hoover appointed him chairman of the conference's Committee on Public Relations.[20]

### Silencing the Sob Sisters

Behind the efforts to restrict cars lay deep and widespread fear, grief, and anger about the bloody toll of traffic accidents. Such emotions were potent, but they were also susceptible to the charge of irrationality. By 1926 the American Automobile Association was ready to make this accusa-tion. The AAA's magazine for members, *American Motorist*, attacked such "popular superstition." More bluntly, the president of the Chicago Motor Club announced "the day of the emotional sob sister campaign has passed."[21]

Auto clubs defined themselves by their service to the motorist, and, as Zack Elkins of the Chicago Motor Club told his colleagues in 1925, "cer-tainly a part of this service should be to protect him from being classed with the thugs and criminals of the community." Auto clubs, Elkins explained, had long fought the strict regulations proposed in safety's name, but now the clubs should take more positive steps to redefine the safety problem and lead the way to its solution. "You are paying good money to defeat . . . freak legislation" such as speed governors, Elkins told club leaders, "when you should have warded it off by active accident prevention work in advance." The safety councils were rivals: "You can rest assured that some organization in your territory is going to take the lead in acci-dent prevention work," but rightfully traffic safety "belongs to the motor clubs and to no one else. You should be the leader. . . . it is motor club work and it should not be left to any other organization."[22]

By 1926 it was official AAA policy that traffic safety work was "one of the chief duties and functions of automobile clubs of the United States."[23] They and other members of motordom were crafting a new kind of traffic safety effort, one that the councils and reformers of the early 1920s would not have recognized. It was a safety campaign without the fanfare of safety weeks. It claimed that pedestrians were just as responsible as motorists for injuries and accidents. It ignored claims defending the historic rights of pedestrians to the streets—in the new motor age, historic precedents were obsolete. "We are living in a motor age," explained John Hertz of Chicago's

Yellow Cab Company. "And we must have not only motor age education, but a motor age sense of responsibility."[24]

In the motor age model of traffic safety, motorists had an inalienable right to the street. The new model rejected notions of the inherent danger of automobiles, and sought to rid streets of reckless motorists so that the allegedly harmless majority could prove the truth of this claim. And to undermine the premise of the inherent danger of automobiles, motordom attacked the view that speed and safety are mutually exclusive. By the end of the decade, the new model of traffic safety joined motordom's reconstruction of the traffic congestion problem. Motordom used both to buttress its call to rebuild the city for the motor age.

In their struggle to redefine traffic safety, auto clubs were only the most prominent representatives of motordom. The tire manufacturer Harvey Firestone introduced a university scholarship in 1920, awarding it to high school students for essays on assigned "good roads" topics, including "The Influence of Highway Transport upon the Religious Life of My Community."[25] But in 1924, after a record year of traffic death and the Cincinnati speed governor war, the assigned topics were in traffic safety. By then diverse motor interests had united to form the Highway Education Board, which administered the Firestone essay contest. Like the Firestone scholarships, the board had originated as a promoter of good roads, but by 1924 much of its work was in traffic safety. It added an essay contest for elementary school students, offering them many cash prizes (there were 485 winners in 1924). The National Automobile Chamber of Commerce funded the contest. As before, however, the topics were assigned. In 1928 children explained "Why We Have and Practice Traffic Rules."[26] By 1927, as the idea that road design was the best way to make traffic safe gained currency, even the American Road Builders Association was prepared to begin a national traffic safety campaign.[27]

Before motordom could redefine safety, it had to agree to a new definition. This required a level of cohesion its lacked before 1924. Auto interests, diffuse until then, quickly learned to collaborate and compromise among themselves, and to present a fairly united public front. The sales slump of 1923–24 and its attribution to congested city streets and restrictive traffic control helped push auto interests into the field of traffic engineering.[28] The catastrophic traffic casualty toll of 1923, and the restrictive measures safety reformers consequently proposed, similarly pressed motordom into the traffic safety debate. Both of these new roles for auto interests were built upon the foundation laid by Commerce Secretary Herbert Hoover at his first National Conference for Street and Highway Safety.

By bringing together all interest groups with a stake in city traffic in 1924, Hoover helped coalitions of them form and negotiate united public positions. This was particularly advantageous for the relatively poorly organized auto industry, which was well represented at the conference.[29] When the first meeting of the conference adjourned on December 16, 1924, motordom was, for the first time, capable of presenting a clear program. By 1925 it could refer confidently to "the political power of the automobile."[30]

## Speed Can Be Safe

In 1920 strict enforcement of low speed limits seemed the obvious road to safety. Auto clubs, to the extent that they took a stand on safety at all, tended to go along with this view. The Chicago Motor Club fought accidents by posting signs and tying pennants to cars reading "What's Your Hurry?" The club also solicited pledges from 17,000 members to refrain from speeding. The Chicago club's safety slogan was adopted in Detroit and Denver.[31] Yet Cincinnati's speed governor proposition was the ultimate logical consequence of the proposition that speed is the wellspring of danger, and in its fight against governors, motordom first distinctly questioned it. The idea was soon subjected to a fierce combined attack.

Even before the Cincinnati speed governor war, governors found occasional advocates in the press. By the spring of 1922, motordom took notice. In *Motor* magazine, the industry journalist Harold Blanchard attacked advocates of governors for linking accidents to speed. Speed governors were vulnerable on practical grounds, but Blanchard preferred to attack the premise. Lacking accident records, Blanchard nevertheless claimed governors could not prevent many accidents because "most motor disasters occur at moderate speeds, and relatively few are the result directly or indirectly of what is usually called fast driving." Limiting cars to pre-automotive speeds—"twenty or twenty-five miles per hour under all circumstances"—would "rob the automobile of much of its utility." Cars, in other words, would no longer be worth buying.[32]

In 1922 Blanchard was almost alone, but the Cincinnati speed governor war made his position a strategic salient in motordom's campaign to redefine traffic safety. Accident records remained incapable of supplying evidence for settling the speed question one way or the other. This was no obstacle to either side. The speed governor's critics agreed with the *Cincinnati Enquirer* that "very few of these accidents are due to speed," and even that "speed often is essential to safety." The *Enquirer* claimed (with extraordinary but undocumentable precision) that "99 per cent of the accidents now occur when automobiles are going at less speed than the limit in the proposed ordinance"—25 miles per hour.[33]

True to its new principle, motordom campaigned to replace low city speed limits with laws requiring only that the motorist drive at a "reasonable and proper" speed. Such a determination would always be open to dispute in court. A day after the defeat of Cincinnati's speed governor proposition, Roy Britton, president of the auto clubs of Missouri and St. Louis, began such an effort in St. Louis, where the official speed limit was still 10 miles per hour.[34]

Motordom had an important advantage in this campaign. Traffic control engineers' goal was to secure efficient traffic flow. To them, therefore, speed was a positive good, and vehicles should be allowed to travel as fast as safety allowed. Traffic control stood for "the greatest facility of movement consistent with public safety," and therefore engineers opposed arbitrarily low speed limits.[35] The relative success of traffic control in the mid 1920s encouraged an upward trend in speed limits in the mid 1920s.

Yet the prevailing principles of traffic control were also an obstacle to motordom's ambitions. In the mid 1920s, traffic engineers condemned the spatial inefficiency of automobiles and questioned the sense of spending large sums to give them room. The engineers' predilection for low-cost regulatory remedies was not a promising foundation for motordom's campaign to redefine traffic safety. It preferred to redefine the safety problem for itself.

### Recklessness

To exonerate speed, motordom had to propose an alternative culprit. The choice was easy. From the beginning, newspapers and safety reformers loosely linked the words "recklessness" and "carelessness" to the causes of accidents, using them almost as often as "speed" and its cognates, and with little care for distinctions. The St. Louis memorial to child traffic fatalities blamed the deaths on "haste and recklessness," and in the early 1920s these two words were usually treated as synonyms. If motordom could subtract "haste" from "recklessness," it could fight accidents without jeopardizing the automobile's chief virtue: its speed.

There was a price to pay. Motordom would have to agree that some motorists were reckless, and back police efforts to rid the streets of this class. Some balked at this. But there would be rich rewards. By going after some of their own, motordom would demonstrate its commitment to safety. And the plan stood a good chance of working. If it removed reckless motorists from the streets, accidents would surely fall, and the responsible majority of motorists would be vindicated. "The entire army of careful motor car drivers suffer," the *Cincinnati Enquirer* claimed, from the misdeeds of "the lawless element" of reckless drivers.[36] The American Automobile Association agreed: "Law-abiding motorists, who are fortunately in the vast majority, must combine to drive off the streets and highways the small minority of motorists who flagrantly disregard the rules of safety. In doing so they are warding off the stigma and blame which is being placed against motorists generally."[37]

Even before the Cincinnati speed governor war, the AAA sought to redirect stigma and blame to "the driver who terrorizes pedestrians and careful drivers alike."[38] In 1922 the association offered $25—in gold—to the person who submitted an epithet for such motorists equivalent in its sting to "jaywalker." Existing derisive names (such as "joy-rider" and "speed maniac") implicitly linked speed and danger; the AAA needed a word that left open the possibilities of safe speed and slow carelessness. The winning suggestion was "flivverboob."[39] Yet publicity for this word was nowhere near as extensive as that for "jaywalker," and it never caught on.

If recklessness and speed were distinct, then the slow could be reckless. For the first time, pedestrians could be included among the reckless and careless, their innocence would no longer be automatic, and some of the burden of responsibility for street casualties would be lifted from the shoulders of motorists. In the mid and late 1920s, motordom identified and pursued these objectives.

In their fight to defeat the speed governor initiative, Cincinnati auto interests were among the first to propose elements of this new safety model. They agreed that recklessness was the real safety problem, and they distinguished it from speed. Acquitting speed, the *Cincinnati Enquirer* contended that "the great majority" of traffic accidents were "due to carelessness."[40] "The causes of accidents . . . are [reckless] men and not [fast] machines."[41] The day before the referendum, the *Enquirer* addressed voters: "Mechanical devices have not yet been invented that will curb recklessness; and, after all, it is recklessness that causes the greatest proportion of automobile accidents."[42] This position soon became orthodoxy in motordom. When the imbroglio in Cincinnati led Alfred Reeves, general manager of the National Automobile Chamber of Commerce, to travel there to speak to the city's auto dealers, he urged them to back "stern measures against the reckless driver, who is the main cause of accidents."[43] Through Reeves, the Cincinnati auto interests' position went national.

Most in the auto industry accepted the implication of their new safety position. Reckless motorists should be given no quarter. Before 1923 auto clubs rarely recognized reckless motorists' threat to the urban future of the car.[44] With reckless drivers, however, came demands that state and local governments stringently restrict all motorists, or even the automobile itself. The *Enquirer* recognized this danger, and to deflect support from the governor proposal it called for "stiff punishment meted out to those who violate existing laws and ordinances. There should be no compromise on this point."[45] Laws aimed at "eradicating the vicious, the careless and the witless novice," after all, would "not touch the man who drives as he should, except to free him from the constant dread of the reckless minority that has hurt the good name of the more than 12,000,000 American motor car drivers."[46]

Auto interests elsewhere rallied to this position, backing punishment of reckless motorists. In 1925, *Ohio Motorist*, the house organ of the Ohio State Automobile Association, demanded punishment of the minority of "careless, reckless and incompetent operators" of motor vehicles.[47] The National Automobile Chamber of Commerce and its member manufacturers agreed. Already in 1924 an executive could announce that "the industry favors severest punishment for the unskilled, reckless, speed-mad or drunken who cause accidents.[48] In 1926, Alvan Macauley, president of the Packard Motor Car Company, calculated that if the industry could help "rid the streets" of "a minority of unskilled motor-car operators" then "present arbitrary speed laws can be removed."[49]

### The Pamphlet Blunder

Some auto clubs were afraid to take campaigns against the reckless too far. In August 1923, the National Automobile Chamber of Commerce began issuing rather modest safety instructions to motorists in the form of a tag attached to new cars.[50] The combined effect of a stunningly bloody traffic accident toll in 1923, a downturn in sales, and the anxiety bred by the struggle in Cincinnati led the NACC to go much further the following year. To the NACC, it seemed that a few reckless drivers threatened the entire industry.

In 1924, to fight the auto's growing reputation as a scourge of the innocent, the NACC entered the safety lists in a big way. George Graham, chairman of the NACC's Safety Committee, warned reckless drivers that "the fear of God should be put into every murdering criminal." To do this, the committee promised to back judges who sentenced reckless drivers to jail, "no matter how severe may be the penalty."[51] The committee drafted and distributed well over a million copies of a pamphlet for motorists titled Getting the Most from Your Car and sent out a copy with each new 1925 model made by member companies. To protect the reputation of the careful majority of the motoring public, the pamphlet included a bold recommendation: "Let every careless motorist, convicted by due process of law, be deprived of his car for a period to be determined by the court."[52] The NACC's lawyers drafted a model state law with this provision and the organization promoted its adoption in every state.[53] *Automotive Industries* announced that the pamphlet marked "the most definite and specific effort yet taken officially by the industry to combat the traffic accident evil."[54] Its proposals went far to winning the auto industry recognition as a promoter of safety, beginning in the newspapers it influenced. The *Washington Star* reported that the NACC's pamphlet "must be construed as complete proof that the American car manufacturers . . . are not actuated by selfish motives."[55] To write the pamphlet's foreword the NACC secured the services of President Coolidge, who duly commended the organization's record as putting it "among the leaders in efforts for safety on the highways."[56]

But to many auto clubs the NACC had gone too far. The clubs feared that drastic measures against the reckless few would set a precedent that could culminate in regulators "hog-tying the automobile." Ohio's state auto club called the NACC's proposal "the crowning insult." The plan threatened not just the reckless but even "Mr. Average Citizen," whose automobile was already the "victim of perhaps more harassment than any other useful and inanimate object."[57] A more telling criticism came from the National Automobile Dealers Association. Its general manager, C. A. Vane, protested that "the last people in the world to advocate a motor vehicle impounding law should be the National Automobile Chamber of Commerce." To Vane the great danger in the battle over traffic safety was that the automobile itself might be defined as the culprit. As inherently dangerous devices, automobiles would then have to be restricted. They would be automatically suspect in any accident, regardless of the driver's competence or the pedestrian's negligence. Vane concluded that the NACC was "willing," through recommending this law, "to admit the most damnable indictment that even the most radical fanatics among our critics have been careful to avoid, namely that a motor vehicle is an 'inherently dangerous instrumentality.' " This notion "transfers the guilt from the driver to the machine. The legal significance of such a classification is appalling to contemplate."[58]

The NACC backed down. After distributing its 1924 printing of Getting the Most from Your Car, it abandoned this angle of attack. From this nasty dispute within its ranks, motordom learned two valuable lessons. The first was the necessity of unity. Beginning in 1924, auto interests quickly closed ranks. The convening of Secretary Hoover's National Conference on Street and Highway Safety in December 1924 was a crucial step. Auto interests learned to confer on policy positions before they acted publicly. By the late 1920s there were just two clear formulators of the national automotive position on safety and on other traffic problems: the NACC and the AAA. The trend culminated in the industry's establishment of the Automotive Safety Foundation in 1937, by which time motordom had long been speaking quite distinctly in one voice.

There was a second lesson. In taking the lead in traffic safety, motordom must never allow suspicion to fall on the automobile itself. The burden of the safety problem had to be concentrated on individual reckless motorists, pedestrians, and roads. The NACC surely never intended to make the automobile itself the issue in its impoundment proposal; seizure was meant to keep reckless motorists off the road. Yet auto clubs and dealers showed the NACC that impoundment gave at least the appearance that the danger lay in the car itself. To clear up any doubt, Charles Kettering, head of the General Motors Research Corporation, declared the automobile "inherently safe."[59] For the next four decades, manufacturers and clubs did their best to keep the automobile itself out of safety debates.[60]

Motordom needed a way to punish reckless motorists without punishing their vehicles. It found the answer in driver licensing. Revocation of licenses (instead of the impoundment of cars) would keep bad drivers off the roads without casting suspicion on the vehicle itself. From 1924 on, motordom almost unanimously backed proposals to license drivers who passed driving tests, and to revoke licenses for recklessness. At first, bureaucratic inertia and state legislatures impeded the spread of these measures. States were much quicker to issue licenses (and collect fees for them) than to see that drivers were screened for competence. By 1926 only eighteen states and the District of Columbia licensed all drivers, and only ten (including the District of Columbia) required tests. All but one of the states that issued licenses provided for revocation.[61] Yet in August 1926 the National Conference on Street and Highway issued a model Uniform Vehicle Code, with licensing provisions.[62] State legislators knew that it had the backing of a wide spectrum of transportation interest groups, and that it had the Commerce Department's imprimatur. By 1930, 23 states had adopted the code in whole or in part. Driver licensing spread rapidly thereafter.[63]

If recklessness could be distinguished from speed, very slow street users could be reckless. The *Cincinnati Enquirer* promoted this distinction: "The man who drives faster than six or eight miles an hour into a street in which a crowd of children are playing is taking tremendous chances."[64] Yet the redefinition had far more important implications for non-motorists. The pedestrian's innocence was so widely assumed in 1923 that a frustrated motorist was driven to ask the newspaper "Why always pick on the driver? Are there no reckless pedestrians?"[65] By the late 1920s, however, pedestrians and even children could, for the first time, be classified among the reckless.

This change in the sense of the word "reckless" was made far easier by innovations in school safety education, most of them pioneered with the help of safety councils. As a practical way to save lives, reformers, educators, safety councils, and traffic engineers trained children to stay off the streets and gave pedestrians some responsibility for their own safety. Once they had this responsibility, pedestrians, including children, could be reckless.

The shift in responsibility implied a fundamental change in pedestrians' ancient rights. In 1923 the *St. Louis Post-Dispatch* had claimed that even in the case of "a child darting into the street" in "the excitement of play" the "plea of unavoidable accident is the perjury of a murderer."[66] In 1926, however, the AAA's magazine *American Motorist* claimed that motorists who struck children anywhere in the street outside the pedestrian crossings could not be classed among the "careless": "The children who were killed in the middle of the block as well as the children who were playing in the streets were not the victims of careless driving; the drivers never had a chance to stop." Using this premise, the article's author consulted 1924 Chicago accident records and concluded that 62 percent of motorists implicated in child traffic deaths there were innocent.[67] "The pedestrian can not selfishly claim that he alone has all the rights and the motorist none," declared Dodge's chairman of the board in 1926.[68] By then, he and others in the industry had already gone a long way toward reversing this imbalance.

## Pedestrian Responsibility

If pedestrians were to accept some responsibility for their own safety, they would sometimes have to yield to motorists. This infringement of pedestrians' traditional street rights would have to be justified. Motordom had to show that the automobile had rights to the street too. To make this claim, motordom first had to show that the automobile, far from being a needless imposition upon more important street uses, was in fact a necessity. Passenger automobiles were no longer "pleasure cars."[69] "Since automobile traffic is as necessary as foot traffic," one auto advocate asked, "should it not have the same rights as foot traffic?"[70] If pedestrians would have to submit to some degree of regulation so automobiles could travel the streets without killing them, the automobile's importance justified the sacrifice. Therefore, in the 1920s motordom organized to reconstruct the passenger car as a necessity.

In 1921 the National Automobile Chamber of Commerce incorporated in its stationery a quotation from the new president, Warren Harding. Every letter leaving the NACC's offices informed readers that "The Motor Car Has Become an Indispensable Instrument in our Political, Social and Industrial Life."[71] In 1922 a Washington auto dealer recognized that the "pleasure car" idea "has somewhat impeded the progress of the automotive industry." "The automobile," he maintained, "has become an essential rather than a luxury."[72] By then, mass ownership of $400 Fords lent support to the claim, as did the transition from open-air to closed models. Already in 1922, well before he planned the National Conference on Street and Highway Safety, Commerce Secretary Hoover agreed that "one cannot condemn the automobile ... as a perquisite of the rich."[73] In 1923 the NACC surveyed car owners and publicized its finding that only 30 of more than 10,000 "state that their cars are used exclusively for recreation." Cincinnati auto interests used the new survey results during their speed governor war.[74]

But pedestrian control was hard to sell. Safety weeks accomplished very little, and traffic control engineers not much more. Traffic engineers did get pedestrian control on paper, however, in the form of ordinances. These helped motordom assign some of the responsibility for traffic casualties to pedestrians.

### Organized Ridicule

In Los Angeles, Miller McClintock's 1925 city traffic code offered motordom a basis for reconstructing traffic safety. E. B. Lefferts of the Automobile Club of Southern California supplied a technique that turned regulatory principles into pedestrian practice. With help from motordom and from the Commerce Department, its way of pedestrian control spread across the continent.

Admitting that "it is not particularly difficult to sell" pedestrian control to "a population which is 'motor-minded,'" Lefferts nonetheless orchestrated a meticulous sales campaign for the new rules. The auto club persuaded the city to delay enforcement, using the interval to "educate" Angelenos to regard streets as motor thoroughfares.[75] The club distributed printed publicity itself, and more through local oil companies and the police. In the week before enforcement began, every radio station in the city broadcast nightly "informative talks" on pedestrian control, so that "the radio audience had no escape."[76]

Lefferts found that to ensure compliance "the ridicule of their fellow citizens is far more effective than any other means which might be adopted."[77] Rather than fine jaywalkers, police blew their whistles at them, "pointing the finger of scorn" and attracting unwanted attention from passersby.[78] Some reacted with "resentment" or "insisted on retaining their 'personal liberty.'"[79] At least two pedestrians fought back physically and were promptly arrested. But at the sound of a whistle and "an admonition to return to the sidewalk," most "violators grinned sheepishly and scuttled back to the curb."[80] There, Lefferts said, the wrongdoer found himself "facing a large gallery of amused people" and "shamefacedly" returned to the curb.[81] One confessed jaywalker admitted the "shrill blast" of the whistle "pierces my whole system," leaving her "cowering."[82]

In the first two weeks of pedestrian control, police arrested only twelve jaywalkers.[83] Rather than arrest an obstinate jaywalker, one cornerman carried her bodily back to the curb.[84] By refraining from arrests, police avoided antagonizing pedestrians and sidestepped the risk that judges might use common-law precedents to cast the legality of pedestrian control into doubt.[85] Above all, however, the technique turned enforcement into another form of "pedestrian education," not just for the violator but also for those witnessing the violation. After weeks of warnings, police stepped up enforcement. In April, in a campaign the Los Angeles Times endorsed as "guerilla warfare," the police department assigned plainclothesmen to anti-jaywalking duty.[86] By winter 1925–26 a reporter found that pedestrians had "learned to obey the stop and go signals the same as vehicles," and that "jaywalkers are arrested."[87]

Los Angeles pedestrians were not entirely subdued. The 1925 ordinance allowed vehicles turning right to block pedestrians' access to the crosswalks. Cornermen found the rule a "source of a great deal of confusion and dissatisfaction." According to the Times, when pedestrians and motorists both get the "go" signal and the pedestrians begin to cross, "along comes an automobile making a right-hand turn across the pedestrians' path. Another automobile follows right behind, and another, and another, until the signal changes again." Pedestrians trying to catch a streetcar missed their chance, and "the man on foot becomes provoked at what he considers an unfair stunt and the minute his path is cleared of the offending automobiles he ducks across the street in high indignation and defies the ordinance."[88]

The Automobile Club initiated a bond issue proposal to finance pedestrian tunnels that would protect children walking to school. When voters approved it, the club surveyed school neighborhoods and selected the first 40 tunnel sites. The passageways protected children and, according to Lefferts, "expedited vehicular movement" on streets, promoting a new proposition: that pedestrians do not belong in streets.[89]

Later, to boost pedestrians' compliance with signals, McClintock and Lefferts learned from Chicago's experience in signal timing. In Chicago, the timings took no account of pedestrians' needs, leading pedestrians to ignore them and fend for themselves. In Los Angeles, signals were timed to give pedestrians a chance.[90]

Lefferts promoted the Los Angeles way to national audiences. To explain his city's relative success, he noted the importance of the social element in technical problems. To a Chicago convention of the National Safety Council he explained: "We have recognized that in controlling traffic we must take into consideration the study of human psychology, rather than approach it solely as an engineering problem." In pedestrian control this meant extending "ridicule" to pedestrian practices incompatible with the motor age street.[91] Through Lefferts's promotion, the backing of national automotive interest groups, and the incorporation of much of it into the Model Municipal Traffic Ordinance, the Los Angeles Traffic Code of 1925 became—as McClintock had hoped—a standard copied by cities across the country.[92]

### Horsepower versus Agility

Besides the legal power behind traffic codes and the social power behind organized ridicule, another less adaptable form of power was at work. In terms of horsepower, the clash between automobiles and pedestrians was no contest. Challenged by an automobile, most people on foot conceded the roadway (including crosswalks) to motorists as a matter of practical necessity, leaving aside finer matters of custom, right or equity. This change in habits lent support to those who claimed that pedestrians did not belong in the streets. "The custom which made the common law," one journalist observed, "is likely sooner or later to change it."[93]

With their superior horsepower, automobiles could negate pedestrians' rights even at designated crossings. The problem worsened as signals replaced police cornermen. "Policemen are the pedestrian's best friends," said a Chicago waiter, but with the spread of automatic signals these friends grew scarce.[94] At intersections controlled by signals, turning cars often deprived pedestrians of their chance to cross on the green. Sometimes pedestrians found that signals timed for motor vehicles did not give them sufficient time to cross. Signals were "regulated solely to speed up street traffic," a Chicagoan complained. "The pedestrian is given no consideration." Even the pedestrian crossing on the green was "cut off by cars turning corners," then "coming up unnoticed behind him."[95] Others objected to waiting for a green light when traffic was light.[96] Especially along wide, busy thoroughfares, pedestrians who wished to cross in the officially preferred manner found themselves stranded. Under such conditions, jaywalking might be the only recourse. Pedestrians complained of standing "five or ten minutes waiting for a chance to cross" where "streams of motorists" sped past, "utterly regardless of other people's rights."[97] To highlight the injustice of such conditions, another pedestrian characterized such drivers as "men who hog [the road] and refuse to let women and little children cross."[98]

Automobiles' superior horsepower won them dominion over the streets. To many pedestrians, this depredation justified resort to their own advantage in agility. They would cross where and when they could. "Pedestrians would be more inclined to cross at street intersections if autoists would respect their rights at the crossings," wrote one pedestrian.[99] Thus, even in the dawning motor age, many pedestrians remained unsubdued. In Baltimore in 1926, 30 percent of pedestrians were crossing away from the designated crossings—and many of the rest used them by chance rather than compulsion.[100] Even in 1930, observers could justifiably complain that there had been "no progress whatever in pedestrian control."[101]

Pedestrians yielded access to the street less out of obedience than for self-preservation. Enforcers of pedestrians' rights were helpless. Bruce Cobb complained that pedestrians of 1924, facing "superior force in the shape of the omnipresent motor car," had "been compelled to forgo asserting [their] legal rights of substantial equality on the highway, as they have existed almost from time immemorial." "As it now stands," the traffic court magistrate continued, "the motorist has won his contest for the use of the streets over the foot passengers, despite the best efforts of police, courts and motor vehicle authorities to regulate him and his kind. The motorist has inspired fear and the sort of respect that brute force inspires." Horse-power overwhelmed all other considerations. "Though he may hug these rights to his breast," said Cobb, "Mr. Average Citizen, his wife and children, cross the city streets or walk the country roads with much the same assur-ance that a luckless rabbit feels when chased by a pack of hounds." He continued: "Under present conditions there is a deadly competition between pedestrian and motorist for the use of those strips of territory we call streets—a conflict deadly to the wayfarer, with the victory to the motorist."[102] A New York pedestrian concurred: "The automobilist has absorbed the pedestrian's ancient right to cross the street at the intersec-tion. I have yet to meet an automobile driver who will allow me to cross the street prior to his whizzing by in the car, although I have the ancient right of way."[103]

**Right of might**

**Figure 8.1**
Cartoon by Winsor McCay, *New York Herald Tribune*, 1925; reprinted in *The Outlook* 140 (July 29, 1925), p. 445.

## Safety Education for the Motor Age

In the mid 1920s, motordom began to question the presumed innocence of children in streets. An industry journalist wrote of "children . . . killed and injured, due to their own carelessness."[104] Traffic engineers' pedestrian control ordinances made this plausible. So did classroom safety education and the redefinition of recklessness. In 1925, with these trends well established, the Cleveland Automobile Club could claim that "disregard of the new traffic regulations by pedestrians" was "the major cause of traffic accidents."[105] The burden of responsibility was spreading beyond the driver's seat. If pedestrians, including children, were often responsible for their own injuries, then motorists could not be asked to bear the whole burden of accident prevention.

In the mid 1920s motordom began to go further, organizing a vigorous and lasting campaign to keep school children out of the streets except to cross them, and to train them to cross streets carefully. The Automobile Club of Southern California had undertaken such work in 1921, but now the effort was national.[106] Motordom picked up where the safety councils left off. Safety councils were the pioneers of organized school safety efforts. Through "junior safety councils" they organized school children and trained them in traffic safety. In the early and mid 1920s auto clubs sometimes supplemented junior safety councils without replacing them. But late in the decade, the auto clubs' well-funded, centrally controlled school safety campaign eclipsed junior safety councils and other local efforts. The AAA became the leading national supplier of safety education material. With the formation of the association's Safety Division in March 1928, the AAA began a major, semi-centralized national drive for safety education. Teachers could receive entire safety curricula drafted at the AAA's Washington headquarters. By the end of the 1928–29 school year the AAA was supplying 50,000 teachers with monthly instructional materials, thereby reaching 2 million school children.[107]

Through its safety education efforts, the AAA labored to revise the moral drama of traffic safety that was so widely publicized in newspaper cartoons and safety council posters. The old safety imagery put the automobile in the most unflattering light. Still worse (to auto clubs), it released pedestrians from any moral responsibility for the burden of reducing accidents. Auto clubs answered safety reformers' macabre publicity with posters of their own. In the 1928–29 school year, the AAA began issuing posters to schools nationwide. That year, the association distributed 250,000 safety posters to American schools, racing past the safety councils (local and national) as the leading supplier of traffic safety posters to schools.[108]

**Figure 8.2**
This poster by Frank Young, published in *Los Angeles Times* of July 6, 1922, was a product of a classroom program sponsored by the Automobile Club of Southern California.

No AAA school safety poster appealed to pity; none showed an accident. There was no blood, and there were no dead children. All AAA posters promoted a new idea (the responsibility of children for their own safety) with a new method (smiles and good cheer). There was no equivalent poster series for motorists. In the posters, smiling, dutiful child pedestrians replaced limp little corpses.[109] The posters' message was obedience. Children were to obey safety patrols: "School boy patrols are for your protection—obey them." "Keep mother happy—obey the schoolboy patrol." They were to obey traffic police: "The policeman is your friend—always obey him." They were to obey signals: "Wait for the signal—for safe crossings." In one poster, a boy at a corner addressed his dog: "Come on, Sport, let's cross on the green." Posters also taught children that the street is the province of the automobile: "The curb is the limit." "Teach your pets to keep off the street."[110] The AAA slogans were a deliberate reform of earlier child-safety publicity, when children had been advised, for example, "better belated than mutilated."[111]

The AAA saw still greater value in school safety patrols. By the early 1920s local safety councils, in cooperation with school authorities, already

COMMANDMENT No. 1—Don't play on the street. Stay on the side-walk or in your yard. The street is for autos. Autos are increasing. Streets are not. Therefore the street is dangerous at all times, and you must remember to "Always Be Careful."

Copyright 1915 by F. J. Kroulik

**Figure 8.3**

Cover of sheet music for Charles P. Hughes's 1924 song "Beware Little Children," published in 1925 by F. J. Kroulik. Source: National Museum of American History, Archives Center, DeVincent Collection. Courtesy Sam DeVincent Collection of Illustrated American Sheet Music, Transportation, Archives Center, National Museum of American History, Smithsonian Institution.

**Figure 8.4**
A Chicago Motor Club safety poster. Source: Harry W. Gentles and George H. Betts, *Habits for Safety: A Text-Book for Schools* (Bobbs-Merrill, 1932), p. 84. Courtesy Chicago Motor Club.

operated school safety patrols in many cities. The patrols were usually the largest component of the junior safety councils. Yet to compete with the councils in the struggle to define the traffic safety problem, auto clubs began to sponsor safety patrols themselves in the mid 1920s. Some local clubs had gotten an early start. The Detroit Automobile Club organized a school patrol in 1919; the Chicago Motor Club and the Automobile Club of Southern California soon followed.[112] These efforts, however, were few and isolated compared to the numerous safety council patrols.

In 1926, however, the AAA began a coordinated national effort to get its local clubs to sponsor "schoolboy patrols" in cities across the America. Local AAA clubs supplied their schoolboy patrols generously, and the AAA coordinated local efforts from its Washington headquarters. The campaign was remarkably successful. For example, in Cincinnati in the early 1920s the young "Safety Guards" had been affiliated with the local safety council.

**Figure 8.5**
An American Automobile Association safety poster. Source: *American City* 41 (September 1929), p. 185. Courtesy American Automobile Association.

By 1929 they were sponsored by the auto club.[113] Club-sponsored patrols soon outnumbered junior safety councils, and in 1929 the number of cities with AAA-organized patrols passed 500.[114] Eventually the AAA claimed that it "took the lead in forming school safety patrols early in this century."[115] The claim was inaccurate, but memories of the earlier safety council school patrols had faded.

A feature of the new safety education of these years was orchestrated peer pressure. In Detroit, 1,300 school children gathered in 1925 to witness the public trial of a 12-year-old accused of "jay walking"; the student jury convicted the defendant, sentencing him to wash school blackboards for a week.[116] In 1930 a Texas junior high school pupil convicted of jaywalking by his young peers was ordered to write an essay titled "Why I Should Not Jay Walk."[117]

When all the blame lay with the automobile, pedestrians struck by cars were innocent victims. This was clearest in the case of child deaths. Their characterization as public losses to be grieved publicly depended in part on the unquestioned innocence of the children and their parents. By the mid 1920s, however, attributions of responsibility were getting more complex, thanks in part to the new kinds of safety education. The new tone in newspaper coverage beginning in 1924 also promoted this

**Figure 8.6**
An American Automobile Association poster (1927?). Source: AAA Headquarters
Library, Heathrow, Florida. Courtesy American Automobile Association.

redistribution of responsibility. As parents and children themselves were
assigned more responsibility for children's safety, young victims of traffic
accidents became more like the fatalities in other private disasters. They
were mourned behind closed doors, where the "sob sisters" were less
audible.

### The Legacy of the National Conference on Street and Highway Safety

Through the first two meetings of the National Conference on Street and
Highway Safety, in 1924 and 1926, motordom established its positions on
safety questions as the most authoritative in the nation.

Herbert Hoover understood that the very future of the automobile in the
city was at stake. Inviting interest groups to send representatives, Hoover
warned that the accident toll, and some of the local regulatory remedies

it inspired, "if permitted to go unchecked," could "threaten the prosperity of a great industry."[118] But as he planned the conference in the spring and summer of 1924, Hoover remained ignorant of motordom's new model of traffic safety. To work, the conference needed auto interests' participation and support. They used this leverage to introduce Hoover to their new safety position and to persuade him to accept it. Months before the conference first formally convened in December, motordom was shaping it. Hoover began drafting a public statement on the conference in March. His early drafts reflect a fervor to fight the traffic accident horror, and his ideas about how best to do this reflected old assumptions about streets and who belongs in them. Hoover did not yet know motordom's new position on traffic safety. But he would soon learn it.

In early drafts of his statement, Hoover betrayed his ignorance of the industry's new position that a small minority of reckless motorists caused most traffic casualties. "It is not to be assumed," Hoover wrote, "that the reported 17,000 persons killed and some 50,000 injured in accidents involving automobiles were deliberately or even recklessly killed and injured." Only "a very small percentage might be charged to reckless 'taking of chances.'" Hoover all but stated that speed was the true menace, and that the automobile was inherently dangerous. "The automobile going at speed is . . . a tremendous and deadly instrument of power," Hoover wrote, with "a capacity for destruction, equal to any deadly weapon." He tied street hazards to "streets and roads where automobiles are permitted to be driven at high speed."[119]

Hoover sent a draft of this address to Percy Owen for comment. Owen was chief of the Automotive Division in the Commerce Department's Bureau of Foreign and Domestic Commerce, a liaison office with industry. Owen and his staff read Hoover's proposed statement with dismay. "In our opinion the article needs to be toned down somewhat," Owen tactfully replied. "The automobile may be considered an element of construction much more than an element of destruction." Owen recommended that Hoover cut passages that disparaged the role of recklessness, that named speed the chief culprit, and that compared the car to "a deadly weapon." Hoover revised the document accordingly, eliminating all objectionable references. Motordom's position on traffic safety was beginning to work its way into the official position of the most important traffic safety body in the United States.[120]

Hoover's opening address to the conference evolved too. An early draft shows that Hoover planned to say "I approach the subject largely in terms of these pedestrians."[121] Hoover later cut this line. Instead, after noting that pedestrians bore the largest share of the casualties, Hoover added: "The next largest group who suffer are the motorists themselves." As to the source of the accident problem, Hoover had learned motordom's position well. He dropped all mention of the danger of speed, and he took a new view of recklessness. Most motorists involved in accidents, he told the conference, "are the victims of reckless driving upon the part of a small minority of vicious or ignorant" motorists. "Incompetence, carelessness, and recklessness are the largest of the contributors" to the "ghastly death toll." No longer did Hoover characterize the vehicle itself as "deadly." Indeed, in the final version of the speech, he added: "The automobile is no longer a luxury—it is a complete necessity."[122] This last statement was a great achievement for motordom. It was official recognition that the automobile was no longer a "pleasure car," and thus cities would have to accommodate it, even at considerable inconvenience to other street users.

Motordom won control of crucial committees at the conference. One of Hoover's personal secretaries, Harold Stokes, described the conference's traffic control committee as "the heart of the whole business,"[123] and motordom virtually owned it. Among the committee members, its representatives outnumbered other street transportation interests by a substantial majority. Roy Britton, the energetic president of the auto clubs of Missouri and St. Louis, served as chairman.[124]

True to Britton's construction of the safety problem, the committee abandoned the usual association of speed and danger, and instead saw its mission as devising "the best methods for keeping reckless and incompetent drivers off our streets."[125] The committee attacked low speed limits as safety measures, recommending that states forbid their cities from establishing limits under 15 miles per hour. Committee members recommended no maximum speed, except to say that in rural areas motorists exceeding 35 miles per hour should have the burden of proving such speed reasonable. The committee announced its recommendations to the press.[126] In 1926 its positions were incorporated in the widely adopted model Uniform Vehicle Code.[127]

Harold Stokes (Hoover's secretary) noticed the departure from earlier conceptions of traffic safety. Taken aback by the traffic control committee's speed recommendation, he wrote Hoover the day the committee made its report public. The report, Stokes noted, "bears every evidence of being written from the point of view of the motorist." "Excessive speed is a major factor in excessive fatalities," yet a recommendation of a *minimum* speed limit was "about the only mention of speed limits." Stokes objected also to the conditional allowance of speeds over 35 miles per hour: "Personally I do not believe that the law should condone a speed in excess of 35 miles an hour under any circumstances."[128]

Most of all Stokes feared the public reception of such recommendations. Since the traffic control committee was "at the heart of the whole business," the success of the entire conference was at stake. "The main point," Stokes told Hoover, was that "speed is identified in the public mind—at least in the pedestrian public mind—with accidents, and for a report of this kind to recommend minimum speed limits as a safety measure I am afraid will bring its sponsors a great deal of ridicule." Stokes urged Hoover to send the committee back to the drawing board. "I think the conference will have to deal with the whole subject a great deal more vigorously than this report deals with it if the conference is going to command the respect of the man who walks as well as the man who rides."[129]

Hoover's reply to Stokes was probably a spoken one, and in any case is apparently not preserved. It seems likely, however, that it was along the lines of a comment Hoover made in a letter some two years later: "I never deliberately inject myself into a row."[130] To do so would have gone against Hoover's characteristic shyness, but more to the point it would have been inconsistent with his conception of "industrial self-government." Government simply had no place dictating methods to business, and if pedestrians had thus far failed to organize a national trade association this was not grounds for government to step into the breach.

The conference adopted motordom's position on recklessness as its own, including the new idea that some pedestrians are reckless. The conference's Committee on the Causes of Accidents included in its ranks the editor of an auto industry trade journal (*Motor*) who later would publish a book advocating fast driving. In it he argued that even high speeds could be safe, and he coached motorists to drive 80 miles per hour on two-way public roads.[131] The committee made the existence of the reckless pedestrian official. It found that "more than 90% of the accidents" in one state in 1924 "were directly traced to recklessness, carelessness or incompetence of pedestrians or drivers involved."[132] The 1924 conference buttressed motordom's claims that pedestrians' carelessness was at fault in many or most pedestrian traffic deaths. "Pedestrians are often as flagrant offenders against traffic regulations as motorists," reported a conference press release.[133] The 1926 conference advised cities and states that "pedestrians as well as motor vehicle operators should be required to obey the traffic rules and regulations and should be punished by adequate fines for failure to do so."[134]

The significance of the 1924 conference was best captured by a journalist shortly after it adjourned. He commented that after the conference the automobile could no longer justly be attacked as a "juggernaut"—a needless and indiscriminate destroyer of life. "One moral" of the conference, he wrote, "seems to have been that this idea must be discarded, and the burden of the problem put somewhere else."[135]

## A Safety Expert for the Motor Age

Hoover's traffic conference put the prestige of the federal government behind the motordom's new conception of traffic safety. It also promoted the cohesiveness of auto interests, which conferred frequently through conference committees in the years between meetings.

Between the first meeting of the conference (1924) and the second (1926), the Studebaker Corporation established and fully funded an institution of traffic expertise at Harvard, the Albert Russel Erskine Bureau for Street Traffic Research, putting it under the direction of a recognized authority in traffic control, Miller McClintock. McClintock's views on traffic changed markedly after the appointment, and he took up the cause of traffic safety as motordom had redefined it. With Studebaker funds, McClintock and the Erskine Bureau produced the first generation of traffic engineers trained as such from the start. In the 1930s, as Studebaker struggled to survive, the Automobile Manufacturers Association took responsibility for funding the Bureau and paid its students stipends to cover their expenses.

In McClintock motordom had a recognized expert it could cite for authority when it questioned the linking of speed to traffic accidents and when it spread the burden of responsibility for accidents to pedestrians. Motordom played up its expert ally. Paul Hoffman, Studebaker's vice president for sales and McClintock's closest friend in the industry, once prefaced his case for more toleration of speed with the phrase "No less an authority than Dr. Miller McClintock," not mentioning that the Erskine Bureau and McClintock's position in it were Studebaker's creation.[136]

McClintock was a convert to the new model of traffic safety. When still a disciple of traffic control, McClintock had linked speed and safety closely. Though speed should not be restricted below the limit safety required, it nevertheless was the crucial factor. "Fundamentally," he wrote in 1925, "all traffic accidents can be reduced to one cause, that is, too great speed under a given set of conditions."[137] McClintock argued for definite, posted limits. "In some instances," he admitted, "it has been urged that the proposed limitations of speed will be a hardship for drivers of discretion." Yet with a sound speed ordinance "drivers of discretion are not exceeding the limitations prescribed." McClintock condemned "reasonable and proper" speed ordinances, arguing the "impossibility of trusting individuals to use proper discretion." Even motorists obeying posted speed limits ought still to be subject to arrest for speeding when conditions made their speed unsafe.[138]

But as director of the Erskine Bureau, McClintock reversed his position. He heartily endorsed the new speed recommendations promulgated at the 1930 meeting the National Conference on Street and Highway Safety. Under this proposal, "speed per se is not made an offense"; motorists had only to be in control of their vehicles. McClintock thus faulted both fixed speed limits and newer "prima facie" speed limits (exceeding them was prima facie evidence of guilt—evidence the motorist could challenge). A newspaper cartoon accompanying McClintock's article suggested that the complexity of prima facie rules overtaxed the brainpower of police officers and motorists alike. Instead, McClintock and the Hoover traffic conference proposed that speed limits be abolished altogether.[139] The idea dated back at least to Roy Britton of the Missouri and St. Louis auto clubs, who proposed it for St. Louis in 1923.[140] Motorists who kept control of their cars would be safe from the traffic police. The rule (or non-rule) would revive the element of discretion McClintock had once objected to. "It is quite impossible," he later argued, "to set fixed rates of speed which can be equitably and safely applied."[141] Much later, McClintock ridiculed the critics of speed as non-experts in the grip of a superstition: those who made speed the issue were trembling before a bogeyman, "Dat Ole Debbil Speed."[142]

Ultimately, highway engineers devised a solution to speed even better than those recommended at the 1930 highway conference. Instead of allowing motorists to exceed low city speed limits if they could show that their speed was "reasonable and proper" under the traffic conditions, highway engineers designed city thoroughfares that allowed (and even required) far higher speeds under normal circumstances. Practical law enforcement required fixed speed limits, but with these now set at or above 50 or 60 miles per hour the critics of fixed speed limits got more than they had hoped for. By 1939, General Motors was promoting to millions of Americans a future of safe urban highways with segregated speed lanes of 50, 75, and 100 miles per hour.[143]

## Highway Safety: Traffic Safety for the Motor Age

Traffic casualties rose relentlessly through the 1920s. By the end of the decade, growth in annual traffic fatalities even exceeded growth in vehicle registrations. Thus the safety problem still cast a shadow on the automobile's future in cities, enough even to keep speed governors a constant threat.[144]

Motordom's disenchantment with traffic control engineers, and its increasing anxiety about cities' lack of "floor space" for automobiles, led it to back a civil engineering solution. State highway projects, financed through state gasoline taxes, would extend into cities, supplying cars with thoroughfares bought and paid for by motorists and reserved for their exclusive use. By 1930 such projects were under way. Highway engineering thus began to supplement traffic control, and to in places to supplant it. Where it remained, traffic control looked more and more like a makeshift awaiting the expertise of the highway engineer.

Urban highway engineering was introduced primarily as a means of fighting traffic congestion. Yet its proponents soon found in it an answer to the accident problem. Highway engineers in the 1930s fought traffic accidents with highway projects, contending that "the same method for overcoming causes of congestion"—highway engineering—would "also overcome the opportunity for accident."[145]

In the mid 1920s, few thought of road design as an important safety factor. A 1926 report on the accident problem in Connecticut ascribed only one accident in 150 to road design, adding that "defective highways cause fewer accidents each year."[146] But in 1925 Cleveland's auto club speculated that the way to prevent accidents might be to "provide streets on which vehicles can travel at a good rate of speed with the maximum element of safety for other drivers and for pedestrians."[147] Two years later, industry leaders serving on the NACC's Traffic Planning and Safety Committee argued that separating intersecting roadway grades would accomplish this end.[148] At a 1927 meeting of the American Society of Civil Engineers, the engineer William Cox prophesied that soon "we shall no doubt see automobile traffic . . . more and more collected in great arterial ways, from which the pedestrian will be excluded, being overpassed or underpassed, and being kept from playing or wandering in the roadway."[149]

Motordom boldly challenged the majority who blamed motorists alone for the accident toll, and won. It proposed that all street users must share responsibility for safety, and that rooting out the reckless minority would go far to making streets safe. Yet the advantages of this new approach were limited. For one thing, real pedestrian control seemed unachievable. Reckless drivers could be deprived of their licenses, but only after they had already imperiled the safety of others. Licensing and harsh penalties could never eliminate inattention and poor judgment. Though motordom might get rid of 10- and 15-mile-per-hour speed limits, there seemed little prospect that city drivers would be able to exploit the engine power of the new cars. These facts led motordom to adapt its new way to fight traffic accidents to its new way to fight traffic congestion. Urban highways, properly designed for the motor age, could solve both problems. This was a new method of traffic safety. Its proponents, recognizing the change of emphasis, called it "highway safety."

### The Limits of Education and Enforcement

Before highway engineers came to the rescue, motordom's safety program depended upon education and enforcement. In practice, however, both of these necessities were very hard to secure. "Education is not the final answer," *Automotive Industries* announced in 1924.[150] It depended upon the intelligence of the great mass of people. "The general intellectual level is very low," Charles Kettering of General Motors explained; "thinking is something which the majority of people refuse to do."[151] Pedestrians were the hardest to reach. A 1936 traffic safety textbook issued by the NACC's successor, the Automobile Manufacturers Association, cautioned motorists about "foolish people" crossing the street.[152] "The man on foot is sometimes accused of traffic reactions that show anything from negligence to stupidity," the AAA warned student drivers. "No wonder! He may actually be stupid!"[153]

While motordom backed strong measures to keep reckless motorists off the road, in the mid 1920s it began to cast doubt on the possibilities of enforcement, even in combination with strong educational efforts. Motordom feared that an enforcement campaign powerful enough to eliminate most reckless drivers would inevitably hobble the mass of motorists with burdensome regulations. This method would leave the traffic safety problem—and motorists' fate—in the hands of unpredictable legislatures. "Legislation, as a means of bettering the traffic situation, probably has been more abused than any other single method," *Automotive Industries* explained late in 1924.[154] In 1928 the director of the American Road Builders Association concurred: "No enforceable legislation will strike at the root of highway accidents."[155] They were more confident that civil engineers could help them.

## Foolproof Highways

Properly designed highways could prevent accidents where education and enforcement could not. As early as 1921 a journalist forecast a motor age future of "broad boulevards, with no danger from sudden inswerves of traffic, no pedestrians, no children, no dogs, no slow horse vehicles and no possibility of skidding."[156] Such roads promised to relieve motorists of much of the safety burden. By the mid 1920s motordom was fighting to get them. Civil engineers, its leaders said, could make roads and streets safe even for those endowed with only the most modest measures of caution, agility, or intelligence (on foot or behind the wheel), and even for Detroit's most powerful new vehicles. With concrete and steel, they promised, engineers could ensure the safety of pedestrians and motorists. "The driver," Miller McClintock later explained, "must be externally restrained from killing himself."[157] Well-engineered motor highways would do just this. By excluding pedestrians, they could save their lives too. With such physical restraints, safety publicity featuring images of mangled corpses would no longer be needed. A New Jersey state highway engineer explained in 1933: "To cope with the weaknesses of the human mind and body and to bring about safety upon the highways, roadway facilities should be designed and used which are inherently safe; that is, they should be so constructed that it will be difficult for highway users to perform improper practices."[158] By then, in New Jersey and elsewhere, engineers were already putting this idea to work.[159] Such highways, built expressly for motorists, promised finally to allow the automobile to go full throttle in and near cities. In the 1920s motordom had feared that accidents and low speed limits had put a bridle on its urban market, but here was a way to remove it. In 1929, Alvan Macauley of Packard argued that with "proper street design and adequate width" speed limits could be "removed so that the public can use motors to travel rapidly but safely."[160] Paul Hoffman of Studebaker agreed. Hoffman spoke of the need for "fine highways" to "make fast travel safe for modern cars," because "the motoring public is . . . demanding speed and more speed."[161]

The reconstruction Hoffman proposed was to be as much for pedestrians as for motorists. The new highways would divert some motor traffic from the main streets frequented by pedestrians.[162] The absence of pedestrians from motor highways would rule out pedestrian casualties, regardless of the competence of drivers or the carelessness of walkers. Hoffman admired Mussolini's new *autostrade*, not least because along them "pedestrians are barred by strong fences."[163] A Chicago highway engineer explained that "the provision of separate routes for high speed traffic will eliminate all causes of accidents to pedestrians and children because pedestrians will not be permitted on the superhighways."[164]

Overpasses and underpasses exclusively for pedestrians were planned. In the late 1920s Los Angeles invested $350,000 in pedestrian underpasses, most of them intended especially for children.[165] In Radburn, New Jersey, a residential city expressly designed "for the motor age," pedestrian underpasses were part of a network of safe footpaths for children.[166]

In practice, however, most projects for pedestrians remained sketches on drawing boards. In October 1929, as Wall Street brokers were busy devising new euphemisms for financial catastrophe, the members of the American Society for Municipal Improvements (a relic of the Progressive Era) were convening in Philadelphia. They complained that in the new highways growing around America's cities there was "a nearly universal absence of sidewalk provision."[167] As motor highways proliferated, the gap grew worse. "We must give the pedestrian the benefits of modern traffic engineering," one traffic engineer urged.[168] Motordom was prominent in such demands. Paul Hoffman, for example, also urged "adequate facilities for pedestrian use."[169] But Radburn's ample pedestrian facilities remained exceptional, and extensive projects to keep pedestrians out of traffic were seldom undertaken. While it appealed for pedestrian projects, motordom fought to keep gasoline tax revenues dedicated exclusively to motor highways, hoping that general revenues would take care of the pedestrians.

To those anticipating the coming of the motor age, the appearance of a new kind of traffic intersection was an auspicious sign. Intersections were the weakest link in the traffic chain. Congestion and accidents were concentrated there. Traffic circles were an early but cumbersome way to promote steady and safe traffic flow through intersections, but professional traffic engineers found that at most intersections well-timed traffic signals did the job better. Beginning in the mid 1920s, however, some cities began to build a few intersections that passed one street over the other, separating conflicting traffic. Such grade-separated interchanges promised safety *and* speed. Westchester County in New York State was an early leader. Its Bronx River Parkway, built in partnership with New York City, opened in 1925 with grade separations at all major intersections.[170] Soon other cities and suburbs followed Westchester's example.[171]

In 1929, New Jersey broke ground on a new highway project funded by gasoline tax revenues. When it was finished, in 1932, the 54-mile four-lane route joined Trenton to the Holland Tunnel. It became the first link in U.S. Highway 1. After inspecting it, a journalist announced: "Here, surely, is one road in the world where the motorist has rights."[172] The safety methods incorporated in this highway lay quite beyond the imagination of any of the safety reformers of a decade earlier. Where safety reformers had called for restricting autos so that pedestrians could enjoy their old rights, on Highway 1 engineers prevented pedestrian casualties by barring pedestrians. Speed was no longer the enemy; the road's very purpose was to permit high speeds with relative safety. The state police permitted an astonishing 45 miles per hour. For the safe navigation of intersections the highway engineers did not depend upon the uncertain judgment of motorists. Instead, intersections were grade separated. Even an inattentive driver could (it was hoped) get through them safely. The highway was designed to make speed safer without demanding expert driving.[173]

Yet the result was, at first, deadly. Highway engineers were still learning. The road attracted speeders. Surely exaggerating, the *New York Times* reported that "nine speedometers in ten read sixty" on the highway.[174] The opposing traffic lanes were adjacent, with no divider, and head-on collisions were terribly frequent. The highway threatened to prove the truth of the popular belief that speed is danger. But highway engineers learned from their mistake. Where the slabs of concrete pavement joined at the center, workers laboriously drove them 12 feet apart. The newly divided highway was safer.[175] Thereafter, highway engineers designed their roads that way from the start. With Studebaker's patronage, Miller McClintock won leadership of the new highway safety movement. His goal, he said, was to find the "fool-proof highway," the motor thoroughfare "built in such a way that accidents will be impossible."[176] In Chicago, in the early 1930s, McClintock designed roads he hoped would fulfill this high ambition, calling them "limited ways."[177] In designing the 160-mile system McClintock applied New Jersey's hard-won lessons. Paul Hoffman defined a "limited way" as "a city street planned and built exclusively for automobiles in a motor age."[178] "There are no pedestrians on limited ways," McClintock explained, so "no pedestrians can be killed on such a structure."[179] Through this work, McClintock refined the doctrine of high-speed safety. He argued that, with grade-separated intersections, divided opposing traffic streams, distinct lanes for fast and slow traffic, and broad shoulders, highways could be fast and safe, even allowing for much human error. "If we could apply all we know," McClintock guessed, "we could eliminate 98 percent of all accidents."[180]

McClintock's vision of the foolproof highway was urban. "A limited way may be constructed right to the heart of any city," McClintock promised.[181] He intended limited ways for even the "densest parts of urban areas," where he believed they could "provide safer and higher speed operation" than existing rural highways.[182] Such roads would bring the motor age to America's cities, "an entirely new era in automobile transportation in metropolitan centers."[183] With new help from Washington, McClintock's hopes for urban limited ways began to be fulfilled in the early 1930s.

In 1931 the federal government endorsed the use of motor highways to fight traffic accidents. As Secretary of Commerce, Herbert Hoover had confined his role in traffic safety to bringing interest groups and state and local government officials together to work out arrangements. Later, spurred in part by catastrophic unemployment, President Hoover brought the national government in. It is not clear whether Hoover, in endorsing highway safety through highway engineering, was thinking first of employment or of safety. It is hard to imagine him committing substantial federal resources to highways without the prod of the Depression. Yet to the Hoover administration, unemployment relief alone did not suffice to justify federal intervention.[184] The official justification was safety. Until 1931, the Bureau of Public Roads largely confined itself to administering federal highway expenditures in rural areas, especially on projects of value in interstate commerce. The strictly local value of most urban highway projects deterred federal action. Yet to the Bureau of Public Roads the promise of highway safety through highway engineering justified federal commitments to urban highway projects.

According to Paul Hoffman, Thomas MacDonald, chief of the Bureau of Public Roads, soon insisted that "no city can be said to be equipped for the motor age unless all of its express highways are some type of limited-way facility." "We must dream," said MacDonald, "of gashing our way rather ruthlessly through built-up sections of overcrowded cities in order to create traffic ways capable of carrying the traffic with safety, facility, reasonable speed."[185] The federal role grew quickly thereafter, and under President Roosevelt the Bureau of Public Roads lost all its former embarrassment about bringing federal money to local projects in cities.

To establish its way of fighting traffic accidents, motordom had been driven to rewrite the plot of the safety fable and add a new concluding moral. Before motordom began its work, accidents were blamed overwhelmingly on motorists. The automobile itself was condemned for its chief advantage: its speed. Attempts to limit cars to pre-automotive speeds threatened to deprive the auto industry of its urban market. The blood of innocents smeared the automobile's public image, especially in cities. Such publicity was so common in the early 1920s that it threatened to outweigh the effect of all Detroit's advertising. Though the plot of evil automobiles and innocent pedestrians never vanished entirely, by 1930 the prevailing story was more complex and included drivers and pedestrians among both the careless and the cautious. As antagonists, innocence and viciousness receded. The leading roles passed to responsibility and recklessness.

By corralling these threats, motordom secured the future of the automobile in the American city. Higher rates of automobile ownership helped, but success, industry leaders believed, depended also on a more deliberate effort to reconstruct the accident problem. By its actions, motordom showed that it did not believe that Americans' "love affair" with automobiles would suffice.

Gothic newspaper tales of evil drivers and innocent pedestrians, moral dramas of mob justice, and mass funerals faded away. Lurid traffic safety publicity from popular sources continued, but it rarely challenged the legitimacy of the automobile itself. It targeted the careless, teenage, or drunk driver. The popular reaction to traffic casualties in the 1920s threatened to limit the automobile's urban sphere and to make the automobile something quite different from what it has indeed become. Automotive interests took this rhetoric seriously—it drove them to bridge their divisions and develop new ways to fight traffic accidents. Motordom's success at reconstructing the accident problem shortened the road from "Better belated than mutilated" to "Come on, Sport, let's cross on the green."[186]

# 9 The Dawn of the Motor Age

New cities will spring up around the edges of the old and these new sections will rapidly rise in value, to the profit of those who own that realty and to the enormous loss of those with great investments in the now valuable sections of the older communities.

—Alvan Macauley, president, Packard Motor Car Company, 1929[1]

The city of tomorrow will be an automotive city.

—Miller McClintock, 1937[2]

In 1930, across a channel called Arthur Kill from Staten Island, New Jersey's state highway engineers finished the "Clover Leaf," the first complete highway interchange of its kind.[3] It was a new kind of traffic structure, and it heralded the dawn of a new way to fight traffic.

The new highway engineers defied the most cherished principles of the traffic control engineers of the 1920s. Facilities like this interchange were expensive, and they served the mode traffic engineers had singled out as the least efficient and the least equitable. Yet by the time New Jersey began to build it, traffic engineering and the city street itself had been radically redefined. Engineers no longer manipulated traffic demands in the name of efficiency, as their predecessors had sought to do. Their job was to identify demands and supply them.

By 1930 the American city was preparing for the coming motor age. Accidents and congestion were as bad as ever, but highway engineers promised to alleviate both problems. Casualties were no longer the inevitable price of fast vehicles in cities; by then the leading experts said they were either human failures, correctable through education and law enforcement, or design errors, correctable through highway engineering. Safety weeks, with their macabre images and their public displays of mass grief, were gone. Examples of grim publicity persisted, but new, cheerful

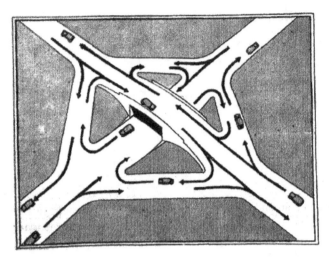

**Figure 9.1**
An early diagram of a cloverleaf. Source: Alvan Macauley, "Over-Passes Overcome Traffic Congestion in Cities," *Motor Age* 52 (September 29, 1927), p. 39. Courtesy *Motor Age*.

**Figure 9.2**
The "Clover Leaf" at Woodbridge, New Jersey, opened in 1929, is considered the first complete example of its type. Source: Norman Bel Geddes, *Magic Motorways* (Random House, 1940), p. 96. Courtesy New Jersey Department of Transportation.

exhortations to caution outnumbered them. The automobile had won a future in the city. The traffic fights of the 1920s shaped this future. In its war on accidents and congestion, motordom in the 1920s began with an attack on safety reformers and traffic control engineers. Long after its victory, motordom fought to keep control of traffic problems. Its highway engineers defined a good thoroughfare as a road with a high capacity for motor vehicles; they did not count the number of persons moved. The industry also fought to control the terms of the safety problem, chiefly by arguing for limited-access divided highways. And it fought to defend its construction of street capacity as a consumer commodity, purchased by motorists and to be supplied to them as demanded. Even today, the American city is substantially a product of this strategy.

## "—And Sudden Death"

By 1930, highway engineers were the leading city traffic experts. Their most ambitious work was still confined to drawing boards and conference papers, but in the following decade many of their plans took shape in poured concrete. As a way of fighting traffic congestion, the new method excited little controversy. Big urban highway projects were still unusual and had obvious employment benefits. Opposition was local and specific. Widespread resistance came decades later, when abundant state and federal gasoline tax revenues funded enormous projects.[4]

Safety was different. City editors gave automobile fatalities front-page headlines from the start, and the frequency and duration of such stories show that they sold papers. The traffic safety problem attracted popular participation. When, after a decline early in the Depression, traffic deaths rose 22 percent in just two years (1932–1934), non-professionals with their updated versions of Danger Land's crippled army came back to haunt highway safety leaders. This frightening apparition taught motordom that though it had won the fight for leadership in traffic safety, it had not won the hearts and minds of many members of the traveling public. In 1935 evidence of motordom's apparent success came in the form of a popular press book, *If You're Going to Drive Fast*, which told motorists they could safely exceed 80 miles per hour on existing undivided two-lane roads.[5] Yet that same year motordom faced a sudden, popular safety rebellion. Its *Common Sense* was a little article in *Reader's Digest*; its pamphleteer was, like Paine in 1776, a young unknown. J. C. Furnas's article "—And Sudden Death" carried an editor's warning that "the realistic details of this article will nauseate some readers" with "sickening facts." Without the help of a single picture, Furnas illustrated in words the daily mangling of human beings in crashing automobiles. In the interests of traffic safety, *Reader's Digest* issued reprints for civic organizations at "1.50 per hundred," and Furnas's piece was probably the most widely read and hotly debated popular press article of the decade.[6] Four months later, *Reader's Digest* described the article's effect as a "sensation" and published a second, even more grisly account of accident casualties—this one by a surgeon who, to prepare fresh corpses for burial, reconstructed their mangled heads.[7]

Readers of "—And Sudden Death" saw the automobile in its ugliest possible light. Motordom had never entirely suppressed gruesome depictions of accident gore, but now they were back in more lurid and less sentimental colors. Worse, Furnas all but declared high speed and safety incompatible. One result was straight out of motordom's worst nightmare. Speed governors were back in style. Five months after "—And Sudden Death" appeared, pollsters found that two out of three Americans favored requiring governors confining automobiles' speed (but this time only to 50 miles per hour, double the limit proposed in Cincinnati in 1923). A year later pollsters found that the public had little faith in highway engineers' chances of reducing traffic casualties. Only 1 percent rated "poor roads" the "biggest cause of automobile accidents."[8]

To motordom this rebellion looked like the Cincinnati speed governor war all over again, and it threatened all they had worked for in the dozen years since. Noting the popular mood, Thomas Henry, president of American Automobile Association, warned auto clubs: "We may expect in the months immediately ahead of us a great amount of crackpot proposals" in the name of safety. "Let us not permit either self-seeking special interests, or well-intentioned long-haired enthusiasts, to hamstring the march of motor transport progress, much less drive us back to the horse and buggy days."[9]

Paradoxically, however, "—And Sudden Death" also confirmed motordom's success in reconstructing the problem of traffic safety. *The Reader's Digest* published the article to "help curb reckless driving."[10] Here was an end that motordom could endorse, even if it did not like the means. The article warned readers of the consequences of "bad motoring judgment," implying that the automobile's dangers were not inherent.[11] There were no suggestions to restrict cars or to promote other modes of transport. Although pedestrians still accounted for about half of traffic fatalities, they were absent from Furnas's article. Furnas also drew attention to dangerous failures in road design—failures which highway engineers, with enough money, could fix.

Nevertheless, motordom took "—And Sudden Death" and the publicity it inspired as a warning. Thereafter it never relaxed its battle for hearts and minds. Outside of classrooms they did not grasp this need sufficiently until *Reader's Digest* forced them to. Just months after "—And Sudden Death" appeared, Paul Hoffman of Studebaker said "The public furor caused by that article frightened us into action."[12] Quite suddenly, beginning in the fall of 1935, highway safety articles penned by representatives of motordom multiplied in the popular press. General Motors bought a full-page advertisement in one Sunday's *New York Times*. In the ad, GM's chairman, Alfred P. Sloan, reassured readers of the safety of the corporation's vehicles, adding that "much progress has been made in advancing the safety of the highway" itself.[13] In the same issue of the *Times*, an assistant director at the Erskine Bureau questioned the " 'shock advertising' " in pieces like Furnas's.[14] *Fortune* summarized the orthodox positions of the leading highway safety experts (especially McClintock), proposing that the automobile has a "right to speed" and that the real problem is that "the road . . . is too slow for the car."[15]

## Selling Highway Safety

The real charm offensive, however, began on January 21, 1936, when the Automobile Manufacturers Association announced "the most comprehensive cooperative educational program for greater safety on streets and highways" ever. The AMA did not intend to investigate means of preventing accidents. Instead it set out to convince the public that the automobile was already safe, and that safer road designs and better driver education and law enforcement would make it still safer. The AMA announced the pointlessness of "attacks emphasizing the morbidity and horror aspects." It published a driver safety textbook for national distribution. It noted that "most of the people killed by motorcars are pedestrians, and the majority of these fatal accidents have been caused by the pedestrian himself."[16] AMA members showed their resolve to fight for their version of traffic safety with a special grant of $54,000 to the Erskine Bureau.[17]

The AMA soon made this new publicity drive permanent and formed a new subsidiary to run it. On June 2, 1937, at a dinner in Detroit, AMA members formed the Automotive Safety Foundation. They named Paul Hoffman as president. The name of the new body dissociated it from the industry group that founded and funded it. Yet the ASF was a lineal descendent of the NACC's Safety and Traffic Planning Department, formed by manufacturers and other auto interests in 1923 (in part to fight the speed governor proposition in Cincinnati). The ASF operated within the AMA, which launched it with an annual budget of half a million dollars.[18]

The Automotive Safety Foundation was above all a publicity agency for the highway safety model. It owed its existence to motordom's new conviction that it would have to get its approach to safety "aggressively supported by an informed public opinion." According to Hoffman, the ASF's programs were "designed to place the utmost public support behind those public officials whose duties and responsibilities include the construction and maintenance of streets and highways." All the ASF's efforts, Hoffman wrote, "aim ultimately at enthusiastic public support of intelligent official action," so that "public opinion is aroused to vitalize official activity."[19] Hoffman promoted this mission by writing for and speaking to mass audiences as the ASF's president.[20] Meanwhile, Thomas Henry of the American Automobile Association, who had just tried out Germany's new autobahns for himself, supplemented the ASF's campaign by promoting a new "Bill of Rights for Motorists." Its first amendment included a right to "arterial routes through cities," in part as a safety measure.[21]

### Revelations of the Motor Millennium

Like any successful public relations campaign, motordom's new battle against bad safety publicity could not merely attack others' positions. It had to offer a positive vision of its own. In the mid 1930s, motordom conceived just such a vision. Like a twentieth-century Saint Augustine, it contrasted a grim city of the present against a vision of a perfect city of the future. Visionary city plans were not new, but now, for the first time, they came from a source powerful enough to present them in spectacular Hollywood-style mock-ups, promote them to mass audiences, lobby for them, and begin building them. Motordom sought to channel popular anxiety about safety into support for bold, new, foolproof highways. The new plans prophesied an automotive millennium without accidents, congestion, or delays. They were visions of the coming motor age.

Road builders promised that they could physically reconstruct the city for the automobile and tie motor age cities together with a web of cement motor highways. Styling himself a "prophet," Frank Sheets of the Portland Cement Association estimated the project would cost $57 billion over 25 years. Admitting this was a "stupendous" sum, Sheets nevertheless claimed that his was a practical vision. With gasoline taxes dedicated solely to roads, plus some highway tolls, the job could be done. "It will be interesting," Sheets mused, "to compare this paper's predictions with the realization of thirty years hence."[22]

Using Sheets's cost accounting, Miller McClintock promised to deliver cities from congestion with motor highways. He was the impresario of a new kind of highway road show. In the spring of 1937, the Shell Oil Company combined McClintock's traffic expertise with the talents of the stage designer Norman Bel Geddes to build a scale model of "the automotive city of tomorrow."[23] It was "built to be photographed," and millions would see it.[24] McClintock presented the model city in slide shows in New York, in Detroit, and elsewhere. When he and Bel Geddes showed their slides to New York City traffic officials, the New York Times described their vision of motor highways and elevated sidewalks as a "Pedestrian Heaven," a "city of the future where the pedestrian can walk without fear of sudden death and the motorist can always find a parking space."[25]

Shell promoted the model in magazine advertisements, calling it "the City of Tomorrow." They presented it as Norman Bel Geddes's prophecy, the work of an "authority on future trends." In the motor age, Bel Geddes promised, "you'll breeze right over cross-town traffic." "By 1960, express traffic will speed along at 50 miles an hour over high-speed boulevards, reached by ramps every ten blocks. No stop lights . . . no intersections."[26]

Bel Geddes's model became motordom's manifesto. Those working to hasten the dawn of the motor age rallied behind it. McClintock presented slide shows of the model for another year, at some point dubbing it "Matronia" and holding it up as the ultimate answer to the safety problem.[27] He was still heading up the Bureau of Street Traffic Research, though "Albert Russel Erskine" was dropped from its name after the Studebaker president committed suicide in 1933, and as other auto manufacturers contributed to its funding. "American cities," he predicted, "will be rebuilt in the next twenty-five to fifty years." To a country recovering slowly from depression, McClintock claimed that with rebuilt cities Americans "could buy 5,000,000 to 10,000,000 more automobiles."[28] He held up Germany's new motor highways as an example for emulation, not mentioning that they did not enter cities. "We have much to learn from what Germany has done with her magnificent Autobahnen." "The Reich," he added, was building "without compromise on the essential principles," and "only such an approach will bring about a permanent solution of the problem of accidents and congestion."[29]

Others interested in the rebuilding the city for the motor age adopted Shell's technique. At the 1939 Golden Gate International Exposition, United States Steel displayed its vision of San Francisco in 1999, with wider streets,

**Figure 9.3**
In 1937, as part of an advertising campaign for the Shell Oil Company, the stage designer Norman Bel Geddes built a model of the city of the new motor age. Bel Geddes's technical consultant was Miller McClintock. This advertisement appeared in the September 18 issue of the *Saturday Evening Post*.

cloverleaf intersections, and an elevated highway.[30] But the crown jewel of this new kind of publicity campaign was General Motors' spectacular exhibit at the New York World's Fair of 1939–40. GM commissioned Norman Bel Geddes to expand his model to gigantic proportions, renaming it "Futurama," for display in its "Highways and Horizons" pavilion. In the first summer of the fair, 5 million people viewed the model from the comfort of cushioned chairs traveling on a track three-tenths of a mile long. As they traveled, the exhibit's "Voice" described the "wonderland" of 1960.[31] As if from an airplane window, visitors saw the vast urban highways of 1960, with safety built into them. Thanks to highway engineers, the motorists of 1960 "still blunder, of course, but when they do, they are harmless"—even at 75 or 100 miles an hour.[32] Raised walkways kept pedestrians safely above motor traffic. The pavilion included a full-scale mock-up of such an intersection; fairgoers looked down from elevated sidewalks to a street intersection crowded with stationary GM cars and trucks.[33]

**Figure 9.4**
Norman Bel Geddes's model for Shell included traffic intersections designed for high vehicular capacity with safety. Source: Norman Bel Geddes, "City 1960," *Architectural Forum* 67 (July 1937), p. 60.

## After Futurama

For 30 years the Automotive Safety Foundation was the leading national institution with authority in traffic safety and the only one with pretensions to disinterested, professional expertise. The American Automobile Association and insurance companies funded separate traffic safety organizations, but these, unlike the ASF, made their interest group sponsorship obvious in their very names.[34] By 1939, Paul Hoffman—now president of Studebaker—was passing himself off in the popular press as an impartial safety expert by styling himself "President, The Automotive Safety Foundation."[35] By then no other coherent or organized safety approach rivaled the ASF's highway safety model. There were a few dissenters. Insurers retained a taste for vivid descriptions of accident casualties, and popular press attacks on speed in general persisted. But there were no important, recognized, ostensibly disinterested traffic safety experts outside of the ASF.

**Figure 9.5**
At the 1939–40 World's Fair in New York, General Motors presented Americans with "Futurama," a vision of the city of 1960. Norman Bel Geddes designed this enormous model, using his model for Shell as a starting point. It was a city rebuilt for the motor age. Source: Norman Bel Geddes, *Magic Motorways* (Random House, 1940), p. 240.

The highway safety model's pinnacle of success was surely reached on February 22, 1955. On that day the use of highway engineering to fight accidents, which President Hoover had inconspicuously raised to official doctrine in 1931, received the highest possible recognition in the most public of all arenas. In his message to Congress that day, President Eisenhower called for a new national highway program. As his first justification for spending the enormous sums required, the president cited the need to save lives.[36] After Congress passed the act creating the National System of Interstate and Defense Highways in 1956, Eisenhower promoted it by repeatedly citing a claim made by the Automotive Safety Foundation. The system, Eisenhower said, would "save four thousand American lives a year."[37]

The 1956 act bore a family resemblance to the Hoover traffic conferences of a generation before. Like them it was the product of interest group politics. In 1927, A. B. Barber had advised Commerce Secretary Hoover to get "automobile dealers and automobile club secretaries" behind him if he wanted results.[38] In a 1955 memorandum, presidential advisor Noorbar Danielian counseled Eisenhower in similar terms: "The primary objective of a new program of highways" should be "to hold together the natural friends of an expanded Federal highway program." The president must not overstep "the tolerance of the friends of the highway system," instead offering them "concessions" to "increase and consolidate the strength of the pro-highway forces."[39] Eisenhower probably didn't need the advice. By then he had already given the job of planning the new interstate highway program to a formidable coalition of "friends of the highway system" under the direction of a personal friend, General Lucius Clay. The Clay Committee's report was the basis for the well-funded interstate highway program, which did not shrink (as European highways did) from venturing into the hearts of cities.[40]

Applying the Automotive Safety Foundation's highway safety doctrines, highway engineers succeeded in making speed relatively safe. In 1938, Paul Hoffman expected that, with highway engineering, "within thirty years we can cut the highway fatality rate from 15.9 deaths per 100,000,000 vehicle miles to five or less."[41] Thirty years later, just as the National Traffic and Motor Vehicle Safety Act of 1966 was going into effect, the death rate was indeed down to 5.4.[42] This accomplishment was overwhelmingly the work of automotive and insurance interests. They achieved it through their promotion of school safety education, driver education and licensing, enforcement of reckless driving laws, and highway engineering. Modest improvements in safe vehicle design, however, were more than offset by prodigious gains in horsepower and average speed.

But the conversion of surface passenger transportation to motor vehicles was so nearly complete that even though each mile of travel carried only one-third the risk, total death tolls kept rising. The conversion of surface transportation to the most dangerous mode robbed American transportation of the fruits of making that mode safer. The last years in which the Automotive Safety Foundation's methods alone ruled in traffic safety were therefore also the worst. The annual death tolls of the late 1960s, at more than 55,000, were high enough to force Detroit to share authority with the National Highway Traffic Safety Administration, a division of the new federal Department of Transportation. The ASF was also forced to retract one of the most basic of its old safety doctrines: the inherent safety of its product. By the late 1960s, the ASF had failed to stop the spreading conviction that Detroit's motor vehicles were indeed inherently dangerous, if not for their speed, then for their inability to protect their occupants in a crash.[43] One major tenet of the 40-year-old highway safety orthodoxy was overturned, but another survived. After 1930, for 50 years, traffic deaths remained private losses, borne behind closed doors. In 1980, however, two grieving mothers of children killed in accidents united to make mothers' grief visible again. Candace Lightner and Cindi Lamb founded Mothers Against Drunk Drivers, believing that action could come only if they made the national loss conspicuous. MADD mobilized the grief and anger of parents nationwide to raise drinking ages, reform drunk driving laws, and step up enforcement. Since then, many parents have erected roadside memorials to accident victims. Their work was a late revival of the grander but forgotten public memorials of the 1920s.[44]

As a symbol of the transformation of the American city for the sake of the automobile, the 1956 highway act was far more prominent than any report of the Hoover traffic conferences 30 years before. But perhaps it was less important. The Model Municipal Traffic Ordinance of 1927 codified a new social construction of the city street. Once a public space for mixed uses, and ruled by informal customs, the street was then becoming a motor thoroughfare for the nearly exclusive use of fast vehicles—especially automobiles. The Hoover conferences, and especially the model traffic ordinance, recognized, legitimized, and promoted a revolution in the perception of the city street. Though old perceptions persisted, thenceforward most streets were chiefly motor thoroughfares. As such, streets were suddenly woefully inadequate in a new way. They were too narrow and too poorly paved, with too many points of access, conflicting paths, and grade intersections—and too many non-automotive street users. Under this new social construction of the city street, the physical construction of a new kind of urban thoroughfare that addressed all these problems was striking, but not revolutionary.

# Conclusion   History, Technology, and the Dawn of the Motor Age

The formulation of a problem already contains half its solution.
—Ludwik Fleck, 1935[1]

Motordom socially reconstructed city streets as motor thoroughfares—places where cars preeminently belong. Their success was never total, for society eludes those who would master it. Nevertheless, motordom's struggle and relative success reveal much about how social groups succeed or fail at remaking the world to suit their needs.

I hope this book demonstrates the benefits of combining theoretical and empirical work to scholarship of both kinds. In 1996 Trevor Pinch observed that "the combination of detailed empirical research with growing theoretical sophistication about science and technology offers genuine new insights into technical change."[2] Yet empirical historians still fault theoreticians for their limited research and heavy theoretical apparatus, while theoreticians often accuse historians of superficiality.[3] Theoretical investigators of technology and society have tended to draw on the work of historians and other empirical researchers for much of the evidence their theories need. This was the proper way to begin. The burdens of developing a new theory should not be compounded by the burdens of new empirical research. Now that theoretical work in technology and society is well developed, however, the advantages of combining theoretical and empirical approaches have grown. Acclimated to theoretical perspectives in technology and society, historians will be less likely (for example) to miss evidence from social groups that lost the struggle to shape a technological system. In turn, the resulting historical scholarship will be more useful to theorists. The value of historians and social theorists to each other will then be clearer, and mutual misunderstanding less frequent.

## Shared Technologies

Most historical studies in the social construction of technology examine distinct artifacts that can be produced in quantity and that need not be shared. Streets are different. A mother cannot conceive of a street as a playground for her children while a motorist thinks of it as a path for driving at speed—at least not for long. The incompatibility of different constructions of a shared technology raises the stakes for relevant social groups. In a shared system, when a new construction becomes dominant, one group cannot easily secede from the prevailing denomination into a dissenter group where the minority construction is preserved. Thus, as the preceding chapters have documented, social groups will use whatever power they have to fight for their construction. These problems are further complicated wherever (as in the United States) a high cultural value of individualism can make the logic of shared systems harder to admit or easier to deny. Such problems have been investigated, particularly in the case of other urban networks. Thomas Hughes's work on electrification showed how such systems tend to develop a "momentum" that is hard to divert.[4] In the fight for the American city street, such momentum is clear. Pedestrians clung to old street uses despite growing threats. Auto clubs and dealers worked within safety councils, giving up only when their disagreements grew rancorous. And in recent decades, policymakers seeking to promote alternatives to driving alone find that decades of physical and social infrastructure make their task almost hopeless. As we have seen, however, the long-standing construction of the street as a public space *was* diverted, despite tremendous momentum. The case of American city streets can therefore help us see how substantial momentum can be overcome so that interpretive flexibility is reintroduced.

The social groups that had the most to lose from the old construction of the street were quite slow to attempt a program to change old constructions. Although automobiles were abundant on American city streets by 1910 or 1915, there was then little sign of any effort to challenge the prevailing construction. Even in 1920, when casualties were causing alarm, automotive interests joined with safety reformers to seek "safety first" solutions. They still did not clearly perceive that there was no solution to their problem that both preserved old constructions of the street and gave automobiles a bright future in cities. How did this change?

A crisis—especially a visible, bloody, public crisis—can create enough social energy to compel a questioning of old assumptions about technologies—even shared ones with substantial momentum. Old, small water distribution networks consisting of wells and public pumps gave way to larger networks of treatment plants, water mains, and sewers when this appeared to be the only alternative to mass death. Edison appears to have grasped the fear of death as a factor in urban network selection in the 1890s, when he attempted to paint AC electric power as deadly to protect his DC network. Faced with a threat from distributors of natural gas in the 1920s, coal merchants tried to give the same reputation to gas.[5] The bloody dawn of the motor age confirms the lesson of these cases. Fear of catastrophe can drive social groups to challenge the prodigious momentum of shared systems. How they will do so remains an open question. In the streets, strict regulation of automobiles and the use of mechanical speed governors might have diminished the death toll. Instead, streets were eventually turned over to automobiles, leaving pedestrians with very limited access. Why?

Water mains and sewers, AC power, and gas replaced wells and pumps, DC power, and coal in large part because of the substantial practical advantages of the former over the latter. Questions, of course remain. Just how and when, and under what circumstances, would the new system replace the old?[6] To many, the relative advantages of the old and new constructions of the city street were harder to sort out. Though pumps, DC, and coal were entrenched by substantial investments, they lacked the mass loyalty that underlay the old construction of city streets. Once water systems, AC, and gas had prevailed, few called bitterly for a return to the old systems. Yet long after the motor age construction of the street prevailed, bitter attacks on it (for example, in letters to newspaper editors) persisted in abundance. The old construction of the city street retained a constituency even after the motor age street eclipsed it. Thus the relative practical advantages of the shared technologies cannot be determined apart from the particular perspectives of the relevant social groups, and cannot settle which construction of a shared system will prevail.

## Organization

Another lesson of the dawn of the motor age is that it is not enough that a social group perceive its interests. If its fight is a hard one, it must organize. Faced with a shared system that had a threatening and obdurate construction, social groups backing the automobile coordinated their efforts. Prevailing constructions of traffic accidents and traffic congestion (both as the fault of cars) threatened cars' future in cities. Promoters of cars were slow to recognize the threat, but when they did they organized to reconceive the safety and congestion problems, and ultimately to reconceive the street itself.

Indeed, auto clubs, dealers, and manufacturers were hardly one social group before 1924, when, at the Hoover traffic conference, they met, organized, and developed a common strategy. They themselves recognized this development as a turning point by christening themselves "motordom." Organizational success required a coherent rhetorical stance, which motordom developed in the mid 1920s. It took money, and motordom found it. The National Automobile Chamber of Commerce organized and funded a new traffic and safety committee. Studebaker spent money on a new, motor age institution of traffic expertise, the Erskine Bureau. The American Automobile Association invested in a national school safety campaign. And motorists themselves paid for it through gasoline taxes, once motordom saw the advantages of tapping this stream of wealth.

Defenders of the old street organized too, especially in local safety councils. But they neither found comparable funds nor developed comparable institutions. To make their voices heard, safety reformers joined local safety councils. But there they competed with a diverse array of safety promoters, including insurance companies and operators of motor fleets, who were not committed to the same construction of the street. Electric railways were more alarmed by the trend toward the motor age, and they did organize to stop it. But their slide into bankruptcy in the 1920s left them too weak to compete with motordom. While motorists made financial contributions to the motor age every time they bought taxed gasoline, streetcar riders were resisting efforts to raise fares, even to the point of covering the costs they incurred by riding. Thus, when Hoover in 1927 wanted his traffic conference to produce a model municipal traffic ordinance that could earn backing energetic enough to win adoption in cities nationwide, he agreed to turn the job of drafting it over to motordom.

Motordom's organizational successes left other social groups closed out of the most influential bodies of expertise and policy formulation and promotion. The Erskine Bureau's growing reputation as the only national body of traffic expertise not only boosted the reputation of the engineers who graduated from it, it left other engineers looking small by comparison. The diminished professional horizons of the efficiency-minded engineers (such as George Herrold of St. Paul) cast their alternative solutions to urban transportation problems into obscurity. As the American Automobile Association took over school safety education, safety councils retreated, taking with them their unflattering portrayals of cars and drivers. And as motordom promoted "highway safety" and "highway engineering," rebuilding cities for the motor age became the chief way to fight traffic accidents and congestion. To change a well-entrenched, shared technological system, and to preserve the change, social groups must organize. Motordom never forgot this lesson.

## Power and Rhetoric

The dawn of the motor age has something to tell us about power. Like money, power is a medium of exchange between social groups.[7] Because it comes in many currencies, it is hard to measure by any one standard. Motordom had substantial and growing financial wealth. By the mid 1920s it was organized enough to dispense this wealth to promote a social reconstruction of the street, through a well-funded rhetorical campaign and through gasoline taxes linked to road construction (including construction of county and state roads in cities). By then it was also exercising direct political power, especially through its influence in the Commerce Department. But drivers themselves exercised power every time they traveled at speed in streets, resorting to the horn instead of the brake to proceed. This exercise of power drove pedestrians from streets and sometimes barred them from access to streets, even at designated crossings. Horsepower gave motorists a literal, physical form of momentum that collided with the social momentum of old constructions of the street, changing their trajectories.

Yet pedestrians exercised power too. Because of their numbers and agility, pedestrians could not be forced to submit to control. When Chicago's coordinated signals did not sufficiently accommodate pedestrians' needs, many pedestrians ignored them with impunity. Los Angeles learned from Chicago that pedestrians' compliance had to be bought. The asking price included signals timed to meet pedestrians' needs. Even then, pedestrians were never subdued. Some turned motorists' unwillingness to hit pedestrians into negotiable currency they were willing to spend. Pedestrians still call motorists' bluffs. Children still sometimes play in side streets, and motorists still assume some of the responsibility for their safety.

But adversaries not only exercise power, they clothe it in a rhetorical dress that legitimizes it and sometimes conceals its uglier aspects. The dress may fit well. In such cases, those who use it conceive of their motives and their rhetoric as indistinguishable. Sometimes the rhetorical stance is more like a sales pitch. In these cases, a group sees its goals first and later finds a respectable rhetorical robe to wrap it in. In competing social groups' struggle for the street we can plainly see the constant companionship of power and rhetoric. In the contest for streets, the rhetorical stances varied—justice, order, efficiency, freedom, modernity—but together they support a common conclusion: that social groups compete not merely through the raw exercise of power, but also through diverse rhetorical plays. These methods can win recruits to a social group from bystanders, and can strengthen the commitment of those already within them.

## Use and Misuse

Social constructivists have shown that users are "agents of technological change," and in recent years they have examined how users change (and are changed by) the artifacts they use.[8] The preceding chapters confirm a fact already implicit in these earlier studies of technologies and users: that misuse shapes artifacts as much as use, and that struggles between rival social groups to fix the meaning of an artifact in ways they prefer often take the form of struggles to define use and misuse. It is time to consider the problem of use versus misuse explicitly.[9]

Ronald Kline and Trevor Pinch showed that rural owners of early automobiles found many uses for them not intended by manufacturers or dealers.[10] Some of these uses were accepted by all relevant social groups as proper, but others were disapproved by some as misuses. Adaptive uses can physically modify artifacts, sometimes by adding or subtracting components. In the 1970s, such adaptations led to a complete redesign of an existing artifact when California youths adapted conventional bicycles for use in mountain races, thereby creating the mountain bike.[11] In so doing, they transformed a misuse (from outsiders' point of view) of conventional bicycles into the most proper use of a new artifact.

Street users also struggled to define use and misuse, transforming streets—and themselves. Prevailing constructions of automobiles clashed with prevailing constructions of the street. Most manufacturers, dealers, and drivers of automobiles conceived of them as vehicles capable of higher sustained speeds than horses and more versatile than crowded and track-bound streetcars. These attributes were not incidental to cars. They made cars what they were; they made cars worth buying. But under prevailing constructions of the street, all of these attributes of cars made them misusers of streets. Custom-bound police responded by requiring motorists to conform. Speed limits and right-angle turning rules were attempts to make cars essentially not cars. Traffic engineers tried to deny cars another essential necessity: access to parking space. Thus the car's future in the city depended upon a reconfiguration of use and misuse. Motordom understood this fact by 1924 and worked actively to achieve it.

Even earlier, however, safety reformers and pedestrians struggled to set the line between street use and misuse by pedestrians and children. Reformers who proposed what they thought of as simple, practical, common-sense rules (such as "Look to the left when you begin crossing, then look to the right") quickly discovered that these offended pedestrians who clung to older street customs. Thus, just as motorists ignored speed limits intended to make them conform to pre-automotive conditions, pedestrians ignored street crossing rules intended to conform them to new street conditions.

In this failure of rules to change practices lies another lesson of the dawn of the motor age for the history of technology. Social groups will try to use rules and low-intensity publicity to change the users' practices. But if these practices are entrenched by users' construction of their systems, more will be required. In struggles over use and misuse, all sides are likely to resort to moral sanctions through shame and ridicule. When cars would not conform to low speed limits (and similar rules), their critics turned to moral attacks that demonized cars and their drivers. For similar reasons, "jaywalker," once an indefinite and obscure Midwestern term of derision, became a universally known word specifically intended to ridicule unreconstructed pedestrians. The pattern is not unusual. Where rules failed, moral outrage and ridicule have been used—for example, in attempts to transform liquor from a catalyst of sociability into an agent of personal and family destruction. Dale Rose and Stuart Blume similarly found that, to discourage resistance to vaccination, governments censured resistors as "bad citizens."[12]

Thus, the city of the motor age was a new kind of city: a city redefined as hospitable to cars. It was heralded by a rhetorical festival celebrating freedom. Yet the automotive city took back much of the freedom it promised. Through technological innovation, a society that prizes individual liberty can unintentionally curtail it. For example, when street users are free to use cars, the freedom of all street users (including motorists) to use anything else is diminished. A city rebuilt (socially and physically) to accommodate cars cannot give street users the good choices a truly free market can provide. In 1920, when Bessie Buckley asked whether streets were "for commercial and pleasure traffic alone," the question was rhetorical.[13] Of course they were not only for motor vehicles. Revolutions, however, turn worlds upside down. In the motor age, Buckley's question became absurd and its answer (a more or less emphatic Yes) obvious to any child. The street is a motor thoroughfare.

# Notes

## Introduction

1. "The Parking Problem in St. Paul," *Nation's Traffic*, July 1927, 28–30, 47–48 (29).

2. "Motor Killings and the Engineer" (editorial), *Engineering News-Record* 89 (Nov. 9, 1922), 775.

3. "Jay Walker Problem," *Providence Sunday Journal*, June 26, 1921.

4. Clay McShane was the first historian to begin to appreciate the scale and significance of urban traffic casualties at the dawn of the motor age, though he nevertheless assigned it a much less important role than it is given here. See Clay McShane, *Down the Asphalt Path: The Automobile and the American City* (Columbia University Press, 1994), esp. 173–179.

5. T. Pinch and W. Bijker, "The Social Construction of Facts and Artifacts, or How the Sociology of Science and the Sociology of Technology Might Benefit Each Other," in *The Social Construction of Technological Systems*, ed. W. Bijker et al. (MIT Press, 1987), 24.

6. T. Pinch and W. Bijker, "The Social Construction of Facts and Artefacts, or How the Sociology of Science and the Sociology of Technology Might Benefit Each Other," *Social Studies of Science* 14 (Aug 1984), 399–441; Pinch and Bijker, "Facts and Artifacts"; Bijker, *Of Bicycles, Bakelites, and Bulbs: Toward a Theory of Sociotechnical Change* (MIT Press, 1995).

7. Ibid.

8. Bijker, *Of Bicycles, Bakelites, and Bulbs*, 122–127.

9. This example corroborates Anique Hommels's observation that groups excluded from prevailing frames "typically propose radical alternative technological designs." See Hommels, "Obduracy and Urban Sociotechnical Change: Changing Plan Hoog Catherijne," *Urban Affairs Review* 35 (2000), 669.

10. Misa, "Controversy and Closure in Technological Change: Constructing 'Steel,' " in *Shaping Technology/Building Society*, ed. W. Bijker and J. Law (MIT Press, 1992), 111.

11. Pinch and Bijker, "Facts and Artifacts," 44–46.

12. "Hog-Tying the Automobile," *Ohio Motorist*, Oct. 1924, 8.

13. Charles L. Wright, *Fast Wheels, Slow Traffic: Urban Transport Choices* (Temple University Press, 1992), 167. See also the comments of Satoshi Fujita, a Japanese visitor in the United States, in "Observations on Literacy and Morality," in *Compulsory Schooling and Human Learning*, ed. D. Bethel (Caddo Gap Press, 1994). Fujita was surprised to find that in American cities it is "necessary for everyone to drive a car in order to survive" (28).

14. The dean of historians of the automobile in America, John Rae, assigns an important role to Americans' "love affair with the motor vehicle," which was "welcomed in American life" (Rae, *The American Automobile Industry*, Twayne, 1984, 59, 69).

15. Scott L. Bottles, *Los Angeles and the Automobile: The Making of the Modern City* (University of California Press, 1987), 254.

16. Ibid., 253–254.

17. Bradford Snell, "American Ground Transport," in Senate, Committee on the Judiciary, The Industrial Reorganization Act: Hearings Before a Subcommittee on S. 1167, 93rd Cong., 2d sess. (1974), 26–49. For a convenient abridgement, see Snell, "American Ground Transport," in *Crisis in American Institutions*, sixth edition, ed. J. Skolnick and E. Currie (Little, Brown, 1985), esp. 322–328.

18. Mark Foster, *From Streetcar to Superhighway: American City Planners and Urban Transportation, 1900–1940* (Temple University Press, 1981), 177.

19. The most important work crediting city planners with major influence is John D. Fairfield, *The Mysteries of the Great City: The Politics of Urban Design, 1877–1937* (Ohio State University Press, 1993). Fairfield's book closely follows the model of elite subversion of the interests of city people.

20. McShane, *Down the Asphalt Path*.

21. Ibid., 212, 203.

22. Almost alone, McShane has noted the distinctly urban hostility to motorists on grounds of safety. See *Down the Asphalt Path*, 176–177.

23. Harland Bartholomew, an engineer for the St. Louis City Plan Commission, expressed a typical expert view in 1924: "It would be financially impossible to provide for unlimited accommodations of all kinds of traffic. The first step toward

a solution of this problem is the reduction or elimination of unnecessary traffic movements." See *Proceedings of the American Society of Civil Engineers*, May 1924, and *Electric Railway Journal* 63 (May 31, 1924), 857–858.

24. The change of terms occurred gradually in the 1920s, with some prodding by auto interests. In 1922 a Washington auto dealer recognized that the "pleasure car" idea "has somewhat impeded the progress of the automotive industry." His reply was to claim that "the automobile itself has become an essential rather than a luxury." R. H. Harper, quoted in "Auto Overcomes Expensive Toy Idea," *Washington Post*, Dec. 3, 1922.

25. Bottles, *Los Angeles and the Automobile*, 249. Early in the century the greater Los Angeles area was exceptionally well suited to automotive transportation and poorly suited to mass transportation. The dense, congested downtown, where mass transit made most sense, was small by comparison to the diffuse surrounding region. The climate was the best in America for early motorists, and an unusually large proportion of Southern Californians could afford automobiles. Los Angeles therefore led the nation in automobile ownership rates and began to accommodate automobiles early. No city as large or larger was nearly so conducive to the automobile. In urban transportation history, Los Angeles is the least representative city in America. Nevertheless, Bottles uses transportation in Los Angeles to represent urban transportation in America, as his book's subtitle and the boldly stated conclusions in his epilogue show.

26. The street, in other words, is an instance of a "commons." See Garrett Hardin, "The Tragedy of the Commons," *Science* 162 (Dec. 13, 1968), 1243–1248. In medieval England a commons was a place of exceptional economic laws, unlike those that obtain in private property. Like a commons, roads and streets are a shared good; individual users cannot be charged for each use. This sense of the metaphor is clearer in its original formulation; see William Forster Lloyd, "Two Lectures on the Checks to Population," in *Lectures on Population Value, Poor-Laws and Rent* (1837; reprint: Augustus M. Kelley, 1968), 30–31. See also Mancur Olson, *The Logic of Collective Action: Public Goods and the Theory of Groups* (Harvard University Press, 1965, 1971), 2: "Unless the number of individuals in a group is quite small, or unless there is coercion or some other special device to make individuals act in their common interest, rational, self-interested individuals will not act to achieve their common or group interests." Motorists have paid for much of their consumption of road and street capacity through gasoline taxes. In the 1920s, however, the tax was a very minor source of funds for streets. The gasoline tax is also not equivalent to a charge for a commodity of trade in a free market, since street capacity varies widely in value. A government cannot use gasoline taxes to charge a motorist for road use with any more success than a department store can charge customers for purchases by the pound.

27. William D. Hudson, "Grade Separations at Intersections," *City Planning* 2 (Jan. 1926), 36–41 (37).

28. Daniel Rodgers documents Haussmann's extensive intellectual influence on American urban design but finds that "few of the Haussman-inspired designs came to fruition." Rodgers also notes that the show boulevards that were built in the United States were "destined to be overwhelmed by automobile traffic" because of "their excessive centralizing tendencies, their étoiles and converging diagonals." See Rodgers, *Atlantic Crossings: Social Politics in a Progressive Age* (Belknap, 1998), 164–174, 181 (173).

29. Paul Barrett, *The Automobile and Urban Transit: The Formation of Public Policy in Chicago, 1900–1930* (Temple University Press, 1983).

30. Ibid., 3, 6. According to Barrett, the reasons for this difference were both cultural and practical. Most importantly, street railways, into the 1920s, often could and did pay their own way; this reinforced the idea that they should and must do so (see esp. 4). A similar disjuncture of definition is apparent today, when passenger rail service is again expected to pay its own way, while roads and highways (though largely funded by gasoline tax revenues) are treated as a public responsibility. The problem of charging users is perhaps the chief reason why the street was traditionally a public responsibility. Street railway entrepreneurs, however, could easily charge users, and therefore willingly built lines. Until early in the twentieth century, city streets were financed primarily by assessments on the owners of abutting property, since they were among the chief beneficiaries of improvements. Clearly, however, many others who were not charged benefited. Note also that a store owner (for example) paid for the benefit accruing from a street improvement, but did not pay for equivalent benefits from street railway extensions, except indirectly (through increased property assessments or rents). Increasingly, bond issues against cities' general revenues paid for improvements; this method corrected the deficiencies of assessments, but clearly benefitted motorists disproportionately. State gasoline taxes grew important in city thoroughfares as state roads more often entered cities in the second quarter of the century. This method had the salutary effect of charging motorists for their use of the roads but also became the basis for claims that roads and streets belong exclusively to motorists.

31. Barrett, *The Automobile and Urban Transit*, 215, 210.

32. Baldwin, *Domesticating the Street: The Reform of Public Space in Hartford, 1850–1930* (Ohio State University Press, 1999).

33. On problem definition, see selections in *The Politics of Problem Definition*, ed. D. Rochefort and R. Cobb (University Press of Kansas, 1994), esp. Rochefort and Cobb, "Problem Definition: An Emerging Perspective," 1–31. For a brief, incisive case study in problem definition and how it can change, see Jameson M. Wetmore, "Redefining

Risks and Redistributing Responsibilities: Building Networks to Increase Automobile Safety," *Science, Technology and Human Values* 28 (summer 2004), 377–405. On social and cultural factors in technological problems, see the growing work of historians studying the social construction of technology, the standard introduction to which is the essays collected in *The Social Construction of Technological Systems*, ed. W. Bijker et al. (MIT Press, 1987).

34. See the growing body of scholarship from the "social construction of technology" school, exemplified by Bijker, *Of Bicycles, Bakelites, and Bulbs*.

## Chapter 1

1. Cobb, in "Nation Roused against Motor Killings," *New York Times*, Nov. 23, 1924.

2. Barnett Wartell to Herbert Hoover, Dec. 15, 1924, file 02767, box 160, Commerce Papers, Herbert Hoover Presidential Library, West Branch, Iowa; "Motorcar Kills Boy in Avoiding Crash," Philadelphia *Public Ledger*, May 31, 1920. Wartell gave his son's age as 9; the *Ledger* said 10.

3. Wartell to Hoover; "Blood-Stained Auto Sought in Killing of 18-Year-Old Youth," *Philadelphia Inquirer*, Oct. 28, 1924 (quotation); "Hunt for Driver Who Killed Boy," *Evening Bulletin*, Oct. 28, 1924; "Bride Brings Back Death Car Driver," *Evening Bulletin*, Nov. 8, 1924; "City-Wide War Is Declared on Death Drivers," *Public Ledger*, Nov. 9, 1924.

4. Casualty figures from the 1920s are unreliable, and their errors are most likely to take the form of underestimates. It is probable that more than 210,000 were killed in traffic accidents in the period 1920–1929 (three or four times the death toll of the previous decade). See Committee on Public Accident Statistics, "Public Accidents in the United States," *Proceedings of the National Safety Council* 13 (1924), 840ff. (841); Bureau of the Census, *Historical Statistics of the United States, Colonial Times to 1970*, part 2 (U.S. Government Printing Office, 1975), 720. For a careful analysis of the problem of estimating accident figures, see H. P. Stellwagen, "Automobile Accidents in the United States," *Safety Engineering* 51 (Feb. 1926), 71–74.

5. The National Conference on Street and Highway Safety found that the death rate in 67 cities with populations over 100,000 was "more than twice as high as the rate in rural districts," but this is probably an overstatement; see Department of Commerce, "More Deaths from Automobile Accidents" (press release for August 17, 1924), file 02781, box 162, Commerce Papers, Hoover Library; see also "Graphical Presentation: Traffic Accident Records" (report), National Conference on Street and Highway Safety, Washington, March 1926, charts 1, 3, 6–9, file 02770, box 161, Commerce Papers, Hoover Library.

6. National Safety Council figures; see Sidney J. Williams, "Getting at the Facts of Public Accidents," *American City* 34 (May 1926), 497–501 (501); "Seventy Lives a Day," *The Outlook* 144 (Nov. 10, 1926), 827.

7. Mitten Management, Inc., "Accidents and the Street Traffic Situation" (Philadelphia Traffic Survey Report no. 4), July, 1929, 10, 18–19, Philadelphia Rapid Transit Company box 1, John F. Tucker Collection, accession 2046, Hagley Museum and Library, Wilmington, Delaware. The same proportion obtained in Chicago in 1925, when 523 of 682 motor deaths were pedestrians; see "Automobile Fatalities Becoming One of Our Deadliest Scourges, *American City* 35 (Sept. 1926), 341–342 (341).

8. Figure cited by Aida de Costa Root of the American Child Health Association in press release, National Conference on Street and Highway Safety, Washington, March 24, 1926, file 02784, box 162, Commerce Papers, Hoover Library.

9. "A Motor Mind for the Motor Age," *The Outlook* 138 (Dec. 31, 1924), 711–712 (711).

10. Wartell to Hoover, Dec. 15, 1924.

11. "Bride Brings Back Death Car Driver," Philadelphia *Evening Bulletin*, Nov. 8, 1924.

12. Jameson M. Wetmore has shown how regulators and automobile manufacturers in the 1990s redefined the problem of child airbag casualties, so that no technological fix was necessary. See Wetmore, "Redefining Risks and Redistributing Responsibilities: Building Networks to Increase Automobile Safety," *Science Technology, and Human Values* 28 (summer 2004), 377–405. On problem definition in general, see selections in David A. Rochefort and Roger W. Cobb, eds., *The Politics of Problem Definition: Shaping the Policy Agenda* (University Press of Kansas, 1994), esp. Rochefort and Cobb, "Problem Definition: An Emerging Perspective," 1–31.

13. Census Bureau figures show 52,900 dead in automobile accidents in the years 1919–1922 (Bureau of the Census, *Historical Statistics of the United States, Colonial Times to 1970*, part 2, U.S. Government Printing Office, 1975, 720); official losses in the American Expeditionary Force were 48,909 dead from enemy action. For examples of the many contemporary comparisons of the two casualty tolls see "1,500,000 Killed or Injured in 1921 in Automobile Mishaps," Baltimore *Sun*, June 11, 1922; "Waste of Human Life" (editorial), Pittsburgh *Gazette-Times*, Oct. 22, 1922; "U.S. Auto Deaths Exceed War Toll, Pastor Declares," *St. Louis Star*, Nov. 19, 1923; Arthur C. Carruthers, "Automobile Accidents to Children," *Safety Engineering* 49 (May 1925), 189–193 (189); "The Murderous Motor," *New Republic* 47 (July 7, 1926), 189–190 (189); "The War after the War" (editorial), *Detroit Free Press*, August 23, 1927, 6; "The Motor More Deadly Than War," *Literary Digest* 94 (August 27, 1927), 12.

14. Ralph Nader and Joel W. Eastman have concentrated attention on vehicle design, extending a critique of safety in the 1960s back to the 1920s; see Nader, *Unsafe at Any Speed: The Designed-In Dangers of the American Automobile* (Grossman, 1965); Eastman, *Styling vs. Safety: The American Automobile Industry and the Development of Automotive Safety, 1900–1966* (University Press of America, 1984). Finding little concern for safe vehicle design in the the 1920s, they concluded that safety was not an important issue. James J. Flink is probably the leading historian of the automobile in America; he has deferred to Eastman on matters of safety. See Flink, *The Automobile Age* (MIT Press, 1988), 290. The popular traffic safety movement has been overlooked also because it was local in character, with a low national profile. A dissertation writer (Anedith Jo Bond Nash), searching the *Readers' Guide to Periodical Literature*, found that "widespread national concern about automobile accidents became evident in many media" only "In the early years of the Depression." Nash, "Death on the Highway: The Automobile Wreck in American Culture, 1920–1940" (Ph.D. dissertation, University of Minnesota, 1983), 5, see also 5–6, 37–38. Another dissertation writer agreed, finding that the "Public Outcry Over Accidents" (7) began only in the 1930s. See Daniel M. Albert, "Order out of Chaos: Automobile Safety, Technology and Society 1925 to 1965" (Ph.D. dissertation, University of Michigan, 1997), esp. 6–12.

15. Miller, *Manslaughter* (Dodd, Mead, 1921), see esp. 119–123; *Manslaughter* (film), directed by Cecil B. De Mille, Paramount Pictures, 1922.

16. *The Crowd*, directed by King Vidor, screenplay by Vidor and John V. A. Weaver, Metro-Goldwyn-Mayer, 1928.

17. Miller, *Manslaughter*, 123.

18. In *The Crowd* the truck driver is shown sympathetically, but he cannot stop his truck. In the 1920s motor vehicles were often depicted as almost wilfully destructive, despite their drivers' efforts.

19. The expression "death drivers" (and variants) was common in contemporary newspapers, at least in Philadelphia. See "Bride Brings Back Death Car Driver," *Evening Bulletin*, Nov. 8, 1924; "City-Wide War Is Declared on Death Drivers," *Public Ledger*, Nov. 9, 1924; "Death Drivers" (editorial), *Evening Bulletin*, Nov. 10, 1924; "Calls Death-Truck Driver A 'Disgrace,' " *Evening Bulletin*, Nov. 11, 1924.

20. Philadelphia *Public Ledger*, 1920, as quoted in "Auto-Killings Now at the Rate of Two an Hour," *Literary Digest* 67 (Nov. 6, 1920), 80, 82, 86, 88 (86). I thank Erik Uttermann of the University of Virginia for bringing this article to my attention. Clay McShane found that in New York, by 1914, there had already been several violent assaults on drivers who struck children; one driver was killed; see McShane, *Down the Asphalt Path*, 176–177.

21. "Choir Singer Killed; Mob Menaces Driver" *Cincinnati Enquirer*, Nov. 5, 1923. The victim was Marietta D. V. O'Donnell.

22. "Eight Children Die in Auto Accidents," *New York Times*, April 13, 1927.

23. The author has taken a small sample of letters written to St. Louis papers Nov. 19–30, 1923, during and just after its Safety Week (Nov. 19–26). Twenty-four letters on traffic safety matters were written to two major city newspapers, the *Star* (19 letters) and the *Post-Dispatch* (5 letters); the *Globe-Democrat*, St. Louis's other major daily, did not then publish letters. Of the 24 letters, 2 found the problem complex, with enough blame to go around. Twenty-two took an unambiguous position. Twenty-one of these 22 letters faulted motorists; only one faulted pedestrians. Of course neither newspaper published all letters it received. While papers also tended to fault motorists in its editorial columns (and so may have selected letters from like-minded writers), this would also indicate that the papers suspected that most St. Louisans faulted the automobile for the accident problem.

24. Letter to editor ("The Criminally Reckless"), signed " 'Very Indignant,' " *St. Louis Post-Dispatch*, Nov. 24, 1923.

25. " 'Eyes Left at Curb, Right in Mid-Street,' Is New Crossing Tip," *St. Louis Star*, Nov. 19, 1923; letter to editor ("Cannon to Right, Cannon to Left," signed "Sic Semper Tyrannus," *St. Louis Star*, Nov. 24, 1923).

26. James R. Doolittle, ed., *The Romance of the Automobile Industry* (Klebold, 1916), 440.

27. Bessie E. Buckley to editor, Sept. 1920, in "Where Can the Kiddies Play? Send Views to *Journal*," *Milwaukee Journal*, Sept. 29, 1920.

28. Lew R. Palmer, "History of the Safety Movement," *Annals* 123 (Jan. 1926), 9–19 (17). For four more juggernaut comparisons, see "Nation Roused against Motor Killings," *New York Times*, Nov. 23, 1924; editorial cartoon, *Norwalk Sentinel*, Dec. 13 1924; "Making the Automobile Safe for Everybody," *Literary Digest* 84 (Jan. 10, 1925), 56–58 (56); Don C. Seitz, "Murder by Motor," *The Outlook* 144 (Sept. 22, 1926), 113–114 (113); Jacksonville *Times-Union*, 1928, quoted in "Taking Arms against the Auto Killer," *Literary Digest* 97 (April 7, 1928), 70–73 (70).

29. James, "Sacrifices to the Modern Moloch" (cartoon), *St. Louis Star*, Nov. 6, 1923.

30. *Cincinnati Times-Star*; see "Our Modern Moloch," *American City* 18 (June 1918), 515. The comparison crossed the Atlantic; see "The Dangerous Art of Crossing the Road," *Public Opinion* 128 (August 14, 1925), 151–152 (152).

31. William Ullman, *Washington Star*, Dec. 1925, quoted in "Making the Automobile Safe for Everybody," *Literary Digest* 84 (Jan. 10, 1925) 56–58 (56).

32. *Post-Dispatch*, "Safety and the Motors" (editorial), Nov. 19, 1923. See also "The Murderous Motor," *New Republic* 47 (July 7, 1926), 189–190, and Don C. Seitz, "Murder by Motor," *The Outlook* 144 (Sept. 22, 1926), 113–114.

33. "Safety Week Is Killing Week" (editorial), *St. Louis Star*, Nov. 24, 1923.

34. " 'Municipal Murder Maps' Effectively Used," *American City* 39 (Oct. 1928), 87.

35. J. L. Warhover to editor ("Sunday's Death Toll"), *St. Louis Post-Dispatch*, Nov. 19, 1923.

36. Letter to editor ("Consider One Another," signed "Driver and Walker"), *St. Louis Star*, Nov. 21, 1923.

37. Letter to editor ("Jail for Reckless Drivers," signed "Soak Them"), *St. Louis Star*, Nov. 27, 1923.

38. Robert A. Crabb to editor ("Penitentiary for Speeders"), *St. Louis Star*, Nov. 27, 1923.

39. "Thousands March in Safety Parade," *Cleveland Plain Dealer*, Oct. 3, 1919. The woman was Leonora Bortz. See also Viviana Zelizer, *Pricing the Priceless Child: The Changing Social Value of Children* (Basic Books, 1985). In *Taking the Wheel: Women and the Coming of the Motor Age* (Free Press, 1991), Virginia Scharff shows that many women—especially rural women—welcomed the automobile as a means of liberation, giving them greater independence and access to the public sphere. As the present chapter shows, urban women who did not own automobiles more often constructed them as intruders and threats.

40. "Safety Congress to Open Here Tomorrow," *St. Louis Post-Dispatch*, Sept. 15, 1918.

41. "10,000 Take Part in Drive to Make Life Safer Here," *St. Louis Star*, Oct. 23, 1922; "Safety First," *St. Louis Star*, Oct. 24, 1922 ("Daily Page for Women").

42. Mrs. John Thomy, quoted in "Safety First," *St. Louis Star*, Oct. 24, 1922.

43. "Women to Decide on Safety Service," *Washington Post*, Nov. 24, 1922.

44. "Mothers Join for Safety," Philadelphia *Evening Bulletin*, Oct. 6, 1927; "Accident Victims' Mothers Organize," *Philadelphia Inquirer*, Oct. 7, 1927; "Victims' Mothers Seek Safe Traffic," Philadelphia *Public Ledger*, Oct. 7, 1927; K. H. Lansing, "First Mothers' Safety Council Formed in Philadelphia," *American City* 37 (Dec. 1927), 745–746.

45. "10,000 Children Join Safety March," *New York Times*, Oct. 10, 1922 (white star mothers). "Accident Victims' Mothers Organize," *Philadelphia Inquirer*, Oct. 7, 1927; "Victims' Mothers Seek Safe Traffic," Philadelphia *Public Ledger*, Oct. 7, 1927 (gold

star mothers). Such remarkable public expressions of grief over the premature death of children have been overlooked by historians. The only exception I know of is Zelizer's *Pricing the Priceless Child*. Zelizer documents the rise of a sense of the immeasurable preciousness of child life at the beginning of the twentieth century, which she contrasts to more practical valuations that prevailed earlier. By 1900 children were newly regarded as "emotionally 'priceless' " beings (p. 3), whose loss (through any misfortune) "became a national disgrace" (p. 12). An effect of the transition was recognition of child death as a public loss through ceremonies and monuments. Zelizer (40–41) briefly describes one such case in New York. Just as remarkable, however, is the rapid return of child loss to a private matter as streets were redefined for the motor age (see chapter 8 of this book).

46. "Autocrats of the Highway," *The Outlook* 140 (August 12, 1925), 507.

47. Quoted in "Auto-Killings Now at the Rate of Two an Hour," *Literary Digest* 67 (Nov. 6, 1920), 80, 82, 86, 88 (82).

48. George P. Le Brun, quoted in "New York's Auto Death toll 858 for Year Ended July 31," *St. Louis Star*, Oct. 29, 1922.

49. "The Toll of the Automobile," *The Outlook* 140 (June 3, 1925), 173–174 (174).

50. In 1924 an auto manufacturer used the term 'overspeeding' to characterize recklessly fast drving. See George M. Graham, "Education, Punishment and Traffic Safety" (address, Feb. 14, 1924), p. 9, in bound collection marked "Pamphlets" (New York: National Automobile Chamber of Commerce, 1919–1933), Library, United States Department of Transportation, Washington; reprinted as "Stern Action Is Urged to Banish Death from Highways," *Automotive Industries* 50 (Feb. 28, 1924), 495–498, 511 (498).

51. H. W. Slauson, "A New Plan for Traffic Laws," *Scientific American* 132 (May 1925), 296–298 (297).

52. Letter to editor ("Curbing the Speeders," signed "Slow and Careful"), *St. Louis Star*, Nov. 22, 1923.

53. J. Cower to editor ("Revising a Commandment"), *St. Louis Star*, Nov. 28, 1923.

54. W. G. to editor ("Comparative Caution"), *St. Louis Star*, Nov. 23, 1923.

55. Letter to editor ("Tyrrany of the Automobilists," signed "Pedestrian"), *St. Louis Star*, Nov. 23, 1923.

56. "Safety Week Is Killing Week" (editorial), *St. Louis Star*, Nov. 24, 1923.

57. *Sun*, quoted in "Auto-Killings Now at the Rate of Two an Hour," *Literary Digest* 67 (Nov. 6, 1920), 80, 82, 86, 88 (80); Stephen S. Tuthill, "The Attitude of the Motorist Toward Traffic Regulations," Oct. 2, 1923, Proceedings of the National Safety

Council, Twelfth Annual Safety Congress, Buffalo, Oct. 1–5, 1923, 810–815 (810). See also Don C. Seitz, "Murder by Motor," *The Outlook* 144 (Sept. 22, 1926), 113–114 ("speed madness," 113) and W. Bruce Cobb in "Nation Roused against Motor Killings," *New York Times*, Nov. 23, 1924 ("motor madness").

58. "What Shall Be the Cure for Automobile Speed Mania?" *Illustrated World* 34 (Sept. 1920), 85–86 (85).

59. F. E. Cogswell, summarizing the view of the city manager of Pipestone, Minnesota, in Cogswell, "Requests Succeed Where Other Measures Often Fail," *American City* 21 (July 1919), 34–35 (34).

60. In 1911 the state of Washington imposed the first effective workmen's compensation law. Other states quickly followed Washington's example.

61. Mark Aldrich, *Safety First: Technology, Labor, and Business in the Building of American Work Safety, 1870–1939* (Johns Hopkins University Press, 1997), 96. A second precedent, older and less directly influential, lay in safety campaigns intended to reduce casualties caused by new street railways; see Barbara Young Welke, *Recasting American Liberty: Gender, Race, Law, and the Railroad Revolution, 1865–1920* (Cambridge University Press, 2001), 16–18, 36–41.

62. Credit for the origins of this phrase was contested. See Dianne Bennett and William Graebner, "Safety First: Slogan and Symbol of the Industrial Safety Movement," *Journal of the Illinois State Historical Society* 68 (June 1975), 243–256, esp. 245–246.

63. On the industrial safety movement, see Mark Aldrich, *Safety First: Technology, Labor, and Business in the Building of American Work Safety, 1870–1939* (Johns Hopkins University Press, 1997).

64. Aldrich, *Safety First*. Throughout this book, Aldrich shows convincingly that in industrial accidents, the low damage awards of employer liability rulings encouraged the safety abuses that led to state workmen's compensation laws. This relationship helps to explain the relatively modest role insurance companies played in traffic safety in the the 1920s as well. Because liability laws, not compensation laws, covered injuries in traffic accidents, insurers tended to invest their money in defending claims against their clients, rather than in accident prevention. A few state laws to compensate victims of traffic accidents were proposed, but these were defeated (Herbert L. Towle, "Financial Balm for Motorists," *The Outlook* 142, March 24, 1926, 459–462 (460)). Insurance companies, organized in the National Bureau of Casualty and Surety Underwriters and in the American Mutual Alliance, were nevertheless important elements in the NSC and in the National Conference on Street and Highway Safety. Their cooperation with the NSC sometimes took the form of granting discounts to operators of commercial motor fleets who were members of the NSC or of a local council. See C. W. Price, in discussion, "Wednesday Afternoon

Session," Sept. 28, 1921, Proceedings of the National Safety Council, Tenth Annual Safety Congress, Boston, Sept. 26–30, 1921, 352–364 (363–364).

65. "Safety as a Community Asset," *American City* 12 (April 1915), 329–330.

66. Carl L. Smith, "The Experience of a Large City in Local Council Work," address, Oct. 3, 1919, to the Eighth Annual Safety Congress, Cleveland, Oct. 1–4, 1919 *Proceedings of the National Safety Council* 8 (1919), 204–218 (206).

67. "Growth of the Safety Council Movement," *American City* 31 (Sept. 1924), 269.

68. E. George Payne, "How the St. Louis Schools Have Reduced Accidents," *American City* 27 (Nov. 1922), 437–438.

69. Harry J. Bell, "Working to Make Chicago Safe," *American City* 28 (April 1923), 395.

70. New York's 1922 safety week was "managed throughout by business men"; see "The Nation's Needless Martyrdom," *Literary Digest* 75 (Oct. 28, 1922), 29–31 (29).

71. On the accident that killed Latimer (Sept. 26, 1913) and on Syracuse's safety month, see the following items from the Syracuse *Post-Standard* (1913): "Woman Struck by Racing Auto Dies Instantly," Sept. 27; "Hibbard to Be Arraigned on Manslaughter Charge," Sept. 29; "Safety in the Streets" (editorial); Syracuse Chamber of Commerce, Automobile Club of Syracuse, and Syracuse Police Department, "Safety First: A Public Spirited Campaign to Reduce Accidents" (advertisement), Dec. 1; "Boy Scouts Will Patrol Busy Streets Entire Day," Dec. 6; "Will March to Safety Rally," Dec. 18; "Stop, Look and Listen—Slogan against Danger," Dec. 20. See also Ivan H. Wise, "How a 'Safety First' Campaign Was Conducted," *American City* 10 (April 1914), 322–326.

72. Safety First was originally an industrial safety slogan (see above). It appeared in organized public safety campaigns by 1913. The industrial safety movement was a model for the public safety reformers. Marcus A. Dow, general safety agent of the New York Central Railroad and later president of the National Safety Council, was a guest of honor at Syracuse's Safety Month on Dec. 19, 1913; see "Will March to Safety Rally," *Post-Standard*, Dec. 18, 1913, and "Stop, Look and Listen—Slogan against Danger," *Post-Standard*, Dec. 20, 1913. See also Dow, "The Public Conscience and Accidents" *American City* 28 (June 1923), 556–558. For a longer-range view on various social groups' use of parades as publicity techniques, see Susan G. Davis, *Parades and Power: Street Theatre in Nineteenth-Century Philadelphia* (Temple University Press, 1986).

73. C. W. Price, in discussion, "Wednesday Afternoon Session," Sept. 28, 1921, Proceedings of the National Safety Council, Tenth Annual Safety Congress, Boston,

Sept. 26–30, 1921, 352–364 (355–357); "Campaigns to Reduce Automobile Accidents," *American City* 28 (May 1923), 507.

74. P. S. Case, in discussion, pp. 222–233 (223), following Case, "The Experience of a Small City in Local Council Work," address, Oct. 3, 1919, to the Eighth Annual Safety Congress, Cleveland, Oct. 1–4, 1919, *Proceedings of the National Safety Council* 8 (1919), 218–222.

75. William A. Searle, " 'Safety First' in Precept and Policy," *American City* 11 (Sept. 1914), 227–230; on a more modest 1915 campaign in a west coast city, see "Safety-First Traffic Signals in Portland, Ore.," *American City* 13 (Sept. 1915), 180–181.

76. O. R. Geyer, "Carles W. Price," in "Men and Methods," *System* 30 (Nov. 1916), 506–512 (510–512).

77. See "Fire Prevention Day," Rochester *Union and Advertiser*, Oct. 9, 1916.

78. "Rochester's Notable Public Safety Campaign," *American City* 19 (Nov. 1918), 363–366.

79. "Safety Congress to Open Here Tomorrow," *St. Louis Post-Dispatch*, Sept. 15, 1918; "Convention of National Safety Council Opens," *St. Louis Post-Dispatch*, Sept. 16, 1918.

80. Price, in discussion, Oct. 3, 1919, Proceedings of the National Safety Council, Eighth Annual Safety Congress, Cleveland, Oct. 1–4, 1919, 222–233 (225).

81. From an address of Charles M. Talbert to the NSC as reported in "Aid of Public Vital to Any Safety Move," *Milwaukee Sentinel*, Sept. 29, 1920.

82. Reporter's paraphrase of remarks of Charles M. Talbert to NSC, in "Brand Auto Chief Peril to Safety," *Milwaukee Evening Sentinel*, home edition, Sept. 28, 1920.

83. Price, "Startling Figures for Automobile Fatalities," *American City* 25 (Sept. 1921), 205–206 (205).

84. Price, "Report of the General Manager," Sept. 26, 1921, Proceedings of the National Safety Council, Tenth Annual Safety Congress, Boston, Sept. 26–30, 1921, 11–15 (14).

85. Price, "Startling Figures for Automobile Fatalities," *American City* 25 (Sept. 1921), 205–206.

86. Price, "Startling Figures," 206.

87. Letter to editor ("Didn't Join—Was Fired," signed "A Victim"), *St. Louis Star*, Nov. 3, 1922.

88. A. A. Bureau, quoted in "Boy Scouts Give Safety Drills," *Milwaukee Evening Sentinel*, home edition, Oct. 1, 1920.

89. "Pastors of Churches to Preach on Safety Today," *St. Louis Post-Dispatch*, Oct. 22, 1922; "10,000 Take Part in Drive to Make Life Safer Here," *St. Louis Star*, Oct. 23, 1922.

90. S. M. Lippincott, "The Story of a Volunteer Local Council," *National Safety News* 9 (Jan. 1924), 23–24.

91. "Brand Auto Chief Peril to Safety," *Milwaukee Evening Sentinel*, home edition, Sept. 28, 1920. The words quoted are those of the *Sentinel*'s summary of the NSC's resolution.

92. Safety and Sanitation Division, Milwaukee Association of Commerce, "The Days of 'Old Dobbin' Have Passed" (advertisement), *Milwaukee Sentinel*, Sept. 27, 1920.

93. John J. Boobar, as paraphrased in "Safety Week Nov. 26," *Washington Post*, Nov. 7, 1922.

94. See National Safety Council posters reproduced in Dianne Bennett and William Graebner, "Safety First: Slogan and Symbol of the Industrial Safety Movement," *Journal of the Illinois State Historical Society* 68 (June 1975), 243–256 (252–253).

95. See photograph captioned "Satan vs. Safety," *Milwaukee Evening Sentinel*, home edition, Sept. 28, 1920; "A No-Accident Week," *National Safety News*, as reprinted in *The Outlook* 126 (Dec. 1, 1920), 608–609 (609). It seems likely that the safety week's organizers deliberately avoided having a motor truck (the most disreputable of all vehicles, to traffic safety reformers) pull the float.

96. George Starkey, "Keep the Kiddies off the Street!" (poster), reproduced in "Winning Safety Poster," *Milwaukee Journal*, Sept. 25, 1920, and in "Prize Winning Safety Cartoon," *Milwaukee Leader*, Sept. 25, 1920. By a dreadful coincidence, on the day this poster was publicized as the winner, four-year-old Eugene Struck was "almost decapitated" by a truck in Milwaukee; Struck was one of three children killed within 24 hours in Milwaukee ("Third Child is Motor Victim in Last Twenty-Four Hours," *Milwaukee Journal*, Sept. 25, 1920).

97. Harry P. De Bauffer, "Was Daddy Hurt Much?" (poster), reproduced in "Poster Wins Second Prize," *Milwaukee Journal*, Sept. 28, 1920.

98. Norman Morey, untitled poster, reproduced in "Warn of Perils," *Milwaukee Evening Sentinel*, home edition , Sept. 30, 1920.

99. Walter Cohn, "Don't Take a Chance on the Wheel of Mis-Fortune" (poster), reproduced in "$5 Prize for Safety Poster," *Milwaukee Journal*, Sept. 30, 1920.

100. Albert Mullinix, "The Grim Reaper Takes His Daily Toll" (poster), reproduced in "Journal's Winning Safety Poster," *Milwaukee Journal*, Sept. 29, 1920.

101. "In Most Cases Accidents and Casualties Are Avoidable," *American City* 32 (Jan. 1925), 32.

102. Dudley C. Meyers, " 'Don't Kill a Child,' " *American City* 27 (Oct. 1922), 332; Page Fence and Wire Products Association (Chicago), "Page Protection Fence" (advertisement), *American City* 27 (Sept. 1922), advertising p. 74.

103. "Safety First Campaign Reduces Loss of Life by Half," *American City* 21 (Sept. 1919), 230–232 (231).

104. "To Dedicate Shaft in Safety Campaign," Baltimore *Sun*, June 11, 1922; "Baltimore Puts Over Successful No-Accident Week," *National Safety News* 6 (August 1922), 38.

105. "10,000 Children Join Safety March," *New York Times*, Oct. 10, 1922 (the words quoted from a eulogy are those of Royal S. Copeland, New York City health commissioner); "The Nation's Needless Martyrdom," *Literary Digest* 75 (Oct. 28, 1922), 29–31 (29).

106. "Crippled Children in Safety Parade," *New York Times*, Oct. 13, 1922.

107. " 'No Accident Week' Campaign Opens with Memorial Service for Children Killed in 1921," Pittsburgh *Gazette-Times*, Oct. 23, 1922; "Parade of Floats to Be Held Today as Feature of 'No Accident' Week," *Gazette-Times*, Oct. 26, 1922 (parade of injured children); "Safety Parade Warns Against Carelessness," *Gazette-Times*, Oct. 27, 1922 (float); Marcus A. Dow, "The Public Conscience and Accidents," *American City* 28 (June 1923), 556–558 (558).

108. "Drive Cuts Mishaps," *Washington Post*, Nov. 27, 1922; for a photograph of the monument, see *Post*, Nov. 28, 1922. "Taps" was also played daily during a safety week in Detroit, August 24, 1922; see "Save Children, Is Safety Aim," *Detroit News*, August 24, 1922.

109. "2,000 in Safety Plea," *Washington Post*, Dec. 2, 1922. For a photograph of the Agriculture Department's float see H. T. Baldwin, "Washington's Safety Week Parade," *American City* 28 (March 1923), 305. Washington's safety committee, which organized the safety week in conjunction with the National Safety Council, assigned Col. Clarence O. Sherrill to arrange the government department contingents in the parade. Sherrill was superintendent of buildings and grounds for the departments in Washington, and later served as Cincinnati's city manager. In the latter post, Sherrill was a zealous adherent of the efficiency model of traffic control, and he backed strict regulation of automobiles as wasters of scarce space. See Sherrill, "Congested Streets Are Costly," *Electric Railway Journal* 68 (Oct. 9, 1926), 648–650 (649).

110. "Safety to Be Preached in City's Pulpits," *Louisville Herald*, June 3, 1923.

111. "Thousands See Warnings of Death in Hideous Forms on Floats in Safety Parade," *Louisville Herald*, June 9, 1923.

112. "This Monument to Plead for Careful Auto Driving," *St. Louis Star*, Nov. 17, 1923; "Monument to Children Slain Here by Automobiles Is Erected Today," *St. Louis Star*, Nov. 19, 1923; "Scene at Dedication of Safety Monument," *St. Louis Post-Dispatch*, Nov. 20, 1923; "Monument to Child Auto Victims and Speedometer to Show Deaths," *St. Louis Globe-Democrat*, Nov. 20, 1923; "Four Persons Killed by Automobiles as Safety Week Opens," *Globe-Democrat*, Nov. 20, 1923; "U.S. Auto Deaths Exceed War Toll, Pastor Declares," *Star*, Nov. 19, 1923; "Carelessness Mother of Accidents, Asserts Rev. James H. Smith," *Globe-Democrat*, Nov. 19, 1923; "Municipal Departments Cooperate with Safety Council in St. Louis Safety Week," *American City* 30 (Jan. 1924), 97. The children's memorial was temporary, for the duration of the safety week. Besides its most prominent inscription (quoted above), it seems to have included twelve others, three on each of its four sides. Of these, press photographs (in newspapers cited above) show nine, at least four of which imply driver responsibility ("32 Children Killed by Motor Cars This Year," "Every Mother's Plea: Drive with Care," "Motorists—Drive with Care and Save Our Children," "Don't Destroy Childhood's [illegible] by Reckless Driving"). Four emphasize the gravity of accidental child death without directing blame ("Accidents' Greatest Cost Is the Lives of Our Children," "Nothing Is More Precious Than a Child's Life," "Our Children—The Hope of Tomorrow—[illegible]," "Better Constant Vigilance than Ceaseless Remorse"). Only one assigns responsibility to parents or children ("Parents—Teach Your Children Not to Play in the Streets").

113. "In Most Cases Accidents and Casualties Are Avoidable," *American City* 32 (Jan. 1925), 32.

114. "Mourning Flag Raised When Child Is Killed," *American City* 35 (Nov. 1926), 708; see also "Flag Indicates No Child Deaths from Automobile Accidents," *American City* 36 (April 1927), 545.

## Chapter 2

1. George C. Kelcey, "Traffic and Parking Regulations: As They Affect Public Safety and the Business Man," *City Manager Magazine* 8 (Sept. 1926), 22–27, 65 (27, 65).

2. Lent D. Upson, *Practice of Municipal Administration* (Century, 1926), 343.

3. Frances Fisher McMurdo (1901–1998) remembering the streets of Shelbyville, Indiana (population about 9,000), in conversation with the author, March 18, 1997.

4. Eno, *The Story of Highway Traffic Control* (Eno Foundation for Highway Traffic Control, 1939), 2.

5. Harvey W. Corbett in "Discussion," Transactions of the American Society of Civil Engineers 88 (New York, 1925), 223. For a contrasting view, see Mark Foster, *From Streetcar to Superhighway: American City Planners and Urban Transportation, 1900–1940*

(Temple University Press, 1981). Foster argued that traffic congestion was comparably severe before the arrival of automobiles (see p. 35).

6. F. S. Besson, *City Pavements* (McGraw-Hill, 1923), 63.

7. William H. Connell, "The Highway Business: What Pennsylvania Is Doing," *Annals* 116 (Nov. 1924), 113–127 (113).

8. City Plan Commission of Newark, New Jersey, *Comprehensive Plan of Newark* (Newark City Plan Commission, 1915), 34.

9. *Comprehensive Plan of Newark*, 34.

10. A. S. Brainard, "County Management Study of Milwaukee County, Wisconsin" (unpublished report, Record Group 30, National Archives, College Park, Maryland), Sept. 20–22, 1915, p. 22.

11. Miller McClintock, *Report and Recommendations of the Metropolitan Street Traffic Survey* (Chicago Association of Commerce, 1926), table 9 (24–25).

12. Harland Bartholomew, "Reduction of Street Traffic Congestion by Proper Street Design," *Annals* 116 (Nov. 1924), 244–246 (244).

13. For example, a 1908 article on "Traffic Congestion" in a popular magazine is largely concerned freight terminals; see George Ethelbert Walsh, "Traffic Congestion in New York," *Cassier's Magazine* 31 (June 1908), 151–155. A 1910 article in *The American City* on "Congestion and Its Relief" (3 (Dec. 1910), 285–288) is about population congestion. A 1912 *American City* editorial on "Traffic Congestion" concerns the crowding of transit terminals; see "Relieving Traffic Congestion," *American City* 6 (Feb. 1912), 469. A 1915 article on "Traffic Congestion" in *Electric Railway Journal* is about the crowding of streetcars, not of the streets themselves; see "Relieving Traffic Congestion," *Electric Railway Journal* 45 (Jan. 2, 1915), 30–32.

14. Carl W. Condit, Chicago 1910–1929: *Building, Planning, and Urban Technology* (University of Chicago Press, 1973), appendix, table 7.

15. Census Bureau figures for horses cited in "Use of Horses Becoming Less in Cities," *Scientific American* 125 (August 20, 1921), 138; the eight cities included are New York, Chicago, Philadelphia, Cleveland, Boston, Baltimore, Pittsburgh, and Cincinnati. Between 1909 and 1919, the number of horse-drawn vehicles in Chicago was halved; while in 1909 they had constituted 89 percent of vehicles registered, ten years later they were only 23 percent. See Miller McClintock, *Report and Recommendations of the Metropolitan Street Traffic Survey* (Chicago Association of Commerce, 1926), table 9 (24–25). In 1911, 25 percent of traffic on a Minneapolis street with no streetcars consisted of motor vehicles. Five years later it was 78 percent motor vehicles, and in 1921 the proportion was 96 percent. See Ellis R. Dutton, "Discussion," 250–251 (251), appended to J. D. MacLean, "Results Obtained from an Experimental Treated Wood-Block Pavement after Fifteen Years Service," *Proceedings*

*of the American Society for Municipal Improvements* 28 (Baltimore, 1921), 240–250. I have counted as horse-drawn vehicles those identified in the source as vehicles with "steel tires" (251), as the accompanying text (251) supports this construction; nevertheless, a portion of these may be motor trucks with steel tires. The number of horse-drawn carts that New York City licensed fell more than a third from 1918 to 1922. See Harold A. Littledale, New York *Evening Post*, March 19, 1923, as cited in Miller McClintock, *Street Traffic Control* (McGraw-Hill, 1925), 5.

16. "Fewer Horses, More Motor Trucks," *American City* 22 (March 1920), 256.

17. In Baltimore, 5.3 percent of traffic was horse drawn in 1925 (Kelker, De Leuw and Company, report to the Traffic Survey Commission of Baltimore, abstracted as "Better Car Routing and Traffic Control Proposed for Baltimore," part 1, *Electric Railway Journal* 67, May 22, 1926, 883–889 (887)); 3.8–4.5 percent of Chicago traffic was horse-drawn in 1926 (McClintock, Report and Recommendations of the Metropolitan Street Traffic Survey, Chicago Association of Commerce, 1926, 18).

18. Ira C. Judson, "In the Wake of the Automobile Joy Rider," Wilmington *Sunday Morning Star*, May 2, 1909.

19. Robert H. Wiebe, *The Search for Order, 1877–1920* (Hill and Wang, 1967).

20. Eric H. Monkkonen, *Police in Urban America, 1860–1920* (Cambridge University Press, 1981).

21. Eno, *Story of Highway Traffic Control*, 3.

22. On the trial-and-error application of common sense to problem solving, which Olivier Zunz terms "American tinkerism," see John Higham, "The Matrix of Specialization," in Alexandra Oleson and John Voss, *The Organization of Knowledge in Modern America, 1860–1920* (Johns Hopkins University Press, 1979), 3–18; Olivier Zunz, "Producers, Brokers, and Users of Knowledge: Making the Century American" (paper presented at UCLA, Dec. 5, 1991), 8–9; and Zunz, *Making America Corporate, 1870–1920* (University of Chicago Press, 1990), 79–90, 154–156.

23. On Eno's life and career, see John A. Montgomery, *Eno—The Man and the Foundation: A Chronicle of Transportation* (Eno Foundation for Transportation, 1988); for a brief treatment, see Peter Norton, "William Phelps Eno," *American National Biography*, volume 7 (Oxford University Press, 1999), 535–536. I thank the Eno Foundation (now of Lansdowne, Virginia) for a giving me a copy of Montgomery's book (among other Foundation publications) without charge, and I want particularly to thank archivist Tracy Larkin for her kind help.

24. Eno, *Story of Highway Traffic Control*, 1.

25. Eno, "Reform in Our Street Traffic Most Urgently Needed," *Rider and Driver*, Jan. 20, 1900, 7–8 (8).

26. Eno to B. J. Shoninger, June 25, 1912, reproduced in Eno, *Story of Highway Traffic Control*, 73.

27. Arno Dosch, "The Science of Street Traffic," *World's Work* 27 (Feb. 1914), 398–409. I thank Meg Jacobs of MIT for bringing this article to my attention.

28. Eno, *Story of Highway Traffic Control*, 3; Montgomery, *Eno*, 11.

29. For a facsimile of the complete "Rules for Driving," see Eno, *Story of Highway Traffic Control*, 10–13. Eno first formally proposed the left-turn rule in 1900 ("Reform in Our Street Traffic Most Urgently Needed," 7); Eno repeated the suggestion in "Rules of the Road Revised," *Rider and Driver*, Jan. 5, 1901, 7–8 (7).

30. Montgomery, *Eno*, 26; Eno, *Story of Highway Traffic Control*, 7. In 1902, however, Eno himself had called for a modest 10-mile-per-hour limit for motor vehicles; see his "Rules of the Road Revised," *Rider and Driver*, Jan. 5, 1901, 7–8 (7). On "rotary traffic control" at intersections, see Eno, *Story of Highway Traffic Control*, 8, 24–25, 106–107, 109; Eno, "Standardized Street Traffic Regulation," *American City* 9 (Sept. 1913), 223–226 (225); Arno Dosch, "The Science of Street Traffic," *World's Work* 27 (Feb. 1914), 398–409 (399, 407, 409); Montgomery, *Eno*, 48, 58, 78–79.

31. Eno, *Story of Highway Traffic Control*, 9, 14.

32. Ibid., 96.

33. Bureau of the Census, *Historical Statistics of the United States: Colonial Times to 1970* (Washington: Government Printing Office, 1975), 1:716.

34. Chicago gave the right of way to horse-drawn vehicles over motor vehicles; see A. J. Cunningham, "Traffic Regulations and the Use of Streets by Pedestrians," *National Municipal Review* 2 (April 1913), 329–331 (330). Numerous states required motor vehicles to yield to horse-drawn vehicles early in the century; see Huddy, *Law of Automobiles*, 116–325.

35. "Street Traffic Regulations," *Municipal Journal and Engineer* 23 (Nov. 20, 1907), 577–578 (578).

36. By 1920 references to the outside-left-turn rule were abundant and widespread. See, e.g., "St. Paul All Set for Safety Week," *St. Paul Dispatch*, April 28, 1922.

37. "Regulation of Street Traffic" (editorial), *American City* 13 (Sept. 1915), 174; John P. Fox, "Traffic Regulation in Detroit and Toronto," *American City* 13 (Sept. 1915), 175–179 (176); Thomas Adams and Harland Bartholomew, "City Planning as Related to Public Safety," *Proceedings of the National Safety Council* 7 (St. Louis, 1918), 355–364 (363); Peter Kline, "Newspaper Publicity for Traffic Rules," *American City* 22 (Feb. 1920), 126–127; "Marking Traffic Zones," *American City* 25 (Dec. 1921), 531, 533; R. D. Ballew, "Solving the Parking Problem in a Small City," *American City* 27 (July 1922), 31; A. R. Hirst, "You Can Say 'Danger' Once Too Often," *American City* 27

(August 1922), 122; William M. Myers, "Traffic Lights and Zone Markers in Richmond," *American City* 28 (May 1923), 454; Miller McClintock, *Street Traffic Control* (McGraw-Hill, 1925), 212–213.

38. By the early 1920s most of the downtown streets of Philadelphia and Pittsburgh were one-way. See George C. Kelcey, "Traffic and Parking Regulations: As They Affect Public Safety and the Business Man," *City Manager Magazine* 8 (Sept. 1926), 22–27, 65 (25–26); Bartholomew, "City Planning as an Aid to Public Safety," *Proceedings of the National Safety Council: Thirteenth Annual Safety Congress* (1924), 831–840 (839–840).

39. The term "adversary model" is David Nord's. See "The Experts versus the Experts: Conflicting Philosophies of Municipal Utility Regulation in the Progressive Era," *Wisconsin Magazine of History* 58 (Spring 1975), 219–236 (229).

40. "Causes of Traffic Accidents—As Seen By Chiefs of Police," *American City* 34 (May 1926), 542. The survey was conducted by *American City*. Though unprompted, 69 of 480 respondents volunteered their view that a major cause of city traffic accidents was "petting parties."

41. The seal of the International Traffic Officers Association, with its slogan, is reproduced in Electrical and Specialty Supply Company, "Traffic Police Approve Mushroom Traffic Light—Milwaukee Type" (advertisement), *American City* 26 (May 1922), advertising p. 84.

42. Philadelphia's Magistrate Costello, quoted in "Calls Death-Truck a 'Disgrace,' " *Evening Bulletin* (Philadelphia), Nov. 11, 1924, 1.

43. Shepard Bryan, "Who Is Responsible for Public Accidents?" Sept. 30, 1924, *Proceedings of the National Safety Council, Thirteenth Annual Safety Congress*, Louisville, Sept. 29–Oct. 3, 1924, 79–82 (81).

44. Median speed limit based on state codes reproduced in Xenophon P. Huddy, *The Law of Automobiles* (Matthew Bender, 1906), 116–325.

45. South Bend Police Department, "Automobile Drivers Attention," *South Bend Tribune*, Oct. 23, 1919, 11; facsimile in Peter Kline, "Newspaper Publicity for Traffic Rules," *American City* 22 (Feb. 1920), 126–127 (127).

46. See, e.g., "Death Drivers" (editorial), *Evening Bulletin* (Philadelphia), Nov. 10, 1924: "Daily slaughter on the streets of the city calls for Law Enforcement quite as loudly as does the business of the bootlegger and his pals."

47. See, e.g., "No Speed Limit Plea Amplified by Roy Britton," *St. Louis Star*, Nov. 17, 1923.

48. In 1927 about 10,820 pedestrians were killed in auto accidents, while 3,430 motorists were killed in collisions with other motor vehicles; see National Safety

Council, *Accident Facts* (Chicago: the Council, 1974). Throughout the 1920s, a large majority of traffic accident fatalities were pedestrians—and this majority was overwhelming in cities.

49. "Causes of Traffic Accidents—As Seen By Chiefs of Police," *American City* 34 (May 1926), 542. The survey was conducted by *American City*.

50. "Value of Traffic Movement and Accident Maps," *American City* 29 (Nov. 1923), 545; George H. Herrold, "The Parking Problem in St. Paul," *Nation's Traffic*, July 1927, 28–30, 47–48 (47).

51. South Bend Police Department, "Automobile Drivers Attention." Where they existed, early state-designated speed limits at city intersections were very low (4–8 miles per hour); see Huddy, *Law of Automobiles*, 116–325. See also Safety First League, *Safety for Twenty Million Automobile Drivers*, 67, 120, 177.

52. Eno, "Reform in Our Street Traffic Most Urgently Needed," *Rider and Driver*, Jan. 20, 1900, 7–8 (7).

53. Eno, *Story of Highway Traffic Control*, 23.

54. Council of National Defense, Highways Transport Committee, "General Highway Traffic Regulations," Article II, section 5; reproduced in full in "Highways Transport Committee's General Traffic Regulations," *Good Roads*, Sept. 3, 1919, 122–23.

55. Guy Kelcey, "Traffic Regulations to Prevent Accidents and to Expedite Fluid Traffic Movement," *Annals* 133 (Sept. 1927), 161–171 (171).

56. G. C. S. Welzel to editor ("Is Jay-Walking Safe?"), Oct. 22, 1924, *Public Ledger* (Philadelphia), Oct. 28, 1924.

57. Kelcey emphasized the safety of the outside left turn for pedestrians. See Kelcey, "Traffic and Parking Regulations as They Affect Public Safety and the Business Man," *City Manager Magazine* 8 (Sept. 1926), 22–27, 65 (22–23). Yet by 1926, as this article shows, Kelcey was silently abandoning his earlier recommendation that left turners always pass the intersection center point on the right. Compare Kelcey, "How Intersections Figure in Traffic Accidents," *American City* 30 (April 1924), 417–419, and "Traffic Problems of the Large City," *Scientific American* 131 (Oct. 1924), 236–237, 294 (237, 294). Kelcey's early book, *Traffic Engineering* (Elizabeth, New Jersey: American Gas Accumulator Co., 1922), presumably reflects his earlier position on left turns, but repeated searches have failed to uncover a copy of it.

58. Kelcey, "Traffic Regulations to Prevent Accidents and to Expedite Fluid Traffic Movement," *Annals* 133 (Sept. 1927), 161–171 (161). In a footnote Kelcey admits that the precise figures he cites are guesses lacking any empirical basis. Engineers found Kelcey's guesses to be far off the mark, and Kelcey's own guess that 75 percent of accidents were at intersections (161) is hard to square with his view that accidents were infrequent downtown.

59. Kelcey, "How Intersections Figure in Traffic Accidents," *American City* 30 (April 1924), 417–419 (417) (emphasis added). See also Kelcey's *Traffic Engineering* (Elizabeth, New Jersey: American Gas Accumulator Co., 1922), p. 7: "As the angle of the intersection is changed more and more away from a right angle, *and as the speed increases*, the confusion area, or the place in which accidents are more likely to take place, is increased" (emphasis added; quoted in Miller McClintock, *Street Traffic Control*, McGraw-Hill, 1925, 115).

60. Eno to Charles Henry Davis, July 21, 1915, reproduced in Eno, *Story of Highway Traffic Control*, 110–111 (110).

61. Burton W. Marsh, "Traffic Control," *Annals* 133 (Sept. 1927), 90–113 (93). For a closer look at cornermen and their trying work, see "It's Not All Blowing the Whistle with the Traffic Cop," *Illustrated World* 34 (Sept. 1920), 77–78.

62. Marsh, "Traffic Control," 95; "Where the Money Goes," *American City* 23 (Sept. 1920), 278. The budget detailed in the latter article lists "police" and "street intersections" as separate line items; I combine them to arrive at a total police budget, of which I make "street intersections" a portion.

63. The number of traffic officers per 1,000 vehicles registered in Chicago fell 69.2 percent between 1916 and 1925, from 39.6 to 12.2; see McClintock, *Report*, table 30 (170).

64. George Cutter Company, "The Shadow Shows the Purpose It Serves" (advertisement), *American City* 20 (Jan. 1919), 103; see also numerous advertisements in *The American City*, especially c. 1917–1923.

65. George Garrett, "A Simple and Effective Traffic Standard," *American City* 26 (Jan. 1922), 33.

66. "A Center Type Traffic Signal," *American City* 32 (May 1925), 601, 603 (601). An advertisement for a silent policeman in *American City 31* (July 1924) freely admitted that the device damaged cars: "Smashing automobiles? Yes—but saving [pedestrians'] lives and reducing upkeep costs to the minimum"; see Gordon M. Sessions, *Traffic Devices: Historical Aspects Thereof* (Washington: Institute of Traffic Engineers, 1971), 55.

67. "Mechanical Devices for Highway Traffic Regulation," *American City* 29 (Dec. 1923), 634–641 (636); reprinted from a report of the Committee on Mechanical Devices for Highway Traffic Regulation, National Highway Traffic Association.

68. Garrett, "A Simple and Effective Traffic Standard."

69. "Highway Traffic Engineers," *Engineering News-Record* 97 (Nov. 4, 1926), 730.

70. Kelcey, "How Intersections Figure in Traffic Accidents," *American City* 30 (April 1924), 417–419; "Traffic Problems of the Large City," *Scientific American* 131 (Oct.

1924), 236–237, 294; "Traffic and Parking Regulations: As They Affect Public Safety the Business Man," *City Manager Magazine* 8 (Sept. 1926), 22–27, 65; "Traffic Regulations to Prevent Accidents and to Expedite Fluid Traffic Movements," *Annals* 133 (Sept. 1927), 161–171.

71. Eno, "Traffic Regulation," *National Municipal Review* 6 (Jan. 1917), 141–142 (142).

72. Eno, "Reform in Our Street Traffic Most Urgently Needed," *Rider and Driver*, Jan. 20, 1900, 7–8.

73. See articles by Kelcey cited above.

74. See Gordon M. Sessions, *Traffic Devices: Historical Aspects Thereof* (Washington: Institute of Traffic Engineers, 1971).

75. For example, a leading traffic control device manufacturer, the American Gas Accumulator Company of Elizabeth, New Jersey, developed its traffic device line as a supplement to its main product, gas stoves. For a traffic device from American Gas Accumulator, see Frederick W. Sheaf, "The Flashing Light Attracts the Driver's Attention," *American City* 24 (May 1921), 481–482.

76. John P. Fox, "Traffic Regulation in Detroit and Toronto," *American City* 13 (Sept. 1915), 175–179 (176).

77. "Marking Traffic Zones," *American City* 25 (Dec. 1921), 531, 533.

78. Harry H. Jackson, "Flashing Traffic Signal and Information Booth Attract Attention," *American City* 27 (Nov. 1922), 421–422 (421).

79. "Street Traffic Signaling," *American City* 26 (April 1922), 387. See also Lewis E. Moore, "Railway Signaling Methods for Public Roads," *Engineering News-Record* 94 (Jan. 22, 1925), 140–144.

80. See Gordon M. Sessions, *Traffic Devices: Historical Aspects Thereof* (Institute of Traffic Engineers, 1971), 37.

81. See Sessions, *Traffic Devices*, 21–27.

82. "Tries Traffic Signals," *Cleveland Plain Dealer*, August 6, 1914; "Illuminated Signals for Traffic Control at Cleveland," *Electric Railway Journal* 44 (Oct. 3, 1914), 630–631; Alfred A. Benesch, "Regulating Street Traffic in Cleveland," *American City* 13 (Sept. 1915), 182–184; Sessions, *Traffic Devices*, 27–28, 30. Other claims exist for earlier traffic light installations in the United States; see Sessions, *Traffic Devices*, 6–9, 20, 24, 27. The *Plain Dealer* ("Tries Traffiic Signals") claimed this Euclid Avenue installation was "the first of the kind to be used in the United States," and it was at least the first influential use of the traffic light.

83. Eno, *Story of Highway Traffic Control*, 23–24.

84. Ibid., 124–125.

85. McClintock, *Street Traffic Control*, 129.

86. "The Traffic Tower," *Scientific American* 122 (March 6, 1920), 249; W. T. Perry, "A New Way to Control Traffic on Congested Streets," *American City* 22 (May 1920), 476; Ralph Howard, "Fifth Avenue's Traffic Tower," *Scientific American* 124 (Jan. 15, 1921), 45, 57–58; "Tricks of Traffic Control," *Scientific American* 124 (June 25, 1921), 510, 518–519; Henry Collins Brown, *Fifth Avenue Old and New* (New York: Fifth Avenue Association, 1924), 116.

87. "New Traffic Control Successful in Philadelphia," *Electric Railway Journal* 64 (July 12, 1924), 51–52; J. Borton Weeks, "Philadelphia's Traffic Problems and Their Solution," *Annals* 116 (Nov. 1924), 235–240 (238). Traffic engineer Burton Marsh credited Houston with the first centrally controlled simultaneous traffic lights (sans towers), installed at nine intersections in 1922; see Marsh, "Traffic Control," *Annals* 133 (Sept. 1927), 90–113 (97).

88. Harriss, "Untangling Our Traffic Laws," *Scientific American* 126 (Feb. 1922), 92–93; McClintock, *Street Traffic Control* (McGraw-Hill, 1925), 129.

89. Harriss, "Untangling Our Traffic Tangles," 92; Marsh, "Traffic Control," 98–99; Miller McClintock, *Street Traffic Control*, 131–134; "Unified Traffic Control Tried in Los Angeles," *Electric Railway Journal* 63 (April 12, 1924), 579; L. D. Bale, "Synchronized Traffic Control is Costly to Street Railway," *AERA* 14 (Sept. 1925), 163–169; "Street Signalling a New Science," *Engineering News-Record* 95 (Oct. 1, 1925), 534; "Rule of Thumb Methods Will Not Solve Traffic Problem," *Electric Railway Journal* 69 (Feb. 5, 1927), 232.

90. Lewis W. McIntyre, "Mechanical Devices for Highway Traffic Regulation" (excerpts from a report of the National Committee on Mechanical Devices for Highway Traffic Regulation to the National Highway Traffic Association, 1923), *American City* 29 (Dec. 1923), 634–641 (638).

91. William M. Rumsey, in discussion, Sept. 30, 1921, *Proceedings of the National Safety Council*, Tenth Annual Safety Congress, Boston, Sept. 26–30, 1921, 615–616 (615).

92. *Titus v. Town of Bloomfield*, 80 Indiana Court of Appeals 483 (1923). See also " 'Silent Policeman' as Street Obstruction; Notice of Damage Claims," *American City* 30 (April 1924), 425.

93. "A Traffic Light That Stays in Place," *American City* 24 (April 1921), 443.

94. Sewall Paint and Glass Company, "Guiding a City's Traffic with Paint" (advertisement), *American City* 29 (Nov. 1923), advertising p. 83.

95. Mound Traffic Equipment Company, "Designers and Manufacturers of Special Traffic Equipment" (advertisement), *American City* 27 (Dec. 1922), advertising p. 68.

96. Federal Signal Company, "Federal Traffic Signal" (advertisement), *American City* 26 (March 1922), advertising p. 90.

97. Electrical and Specialty Supply Company, "Traffic Police Approve Mushroom Traffic Light—Milwaukee Type" (advertisement), *American City* 26 (May 1922), advertising p. 84.

98. John P. Fox, "Traffic Regulation in Detroit and Toronto," *American City* 13 (Sept. 1915), 175–179 (178). A 1918 reference to a "Traffic Mushroom" by Detroit's police department similarly specifies its function as a protector of streetcar safety zones; see W. P. Rutledge to Eno, July 27, 1918, in Eno, *Story of Highway Traffic Control*, 125–126 (126).

99. A. P. Child, "A Day and Night Street Indicator," *Scientific American* 121 (Sept. 27, 1919), 318.

100. "A Traffic Light That Stays in Place," *American City* 24 (April 1921), 443. Advertising in *The American City* is indicative of the rapid rise and fall of the Milwaukee mushroom; introduced in 1920, the mushroom was first advertised in the national journal in 1922. In 1922 and 1923 mushroom advertisements were abundant, but by the late 1920s they had vanished.

101. "A New Traffic Control Device," *American City* 30 (May 1924), 585, 587 (585).

102. Electrical and Specialty Supply Company, "Resolved . . ." (advertisement), *American City* 26 (Jan. 1922), 70 (advertising p. 76). The indestructibility of the mushroom was its most promoted feature; see also, e.g., Elkhart Foundry and Machine Company, "Don't Put Off" (advertisement), *American City* 27 (August 1922), advertising p. 66; Mound Traffic Equipment Company, "Officials Everywhere . . ." (advertisement), *American City* 27 (August 1922), advertising p. 66; "The Automatic Regulation of Traffic at Street Corners," *American City* 28 (Jan. 1923), 101.

103. "Mound Traffic Markers," *American City* 27 (July 1922), 95, 97 (95).

104. In *American City* 28 (Jan. 1923), see Safety Traffic Light Manufacturing Company, "No Legal Liability" (advertisement), advertising p. 137, and "Disappearing Safety Dome Traffic Light," p. 97.

105. Electrical and Specialty Supply Company, "Resolved . . ." (advertisement), *American City* 26 (Jan. 1922), 70 (advertising p. 76).

106. I have found only one advertisement that suggests that a mushroom-like device improves traffic flow: Line Material Company of Milwaukee claimed its "Traficons" promoted "the Safe and Expeditious Handling of traffic." See "Not a Device but a System," *American City* 30 (May 1924), advertising p. 105. By 1924, however, the mushroom business was already drying up.

107. See, e.g., Electrical and Specialty Supply Company, "Resolved . . ." (advertisement), *American City* 26 (Jan. 1922), 70 (advertising p. 76).

108. McClintock, *Street Traffic Control* (McGraw-Hill, 1925), 208.

109. "Tricks of Traffic Control," *Scientific American* 124 (June 25, 1921), 510, 518–519 (510).

110. "Traffic and the Law," *Scientific American* 130 (Jan. 1924), 18–19, 65–66 (66).

111. Burton W. Marsh, "Traffic Control," *Annals* 133 (Sept. 1927), 90–113 (90).

112. G. C. Schink, quoted in "Needless Handicaps to Police Regulation of Traffic," *American City* 36 (April 1927), 527.

113. W. B. Powell, "Should Police Departments Have Exclusive Control of Street Traffic?" (excerpt of a report to Francis X. Schwab, Mayor of Buffalo, Jan., 1926), *American City* 34 (June 1926), 699.

114. George C. Kelcey, "Traffic and Parking Regulations: As They Affect Public Safety and the Business Man," *City Manager Magazine* 8 (Sept. 1926), 22–27, 65 (27, 65). Kelcey believed that "the ultimate problem" was "to fit streets to traffic" (p. 65). This opinion came late in his career and did not influence his traffic recommendations.

115. George Kelcey, "How Intersections Figure in Traffic Accidents," *American City* 30 (April 1924), 417–419 (417).

116. James W. Bayless (traffic policeman, Venice, California), "Have We Too Many Traffic Laws?" *Scientific American* 133 (Nov. 1925), 318–319 (318).

117. Arthur Woods, "Keeping City Traffic Moving," *World's Work* 31 (April 1916), 621–632 (629).

## Chapter 3

1. Bessie E. Buckley to editor, in "Where Can the Kiddies Play? Send Views to *Journal*," *Milwaukee Journal*, Sept. 29, 1920.

2. "No Accident Week" (editorial), *Milwaukee Leader*, Sept. 25, 1920.

3. "Safety and the Motors" (editorial), *St. Louis Post-Dispatch*, Nov. 19, 1923.

4. Charles Hayes, "Auto Accidents Are Not Always Due to Drivers," *Chicago Tribune*, Jan. 11, 1920.

5. "Regulation of Street Traffic" (editorial), *American City* 13 (Sept. 1915), 174; John P. Fox, "Traffic Regulation in Detroit and Toronto," *American City* 13 (Sept. 1915), 176; Thomas Adams and Harland Bartholomew, "City Planning as Related to Public

Safety," *Proceedings of the National Safety Council* 7 (St. Louis, 1918): 363; Peter Kline, "Newspaper Publicity for Traffic Rules," *American City* 22 (Feb. 1920): 126–27; "Marking Traffic Zones," *American City* 25 (Dec. 1921): 531, 533; R. D. Ballew, "Solving the Parking Problem in a Small City," *American City* 27 (July 1922): 31; A. R. Hirst, "You Can Say 'Danger' Once Too Often," *American City* 27 (August 1922): 122; William M. Myers, "Traffic Lights and Zone Markers in Richmond," *American City* 28 (May 1923): 454; Miller McClintock, *Street Traffic Control* (New York, 1925), 212–13; "Autoists Must Drive Chalk Line in Loop District," *Chicago Tribune*, April 28 1923; George Earl Wallis, "Teaching Father Knickerbocker to Watch His Step," *National Safety News*, Oct. 1923, 11.

6. L. J. Smyth, "The Jay Walker Problem," *National Safety News* 2 (11 Oct. 1920), 11.

7. Stephen S. Tuthill, "The Attitude of the Motorist Toward Traffic Regulations," Oct. 2, 1923, *Proceedings of the National Safety Council, Twelfth Annual Safety Congress*, Buffalo, Oct. 1–5, 1923, 810–815 (811).

8. The traffic expert Miller McClintock summarized the common-law tradition in 1926: "The relation between drivers and pedestrians, where statutory enactments do not provide specifically to the contrary, is still governed by a Common Law principle which developed centuries ago. This ancient rule is that all persons have an equal right in the highway, and that in exercising the right each shall take due care not to injure other users of the way." *Report and Recommendations of the Metropolitan Street Traffic Survey* (Chicago Association of Commerce, 1926), 133. The best summary of the traditional law of roads and streets the author has found is William M. Maltbie, "The Law of the Highway," address, April 9, 1924, in *Proceedings of a Conference on Motor Vehicle Traffic*, New Haven, April 9–11, 1924, ed. R. Kirby (Yale University Press, 1924).

9. Miller McClintock, *Report and Recommendations of the Metropolitan Street Traffic Survey* (Chicago Association of Commerce, 1926), 134.

10. Robbins B. Stoeckel, "Rights of Pedestrian and Motorist," *Roads and Streets* 61 (June 4, 1924), 1243–1244 (1244).

11. Quoted in W. Bruce Cobb, "Automobile Accidents—Their Cause and Prevention," *American City* 21 (August 1919), 125–128 (125).

12. Frederick B. House, quoted in editorial, *New York Herald*, August 23, 1923, reprinted in Stephen S. Tuthill, "The Attitude of the Motorist Toward Traffic Regulations," Oct. 2, 1923, *Proceedings of the National Safety Council, Twelfth Annual Safety Congress*, Buffalo, Oct. 1–5, 1923, 810–815 (811).

13. The 1812 ruling of Britain's Lord Chief Justice Ellenborough that "Every unauthorized obstruction of a highway to the annoyance of the King's subjects is a nuisance" was a lasting common-law precedent still cited by American courts in the

1920s. See John Otey Walker, "The King's Highway," *Municipal Index* (1930), sec. 6, 314–320.

14. "Death Drivers" (editorial), Philadelphia *Evening Bulletin*.

15. Letter to editor ("Tyrrany of the Automobilists"), *St. Louis Star*, Nov. 23, 1923.

16. Letter to editor ("For an Anti-Auto Society," signed "Semaphore"), *St. Louis Star*, Nov. 12, 1923.

17. J. W. Davis to editor ("For a Pedestrians' Union"), *St. Louis Star*, Nov. 23, 1923.

18. W. H. P. to editor ("Calls Pedestrians to Protest"), *St. Louis Star*, Nov. 21, 1923.

19. Cobb, "Automobile Accidents—Their Cause and Prevention," *American City* 21 (August 1919), 125–128 (125).

20. "Motor Car Accidents," *American City* 28 (Feb. 1923), 189.

21. "Speeder Wants All Street," *Chicago Tribune*, May 6, 1913.

22. Magistrate Costello, Philadelphia, Nov. 8, 1924, quoted in "City-Wide War Is Declared on Death Drivers," *Public Ledger* (Philadelphia), Nov. 9, 1924.

23. Quoted in "Auto-Killings Now at the Rate of Two an Hour," *Literary Digest* 67 (Nov. 6, 1920), 80, 82, 86, 88 (82).

24. *Benison v. Dembinsky*, 241 Illinois Court of Appeals 530; see Myron M. Stearns, "Your Right to Cross the Street," *Outlook and Independent* 155 (May 14, 1930), 50–53, 80 (52).

25. Judge Charles L. Bartlett, address to the National Safety Council, Detroit, Oct. 26, 1926, as reported by A. Le Roy Hodges to the Metropolitan Section, American Electric Railway Association, New York, Dec. 3, 1926, in "Metropolitan Section," *AERA* 16 (Jan. 1927), 1070–1084 (1072).

26. "Calls Death-Truck Driver A 'Disgrace,' " Philadelphia *Evening Bulletin*, Nov. 11, 1924.

27. "Discretion Urged upon Traffic Police," *Philadelphia Inquirer*, Oct. 31, 1924.

28. C. W. Price, "Automotive Industry Should Lead in Safety Movement," *Automotive Industries* 49 (Dec. 13, 1923), 1187–1190 (1189).

29. "Joy Rider and Jay Walker" (editorial), *Post-Standard*, Nov. 25, 1913.

30. Synopsis of *New York Times* article in "Auto-Killings Now at the Rate of Two an Hour," *Literary Digest* 67 (Nov. 6, 1920), 80, 82, 86, 88 (82). The words quoted are those of the *Digest*'s synopsis.

31. Herbert L. Towle, "Financial Balm for Motor Victims," *The Outlook* 142 (March 1926), 459–462 (459).

32. William A. Searle, " 'Safety First' in Precept and Policy," *American City* 11 (Sept. 1914), 227–230 (228).

33. Press release, Public Safety Bureau, Westchester County, New York, 1924, quoted in "Cutting Out Accidents by Safety Propaganda," *Literary Digest* 82 (August 16, 1924), 22–23 (22).

34. "Death at the Wheel" (editorial), *Chicago Tribune*, Oct. 31, 1919.

35. "Preaching Self-Preservation," *Cleveland Plain Dealer*, Sept. 29, 1919. See also L. J. Smyth, "The Jay Walker Problem," *National Safety News* 2 (Oct. 11, 1920), 10–11 (10; "Public Safety Crusader"); George A. Davies to editor ("Auto Traffic"), *St. Louis Star*, Oct. 31, 1922 ("crusade").

36. Stephen S. Tuthill, "The Attitude of the Motorist toward Traffic Regulations," Oct. 2, 1923, *Proceedings of the National Safety Council, Twelfth Annual Safety Congress*, Buffalo, Oct. 1–5, 1923, 810–815 (810).

37. For a closer examination of jaywalking and its role in the social reconstruction of the city street, see Peter D. Norton, "Street Rivals: Jaywalking and the Invention of the Motor Age Street," *Technology and Culture* 48 (April 2007), 331–359.

38. Victor Appleton, *Tom Swift and His Motor-Cycle* (Grosset and Dunlap, 1910), 3–4.

39. Ira C. Judson, "In the Wake of the Automobile Joy Rider," *Sunday Morning Star*, Wilmington, May 2, 1909; "Publicity and Penalty" (editorial), *Providence Journal*, July 2, 1921 ("speed maniacs"). "Road hogs" was occasionally favored; in 1925 a British M.P. defined them as those who "run down pedestrians and attempt to escape detection by driving on." See J. M. Kenworthy, *Daily Mail*, quoted in "The Dangerous Art of Crossing the Road," *Public Opinion* 128 (August 14, 1925), 151–152 (152). For a study of "metaphors of aggressive motoring" in this era, see Kurt Möser, "The Dark Side of 'Automobilism,' 1900–30: Violence, War and the Motor Car," *Journal of Transport History* 24 (Sept. 2003), 238–258.

40. Judson, "In the Wake," 13; Allan Hoben, "The City Street," in *The Child in the City*, ed. S. Breckinridge (Chicago School of Civics and Philanthropy, 1912; Arno, 1970), 453.

41. "In Simple, Child-Like New York" *Kansas City Star*, April 30, 1911. The earliest use of the word in print that this author has found is an untitled one-sentence item in *Chicago Tribune*, 7 April 1909; it appeared eight years before the earliest reference noted in the *Oxford English Dictionary*. A 1921 definition retained the word's early sense: a jaywalker was "One who 'shortcuts' from the regular traffic routes,

pedestrian or motor"; see B. H. Lehman, "A Word-List from California," *Dialect Notes* 5, part 4 (1921), 109–114 (109).

42. " 'Jay' Walkers," *Washington Post*, May 18, 1913. For a second 1913 example, see "Joy Rider and Jay Walker" (editorial), Syracuse *Post-Standard*, Nov. 25, 1913. The *Post-Standard* placed most of the blame for pedestrian casualties on motorists.

43. "Police Department," *Washington Post*, Feb. 20, 1916.

44. "Middle-Block Crossing Is Defended," *New York Times*, Dec. 2, 1915; "Crossing at the Corners," *New York Times*, Nov. 11, 1915.

45. "Police Department Notes," *Washington Post*, Dec. 10, 1916.

46. Robert P. Utter, "Our Upstart Speech," *Harper's Magazine* 135 (June 1917), 66–72 (70), cited in *Oxford English Dictionary*, second edition, as the first known use of "jaywalker."

47. "Convention of National Safety Council Opens," *St. Louis Post-Dispatch*, Sept. 16, 1918.

48. Price, in "Discussion on Traffic Hazards," Sept. 29, 1921, *Proceedings of the National Safety Council, Tenth Annual Safety Congress*, Boston, Sept. 26–30, 1921, 591–594 (592). At least several cities (and perhaps many) enacted pedestrian control ordinances in the 1910s, but because few cities attempted enforcement they attracted little notice. See Frederick S. Crum, "Street Traffic Accidents," *Publications of the American Statistical Association* 13 (Sept. 1913), 473–528 (525).

49. "Rules for Pedestrians," *AERA* (Nov. 1919), p. 537.

50. J. A. Dickinson, in "Discussion on Traffic Hazards," Sept. 29, 1921, *Proceedings of the National Safety Council, Tenth Annual Safety Congress*, Boston, Sept. 26–30, 1921, 591–594 (592).

51. J. H. Truett, in "Discussion on Traffic Hazards," Sept. 29, 1921, *Proceedings of the National Safety Council, Tenth Annual Safety Congress*, Boston, Sept. 26–30, 1921, 591–594 (593).

52. Ibid., 594.

53. A. J. Driscoll (Safety First Association of Washington), quoted in " 'Jay-Walking' Banned," *Washington Post*, Oct. 21, 1917.

54. J. A. McRell, quoted in "Fines for Jay Walkers Help Safety in Newark," *New York Times*, August 12, 1923.

55. "Judge Assails Talk of 'Reckless Pedestrians,' " *Detroit News*, August 24, 1922.

56. Walter H. Douglas, a corner traffic policeman, quoted in Nick Woltjer, "Jay-Walking Campaign Is a Big Success," *Grand Rapids Herald*, June 12, 1921.

57. B. H. Lehman, "A Word-List from California," *Dialect Notes* 5, part 4 (1921), 109–114 (109).

58. Ivan H. Wise, "How a 'Safety First' Campaign Was Conducted," *American City* 10 (April 1914): 322–326 (326).

59. "Convention of the National Safety Council Opens," *St. Louis Post-Dispatch,* Sept. 16, 1918.

60. A. McKie Donnan, "The Human-Interest Safety First Campaign in San Francisco," *Ameriucan City* 23 (August 1920), 153–154 (154).

61. "Jay-Walkers Will Be Curbed To-day," *Providence Journal,* July 1, 1921; "Safety Week Still Leads 1920 Record," *Providence Journal,* July 2, 1921.

62. Perhaps the earliest example of this was in Cleveland during its 1919 safety week; see "Make Final Plans for 'Safety Week,'" *Cleveland Plain Dealer,* Sept. 26, 1919.

63. "Did You Get One of These?" *Grand Rapids Herald,* June 7, 1921.

64. "Jaywalkers Scarce as Safety First Crusade Makes Its Impression," *Grand Rapids Herald,* June 10, 1921.

65. "Safety Gospel Will Permeate City This Week," *Cleveland Plain Dealer,* Sept. 28, 1919.

66. Barron Collier, *Stopping Street Accidents: A History of New York City's Bureau of Public Safety* (New York, 1925), 114.

67. " 'Make April Safe' Was Philadelphia's Slogan," *Electric Railway Journal* 69 (May 7, 1927), 806–812 (811, see also 809).

68. "Wins Safety Cup," *Detroit News,* August 31, 1922.

69. "M. O. Eldridge Named Director of Traffic by Commissioners," *Washington Post,* March 21, 1925; "4 New Candidates for Traffic Chief Entered in Field," *Washington Post,* March 11, 1925.

70. "Eldridge to Ban 'Jaywalkers' Here; Backed by Chiefs," *Washington Post,* August 4, 1925.

71. "83 men and Women Pedestrians Seized in Drive by Police," *Washington Post,* Dec. 31, 1925.

72. "Moller Opposes 'Jay Walker' Term for Pedestrians," *Washington Post,* August 19, 1925.

73. "Walkers Ignoring Traffic Rule, Say Police Officials," *Washington Post,* Dec. 29, 1925.

74. 83 men and Women Pedestrians Seized in Drive by Police," *Washington Post*, Dec. 31, 1925.

75. J. C. Townsend, in "Discussion on Traffic Hazards," Sept. 29, 1921, *Proceedings of the National Safety Council, Tenth Annual Safety Congress*, Boston, Sept. 26–30, 1921, 591–594 (593).

76. Nick Woltjer, "Jay-Walking Campaign Is a Big Success," *Grand Rapids Herald*, June 12, 1921.

77. Charles L. Bartlett, speech to the Vortex Club, Detroit, August 23, 1922, quoted in "Judge Assails Talk of 'Reckless Pedestrians,' " *Detroit News*, August 24, 1922. See also "Saving Women, Children Bartlett's Idea of Justice," *Detroit News*, August 31, 1922.

78. George A. Davies to editor ("Auto Traffic"), *St. Louis Star*, Oct. 31, 1922 (emphasis added). For a rebuttal, see E. A. M. to editor ("Traffic a Two-Sided Question"), *St. Louis Star*, Nov. 3, 1922. As early as 1915, the *Kansas City Star* urged its readers: "Don't Be a Jay Driver," but the term did not catch on; see reprint in *Washington Post*, Oct. 10, 1915.

79. "Make the Streets Safe" (editorial), July 31, 1923.

80. "Theater," *Chicago Tribune*, March 2, 1925; H. B. H., letter to editor ("Jay Driving"), Nov. 1, 1926; *Chicago Tribune*, Nov. 10, 1926.

81. The term did persist (through survival or reinvention). In 1936 a New Yorker complained that "jay drivers are more of a menace than the jay-walker" (William Floyd to editor, July 8, 1936, *New York Times*, July 13, 1936).

82. *Webster's New International Dictionary of the English Language*, second edition (Merriam, 1936), 1331.

83. Herbert L. Towle, "The Motor Menace," *Atlantic Monthly* 136 (July 1925), 98–107 (99).

84. *The Practical Standard Dictionary of the English Language* (Funk & Wagnalls, 1924), 620. The editors did not classify the word as slang or colloquial, though some later dictionaries did. In 1936, in the first new edition of its unabridged dictionary since 1895, *Webster's* classified *jaywalk* as colloquial, and defined it as "To cross a street carelessly or at an unusual or inappropriate place, or in a dangerous, or illegal, direction, so as to be endangered by the traffic." *Webster's New International Dictionary of the English Language*, second edition (Merriam, 1936), 1331. The word appeared in several other English dictionaries in the 1930s; all used variations of the new, restricted definition.

85. For an excellent case study of this transition in Hartford, see Peter C. Baldwin, *Domesticating the Street: The Reform of Public Space in Hartford, 1850–1930* (Ohio State University Press, 1999), 214–225.

86. See synopsis (with excerpts) of Resor's address to a joint meeting of the New York University Men in Advertising and the Yale Men in Advertising, New York, in "Advertising Methods Recommended in Accident Prevention," *American City* 38 (May 1928), 92. The words quoted are from the synopsis, and are not necessarily Resor's exact words.

87. Julian H. Harvey quoted in "Rochester's Notable Public Safety Campaign," *American City* 19 (Nov. 1918), 363–366 (366).

88. Roadifer, "Rejecting the Grundys from Safety First," Sept. 27, 1921, *Proceedings of the National Safety Council, Tenth Annual Safety Congress*, Boston, Sept. 26–30, 1921, 249–253 (251). Mrs. Grundy is an off-stage character in the play *Speed the Plough* (1798) by the English playwright Thomas Morton (1764?–1838). Later her name entered the English language (in a small way) as a synonym for a cheerless moralizer. There may also be influence from the hapless Solomon Grundy, a character in a nursery rhyme first published in a collection by James Orchard Halliwell, *The Nursery Rhymes of England* (1842); see Iona and Peter Opie, *The Oxford Dictionary of Nursery Rhymes* (Clarendon, 1951, 1952), 392.

89. Roadifer, "Eject the Grundys" (abstract of address of Sept. 27, 1921), *Electric Railway Journal* 58 (Oct. 22, 1921), 745–746 (746).

90. "Flag Indicates No Child Deaths from Automobile Accidents," *American City* 36 (April 1927), 545.

91. Ruth Streitz, *Safety Education in the Elementary School: A Technique for Developing Subject Matter* (New York: National Bureau of Casualty and Surety Underwriters, 1926), 25.

92. "Safety-First Instruction to Children," *American City* 13 (Oct. 1915), 341.

93. "Children's Chum Is This Policeman," *Cleveland Plain Dealer*, Oct. 1, 1919.

94. "Safety First" (letter to editor, signed "Springfield Republican"), *St. Louis Star*, Sept. 25, 1918.

95. Ruth Streitz, *Safety Education in the Elementary School: A Technique for Developing Subject Matter* (New York: National Bureau of Casualty and Surety Underwriters, 1926), 25; "Cites Value of Training in Safety," *Evening Sentinel* (Milwaukee), home edition, Sept. 29, 1920; "Schools Must Teach Safety, Says Potter," *Milwaukee Journal*, Sept. 29, 1920; "Street Safety Education in the Schools," *American Review of Reviews* 65 (March 1922), 326–327; E. George Payne, "How the St. Louis Schools Have Reduced Accidents," *American City* 27 (Nov. 1922), 437–438.

96. "Adopts Plan for Safety Teaching," *Cleveland Plain Dealer*, Oct. 5, 1919; C. W. Price, "Report of the General Manager," Sept. 26, 1921, *Proceedings of the National Safety Council, Tenth Annual Safety Congress*, Boston, Sept. 26–30, 1921, 11–15 (15).

97. Harriet E. Beard, "Teaching Accident Prevention," *American City* 24 (March 1921), 256–257; "Teaching Safety and Responsibility in Public Schools," *Detroit Educational Bulletin*, reprinted in *American City* 28 (Jan. 1923), 4. See also Beard, *Safety First for School and Home* (Macmillan, 1924).

98. Brown, "Street Safety Education for School Children," Oct. 5, 1923, *Proceedings of the National Safety Council, Twelfth Annual Safety Congress*, Buffalo, Oct. 1–5, 1923, 863–866 (863–864).

99. A. C. Brown quoting his daughter in "Make Final Plans for 'Safety Week,'" *Cleveland Plain Dealer*, Sept. 26, 1919.

100. Waldo, *Safety First for Little Folks* (Charles Scribner's Sons, 1918), esp. chapters 1 and 3 (1–11, 22–28). Waldo's book includes also a true account of an accident (not in traffic) in which a boy lost a leg, complete with a photograph of the boy (25–28). Baum's influence is evident in the name of the book's principle character: Dotty. See also Safety First League, *Safety First for Children: A Book of Story and Verse Intended to Help in the Work of Saving Human Life* (St. Louis: Garrison Wagner, 1915).

101. George I. Brinkerhoff and Celena Rowe, *Safety First Stories* (Longmans, Green and Co., 1928), 2.

102. W. H. Musselman, "Safety Arithmetic Problems," *Bulletin of the Education Section of the National Safety Council* 1 (April 1, 1924), 5.

103. In *Bulletin of the Education Section of the National Safety Council* 1 (April 1, 1924), see "Safety Posters for School Room Use" (advertisement), p. 2, and "One More Contest—But a Different Kind," p. 3.

104. Based on data from Chicago, 1925. Marian L. Telford, "How Children Are Hurt by Motor Vehicles in Chicago," National Safety News 14 (Oct. 1926), 18–19 (18).

105. "Irene Swartz First Winner in Slogan Contest," *Sunday Sentinel* (Milwaukee), Oct. 3, 1920. According to the Sentinel, judges included this slogan among those deemed "excellent," but lacking " 'dignity.' " The winning slogan was "Danger Stalks Where Safety Balks." A St. Louis parent faulted landlords who barred children from playing on lawns; see letter to editor, "Landlords and Auto Deaths," *St. Louis Star*, Oct. 22, 1922.

106. "Street Accidents," *New York Times*, Dec. 2, 1915.

107. C. C. White to editor, "Safety First Rules," *St. Louis Star*, Sept. 23, 1918.

108. Bessie E. Buckley, in "Where Can the Kiddies Play? Send Views to *Journal*," *Milwaukee Journal*, Sept. 29, 1920.

109. Mrs. John Belavek and George A. Walters quoted in "2 Girls Spread Safety Gospel," *Detroit News*, August 25, 1922 (emphasis added).

110. Beard, "Teaching the Child to Be Careful," Oct. 5, 1923, *Proceedings of the National Safety Council, Twelfth Annual Safety Congress*, Buffalo, Oct. 1–5, 1923, 858–861 (859).

111. On the playground movement before it was an element in the traffic safety movement, see Cary Goodman, *Choosing Sides: Playground and Street Life on the Lower East Side* (Schocken Books, 1979) and Dom Cavallo, "Social Reform and the Movement to Organize Children's Play During the Progressive Era," *History of Childhood Quarterly: The Journal of Psychohistory* 3 (Spring 1976), 509–522. For valuable, brief, contemporary introductions to the earlier playground movement from the perspective of its promoters, see three articles in *American City*: Henry S. Curtis, "The Growth, Present Extent and Prospects of the Playground Movement in America" (address to the Conference on Child Welfare, Clark University, 1909), in volume 1 (Sept. 1909), 27–33; Lerbert H. Weir, "The Playground Movement in America," in volume 6 (March 1912), 577–580, and H. S. Braucher, "Growth and Wealth of the Playground Movement" (abstract of a report to the annual convention of the Playground and Recreation Association of America, Grand Rapids, Michigan, Oct., 1916), in volume 15 (Dec. 1916), 645. See also Everett B. Mero, ed., *American Playgrounds* (Civic Press, 1914).

112. Thomas Adams and Harland Bartholomew, "City Planning as Related to Public Safety" (paper read before the Public Safety Division of the National Safety Council, Seventh Safety Congress, Sept. 17, 1918), *Proceedings of the National Safety Council* (St. Louis, 1918), 355–364 (362).

113. For three examples of playgrounds opened by local business organizations in 1922, see T. A. Stevenson, "Klamath Falls Playground Financed by Chamber of Commerce," *American City* (Dec. 1922), 549; O. O. McLeish, "Playgrounds Equipped and Supervised," *American City* 28 (April 1923), 397 (Findlay, Ohio); Walter Harrison, "Recreation Park Developed and Maintained by Rotary Club," *American City* 28 (June 1923), 602–603 (Oklahoma City).

114. "Frolic in Safety of City Playgrounds," *Detroit News*, August 30, 1922.

115. "The equal rights of motor . . .": "Massachusetts Playgrounds Teach Safety," *American City* 25 (Sept. 1921), 224–225 (224); "to observe and obey . . .": Weaver Weddell Pangburn, "Playgrounds, a Factor in Street Safety," *Annals* 133 (Sept. 1927), 178–185 (183); rules: "A Traffic Game for Children" (reprinted from National Bureau of Casualty and Surety Underwriters, Bulletin of Safety Education), *American City* 33, Nov. 1925, 500; Patrick F. Shea, *Accident Prevention Through Education in the Elementary Schools* (Boston: D. C. Heath, 1928), 66–68. A similar game was devised in Cleveland in the late 1910s by Louise C. Wright; see the description by Herbert Buckman quoted in "Schools Teaching Pupils to Avoid Speeding Cars," *Literary Digest* 68 (Jan. 8, 1921), 58, 62 (58).

116. Howard George Co., Philadelphia, "Specify Apex Apparatus" (advertisement), *American City* 26 (Jan. 1922), advertising p. 42.

117. "Public Playgrounds," Everwear Manufacturing Company, Springfield, Ohio, 1922; see also "How to Create Interest in Public Playgrounds," *American City* 27 (Dec. 1922), 573, 575. There was a sharp increase in playground equipment advertisements appealing to traffic safety in 1922; see esp. 1922 issues of *American City*.

118. See esp. Page Fence and Wire Products Association (Chicago), "Page Protection Fence" (advertisement), *American City* 27 (Sept. 1922), advertising p. 74; see also advertisements: Anchor Post Iron Works (New York), "A Vacation-Time Job That Insures School-Time Safety," *American City* 35 (July 1926), 136; Cyclone Fence Company (Waukegan, Ill. ), "Keep the Children Safe!" *American City* 36 (May 1927), 728. Before 1922 fence manufacturers almost never sold their products as traffic safety devices, but beginning that year the method grew suddenly common. A rare exception is a special fence sold by the Cincinnati Iron Fence Company, which in 1915 promoted it for playgrounds as a way "to protect the children from runaway accidents" (that is, from horses); see "A New Type of Iron Fence," *American City* 12 (April 1915), 369.

119. J. H. Chase, "City Streets Reserved for Coasting," *American City* 23 (Dec. 1920), 609.

120. "Traffic Danger at 2 Schools," *Milwaukee Journal*, Sept. 27, 1920: "Police officials say the force is inadequate to have a man at every school every day."

121. Harriet E. Beard, "Teaching Accident Prevention," *American City* 24 (March 1921), 256–257; "Teaching Safety and Responsibility in Public Schools," *Detroit Educational Bulletin*, reprinted in *American City* 28 (Jan. 1923), 4; "Schools Teaching Pupils to Avoid Speeding Cars," *Literary Digest* 68 (Jan. 8, 1921), 58, 62 (58).

122. Jessica P. McCall, "The Safety Crusade in Brooklyn," *American City* 12 (April 1915), 305–308.

123. Roland B. Woodward, "Organizing Rochester's School Boys for Accident Prevention," *American City* 13 (Sept. 1915), 208–209; Idabelle Stevenson, *Safety Education* (A. S. Barnes, 1931), 26.

124. J. P. Freeman, "Report of the Boy Scouts of America to the National Safety Council," Sept. 29, 1921, *Proceedings of the National Safety Council, Tenth Annual Safety Congress*, Boston, Sept. 26–30, 1921, 589–591 (589–590).

125. During Milwaukee's 1920 safety week, for example, 1,800 boys were organized as "Safety Cadets," which were, "In effect, . . . a junior membership of the National Safety Council." See "Schools Enlist Safety Cadets," *Evening Sentinel* (Milwaukee), home edition, Sept. 28, 1920. See also Louis Resnick, "Milwaukee Conducts Successful 'No Accident Week,' " *American City* 23 (Nov. 1920), 531, 533, 535 (533, 535).

126. Montague A. Tancock, "Omaha Volunteer Traffic Officers Prove Valuable," *American City* 27 (August 1922), 169, 171 (169).

127. "Schools Teaching Pupils to Avoid Speeding Cars," *Literary Digest* 68 (Jan. 8, 1921), 58, 62; Harriet Beard in discussion, Sept. 29, 1921, *Proceedings of the National Safety Council, Tenth Annual Safety Congress*, Boston, Sept. 26–30, 1921, 590–591; E. S. Martin, "Educating the Public in Safety," *American City* 36 (Feb. 1927), 179–181 (179–180).

128. "Flint to Educate Boys in Traffic Control," *Michigan Roads and Pavements* 21 (Jan. 3, 1924), 16.

129. Washington's school traffic squads, organized in 1922, appear to have fallen in this class. See "Washington Safer, Campaign Showing," *Washington Post*, Nov. 28, 1922.

130. G. H. McClain, in discussion, Sept. 28, 1921, *Proceedings of the National Safety Council, Tenth Annual Safety Congress*, Boston, Sept. 26–30, 1921, 583–584 (583).

131. Marcus A. Dow, "The Public Conscience and Accidents," *American City* 28 (June 1923), 556–558; Idabelle Stevenson, *Safety Education* (A. S. Barnes, 1931), 5–7, 42–44; Harry W. Gentles and George H. Betts, *Habits for Safety: A Text-Book for Schools* (Indianapolis: Bobbs-Merrill, 1932), 13.

132. Syracuse Chamber of Commerce, Automobile Club of Syracuse, and Syracuse Police Department, "Safety First: a Public Spirited Campaign to Reduce Accidents" (advertisement), *Post-Standard* (Syracuse), Dec. 1, 1913.

133. In a press release issued by the Washington affiliate of the AAA on Nov. 7, 1915, there is evidence of a proposal for traffic vigilantes by the Automobile Club of St. Louis in 1913. The release itself is now probably lost, but its contents appear in the *Washington Post* (under "Stop Careless Driving") and the *Washington Star* (under "Auto Club News: District of Columbia") for Nov. 7, 1915, or see clippings in Seiler Scrap Book, American Automobile Association Headquarters Library, Heathrow, Florida.

134. James E. Wales, "A City's Volunteer Police Force," *American City* 16 (Feb. 1917), 201, 203. The words quoted are Wales's.

135. Pierce Atwater, "A Phase of Public Safety," *American City* 23 (Sept. 1920), 321, 323, 325, 327; R. C. Haven, in discussion, Sept. 30, 1921, in *Proceedings of the National Safety Council*, Tenth Annual Safety Congress, Boston, Sept. 26–30, 1921, 610.

136. T. M. Ford, "Prevention of Street Accidents," *American City* 23 (Dec. 1920), 604–605 (Newark, N.J.); "Citizens to Help Curb Wild Drivers," *Detroit News*, August 27, 1922; "Mayor Enlists Volunteer Traffic Officers," *American City* 31 (July 1924), 67 (Columbia, S.C.).

137. On the organization of traffic vigilantes by local safety councils, see C. W. Price, H. A. Reninger, F. C. Hill, S. M. Lippincott, and D. R. Faries, discussion, Sept. 28, 1921, in *Proceedings of the National Safety Council, Tenth Annual Safety Congress*, Boston, Sept. 26–30, 1921, 357–360.

138. "Campaigns to Reduce Automobile Accidents," *American City* 28 (May 1923), 507.

139. D. R. Faries (Automobile Club of Southern California, Los Angeles), Sept. 28, 1921, *Proceedings of the National Safety Council, Tenth Annual Safety Congress*, Boston, Sept. 26–30, 1921, 359.

140. Pierce Atwater, "A Phase of Public Safety," *American City* 23 (Sept. 1920), 321, 323, 325, 327 (325).

141. Charles M. Anderson, in discussion, Sept. 30, 1921, in *Proceedings of the National Safety Council, Tenth Annual Safety Congress*, Boston, Sept. 26–30, 1921, 609–610 (610).

142. Lee, "Success in Vigilante Work," Sept. 30, 1921, *Proceedings of the National Safety Council, Tenth Annual Safety Congress*, Boston, Sept. 26–30, 1921, 607–609. An abridgement of this address is available as Lee, "Traffic Vigilantes in St. Louis," *American City* 26 (April 1922), 339–340. See also "How Other Cities Regulate Traffic," *Grand Rapids Herald*, June 5, 1921 ("in Omaha . . . 200 business men make up a force of volunteer citizens to help enforce the city's traffic laws"), and "Citizens to Help Curb Wild Drivers," *Detroit News*, August 27, 1922.

143. G. H. McClain, in discussion, Sept. 30, 1921, *Proceedings of the National Safety Council, Tenth Annual Safety Congress*, Boston, Sept. 26–30, 1921, 613.

144. "Stop, Be Careful!—Read the Label," *Milwaukee Journal*, Sept. 26, 1920. Precisely the same slogan was circulated in a 1918 St. Louis safety campaign; see "The Automobile and the Child," *St. Louis Star*, Sept. 18, 1918.

145. "Crippled Children in Safety Parade," *New York Times*, Oct. 13, 1922.

146. See National Safety Council Figures in Bureau of the Census, *Historical Statistics of the United States, Colonial Times to 1970*, part 2 (Washington: Government Printing Office, 1975), 719–720. Figures before 1918 are more the product of guesswork than record keeping, but the 1914–1917 period clearly suffered a higher rate of growth in the death rate as the automobile evolved from expensive novelty to mass consumer good.

147. On the sales decline and the "saturation crisis," see chapter 6.

148. "Auto Dealers of City Want Safe Driving of Cars, Fahrenkrog Asserts," *St. Louis Star*, Oct. 22, 1922.

149. C. H. Hites, "Thomas P. Henry of Detroit Elected President of A.A.A.," *American Motorist* 15 (June 1, 1923), 1–3, 13 (13). The words quoted are from Hites's synopsis the remarks of the speaker, James Inches.

150. "Detroit Newspapers Seek to Solve Traffic Problem," *Automotive Industries* 48 (June 7, 1923), 1260.

151. Price, "Automotive Industry Should Lead in Safety Movement," *Automotive Industries* 49 (Dec. 13, 1923), 1187–1190.

152. Ibid. On short-haul shippers reverting to horse-drawn vehicles, see also "Dobbin Is Coming Back, Say Draymen," *Milwaukee Journal*, August 31, 1920.

153. Undated letter, Clarence R. Basht (president, Masillon Automobile Dealers, Masillon, Ohio) to Coolidge, received April 22, 1924, in file "Automobiles," box 39, Commerce Papers, Hoover Library. Basht was not specific about the source or nature of the "unjust propaganda."

154. Price, "Automotive Industry Should Lead," 1187–1190. See also Price, "Automotive Industry Should Stimulate Safety Campaigns," *Automotive Industries* 50 (Jan. 3, 1924), 17–19. Price's words show that he was, implicitly, a social constructivist. He was telling his audience that street traffic problem was in stage of "interpretive flexibility."

155. Editorial summary of the position of H. W. Slauson, in Slauson, "A New Plan for Traffic Laws," *Scientific American* 132 (May 1925), 296–298 (297).

156. Robert L. Cusick, "Continued Growth of Motor Accidents a Threat to Industry," *Automotive Industries* 52 (April 23, 1925), 730–733 (730).

157. Elkins, "Safety Work for Motor Clubs," May 1, 1925, American Automobile Association, *Conference of Club and Association Secretaries*, Washington, April 30–May 1, 1925, 150–156 (151).

158. Janes, address, April 30, 1925, American Automobile Association, *Conference of Club and Association Secretaries*, Washington, April 30–May 1, 1925, 34–39 (36–37).

159. Ibid., 37.

160. "Kettig Says Auto Killed Only Few of 25 Children," *Louisville Herald*, June 4, 1923; "Monument to Auto's Youthful Toll," *Louisville Herald*, June 3, 1923.

161. Dean R. L. McCready, quoted in "Tragedies to 25 Children Are Recalled," *Louisville Herald*, June 5, 1923.

162. Louis De Armand, "Report of the Publicity Committee," Sept. 29, 1921, *Proceedings of the National Safety Council, Tenth Annual Safety Congress*, Boston, Sept. 26–30, 1921, 600–601 (601).

163. "Born to the U.S.A.—A New and Gigantic A.A.A.," *American Motorist* 16 (August 1924), 24–25, 34, 36 (24).

164. The "gyp club" problem was a frequent topic at AAA conventions; see esp. H. J. Donnelly Jr., "How to Handle 'Gyp' Clubs," May 1, 1925, American Automobile Association, *Conference of Club and Association Secretaries*, Washington, April 30–May 1, 1925, 142–150.

165. "Kiel's Safety Queries Ill-Advised, Auto Club of Missouri Declares, in Submitting Its Own Answers," *St. Louis Star*, Nov. 20, 1923; "Auto Club Officials Take Exception to 'Safe' Questionnaire," *St. Louis Globe-Democrat*, Nov. 21, 1923.

166. Louis Resnick, "Milwaukee Conducts Successful 'No Accident Week,' " *American City* 23 (Nov. 1920), 531, 533, 535 (533); "A No-Accident Week," *The Outlook* 126 (Dec. 1, 1920), 608–609 (609).

167. Charles M. Anderson, in discussion, Sept. 30, 1921, *Proceedings of the National Safety Council, Tenth Annual Safety Congress*, Boston, Sept. 26–30, 1921, 618.

168. "Support Safe Drivers' Club," *Milwaukee Journal*, August 31, 1920 (dealers made members); "Safe Drivers' Club Cordially Indorsed," *Milwaukee Journal*, Sept. 3, 1920 (motorist endorses club).

169. Carl L. Smith, in discussion, Sept. 30, 1921, *Proceedings of the National Safety Council, Tenth Annual Safety Congress*, Boston, Sept. 26–30, 1921, 616–617. On the Safe Drivers' Club in Erie, Pennsylvania, see S. M. Lippincott, "The Story of a Volunteer Local Council," *National Safety News* 9 (Jan. 1924), 23–24.

170. Charles B. Scott, "Report of Committee on Local Councils," August 28, 1922, *Proceedings of the National Safety Council, Eleventh Annual Safety Congress*, Detroit, August 28–Sept. 1, 1922, 21–23 (22).

171. Carl L. Smith, in discussion, Sept. 30, 1921, *Proceedings of the National Safety Council, Tenth Annual Safety Congress*, Boston, Sept. 26–30, 1921, 617.

172. "Kiel's Safety Queries Ill-Advised, Auto Club of Missouri Declares, in Submitting Its Own Answers," *St. Louis Star*, Nov. 20, 1923; "Auto Club Officials Take Exception to 'Safe' Questionnaire," *St. Louis Globe-Democrat*, Nov. 21, 1923.

173. On safety council driver education programs, see esp. C. W. Price's comments in discussion, "Wednesday Afternoon Session," Sept. 28, 1921, *Proceedings of the National Safety Council, Tenth Annual Safety Congress*, Boston, Sept. 26–30, 1921, 352–364 (355–357). See also "National Safety Council Holds Campaign Against Public Accidents," *American City* 28 (March 1923), 250, and "Teaching Safety to Drivers," *American City* 28 (May 1923), 521.

174. See Herbert J. Stack, *History of Driver Education in the United States* (Washington: National Education Association, 1966), esp. 8–9, 15–16, 49–50, 57, 61.

175. See transcript of spoken exchange between Johnstone and Ernest N. Smith, April 30, 1925, in American Automobile Association, *Conference of Club and Association Secretaries*, Washington, April 30–May 1, 1925, 43–44.

176. "No Accident Week" (editorial), *Milwaukee Leader*, Sept. 25, 1920.

177. "Reducing the Railroad Death List," *World's Work* 22 (Sept. 1911), 14797–14799 (14798); Ralph C. Richards, address to the First Co-operative Safety Congress, Milwaukee, Sept. 30–Oct. 5, 1912, quoted in C. W. Price, "Campaigning for Safety," *The Survey* 29 (Nov. 23, 1912), 222–225 (222).

178. Price, "Startling Figures for Automobile Fatalities," *American City* 25 (Sept. 1921), 205–206.

179. Safety and Sanitation Division, Milwaukee Association of Commerce, "The Days of 'Old Dobbin' Have Passed" (advertisement), *Milwaukee Sentinel*, Sept. 27, 1920.

180. "Brand Auto Chief Peril to Safety," *Evening Sentinel* (Milwaukee), home edition, Sept. 28, 1920.

181. *Post-Dispatch*, "Safety and the Motors" (editorial), Nov. 19, 1923. See also "The Murderous Motor," *New Republic* 47 (July 7, 1926), 189–190: "the mounting death toll is an adequate reminder to everyone, including the automobile industry itself, that the time is near at hand when serious and even drastic remedies must be applied" (190).

182. "Safety and the Motors." For an earlier example of a demand for such equipment, see "What Shall Be the Cure for Automobile Speed Mania?" *Illustrated World* 34 (Sept. 1920), 85–86 (86).

183. H. J. Schneider to editor ("Gear Down the Autos"), *St. Louis Star*, Nov. 24, 1923; R. F. to editor ("Governors on Autos"), *Star*, Nov. 19, 1923.

184. "Causes of Traffic Accidents—As Seen By Chiefs of Police," *American City* 34 (May 1926), 542. The survey was conducted by *American City*; 318 chiefs favored governors (speed limit not designated), 161 opposed. Chiefs suggested limits from 20 to 55 miles per hour, 35 was the approximate average.

185. "If You Do Not Help Regulate Traffic, Mr. Dealer, Some Day Traffic Will Regulate Your Bank Account," *Motor Age*, Nov. 1, 1923, 58.

186. Bob Beiser, "Ridicule Would Come to City," *Cincinnati Enquirer*, Oct. 28, 1923 (Automobile Section).

187. The most important later examples of such names are the Bureau for Street Traffic Research, founded by Studebaker in 1925, and the Automotive Safety Foundation, founded by the Motor Vehicle Manufacturers Association in 1937. The Automotive Safety Foundation was the descendent of an earlier manufacturers'

committee on safety that was formed in the wake of the Cincinnati speed governor fight (see below).

188. Cincinnati Automobile Dealers Association to "Owner," Oct. 17, 1923, facsimile in "If You Do Not Help Regulate Traffic, Mr. Dealer, Some Day Traffic Will Regulate Your Bank Account," *Motor Age*, Nov. 1, 1923, 58.

189. "Business Would Suffer," *Cincinnati Enquirer*, Nov. 6, 1923.

190. "Ordinance Defeated 7 to 1," *Cincinnati Enquirer*, Nov. 7, 1923.

191. "If You Do Not Help Regulate Traffic, Mr. Dealer, Some Day Traffic Will Regulate Your Bank Account," *Motor Age*, Nov. 1, 1923, 58.

192. Reeves, address to the Cincinnati Automobile Dealers Association, Cincinnati, Oct. 1923, quoted in "Accidents Can Be Prevented," *Cincinnati Enquirer*, Oct. 21, 1923 (Automobile Section).

193. "Just Among Ourselves," *Automotive Industries* 49 (Nov. 8, 1923), 940–941 (940); "NACC Creates Safety Committee," *Automotive Industries* 49 (Nov. 8, 1923), 977.

194. "Personnel of Safety Committee Appointed," *Automotive Industries* 49 (Nov. 29, 1923), 1122.

195. George E. Mix, "Law Enforcement and Its Application," *Proceedings of the National Safety Council, Ninth Annual Safety Congress*, Chicago, 1920, 337, cited in McClintock, *Street Traffic Control* (McGraw-Hill, 1925), 88n. 1.

196. Ray W. Sherman, *If You're Going to Drive Fast* (Crowell, 1935), 15. See also Miller McClintock, "Readjustment Is Necessary," *Los Angeles Times*, July 20, 1924: ". . . motor vehicles whose most valuable characteristic is their ability to move with speed."

197. J. Maxwell Smith, " 'Be Careful' Campaigns," in National Motorists Association, "Report on the National Convention of Automobile Club Officials," Cleveland, Sept. 20–22, 1923, AAA Headquarters Library.

198. See transcript of spoken exchange between Smith and Norman Johnstone, April 30, 1925, in American Automobile Association, *Conference of Club and Association Secretaries*, Washington, April 30–May 1, 1925, 43–44.

199. Lawrence L. Jewell, "Putting Horse Sense into Control of Traffic," *American Motorist*, June 1925, 8–10, 36 (9).

200. Hoover to Smith, August 12, 1926, in file "American Automobile Association," box 21, Commerce Papers, Hoover Library.

201. Smith, in American Automobile Association, *Third Conference of Club and Association Secretaries*, Washington, March 23–25, 1927, 215, 218.

## Chapter 4

1. *Engineering News-Record* 93 (Oct. 16, 1924), 615.

2. For a good overview of the development of city services, see Joel A. Tarr and Joseph Konvitz, "Patterns in the Development of the Urban Infrastructure," Howard Gillette Jr. and Zane L. Miller, *American Urbanism: A Historiographical Review* (Greenwood, 1987), 195–226

3. Ross, *Social Control* (Macmillan, 1901).

4. On social control and on E. A. Ross, see esp. Dorothy Ross, *The Origins of American Social Science* (Cambridge University Press, 1991), chapter 7 (219–256). Dorothy Ross is far from the first to see Progressive Era social scientists as seeking to impose "social control" on a society unsettled by industrialization, but she returned the phrase to its original progressive setting, apart from the Marxian context it received from later writers; see esp. 247–253.

5. On the development of an "ideology of engineering" applicable to social problems, see Edwin T. Layton Jr., *The Revolt of the Engineers: Social Responsibility and the American Engineering Profession* (Press of Case Western Reserve University, 1971), chapter 3 (53–78).

6. *Engineering News-Record* 93 (Oct. 16, 1924), 615. *Engineering News-Record* was a voice for civil engineers, some of whom opposed regulatory traffic control and favored the accommodation of increasing motor traffic through large reconstruction projects. The editorial quoted faults engineers for seeking to accommodate human nature to physical circumstances, instead of the other way around. The view faulted, however, was the dominant one among engineers involved in traffic control.

7. The term "social organization" was used in this sense in 1915 by Edwin Anderson Alderman, president of the University of Virginia, in an address. See "Can Democracy Be Organized?" in *National Ideals and Problems*, ed. M. Fulton (Macmillan, 1918), 334–335.

8. J. Rowland Bibbins, "Traffic-Transportation Planning and Metropolitan Development," *Annals* 116 (Nov. 1924), 205–214 (212).

9. Editorials in *Engineering News-Record* show understandable resistance among some civil engineers to the substitution of regulatory for building methods; see especially "Motor Killings and the Engineer," volume 89 (Nov. 9, 1922), 775; "Capitalizing Traffic Congestion Cost," volume 91 (Nov. 8, 1923), 747–748; "For a New View of Traffic Control," volume 93 (Oct. 16, 1924), 615. Yet many civil engineers grew accustomed to the use of regulatory measures; see professional biographies in John William Leonard, *Who's Who in Engineering* (Who's Who Publications; first edition 1922, second edition 1925).

10. Charles Evan Fowler, "The Traffic Jam," *Roads and Streets*, Jan. 1924, 58–60 (59).

11. Fred W. Powell, "The Problem of Additional Sources of City Revenue," *American City* 16 (Jan. 1917), 31–34 (31); see also "The Increased Cost of City Government," *American City* 23 (August 1920), 177–179 (177).

12. Powell, "The Problem of Additional Sources of City Revenue," 32.

13. Mark H. Rose and John G. Clark, "Light, Heat, and Power: Energy Choices in Kansas City, Wichita, and Denver, 1900–1935," *Journal of Urban History* 5 (May 1979), 340–364 (359).

14. Business associations were crucial in the history of many aspects of city life in the twentieth century, but they have received relatively little attention from historians. Robert M. Fogelson has gone far to correcting this in his recent book *Downtown: Its Rise and Fall, 1880–1950* (Yale University Press, 2001). Fogelson shows the distinct role of central business district business associations, which saw in the traffic crisis a threat to the economic future of city centers. See esp. 208–213.

15. American City Bureau, "Why Your City Needs the City Planning Exhibition" (advertisement), *American City* 10 (March 1914), advertising p. 80; American City Bureau, "Another National Record in Chamber of Commerce Development" (advertisement), *American City* 13 (Dec. 1915), advertising p. 30.

16. "The Reorganization of the Syracuse Chamber of Commerce," *American City* 10 (Jan. 1914), 47–49 (47).

17. The quotation refers to state municipal leagues; Richard R. Price, "Scope and Proper Limitations of Leagues of Municipalities," *American City* 10 (Jan. 1914), 29–31 (29).

18. Ralph Howard, "Fifth Avenue's Traffic Tower," *Scientific American* 124 (Jan. 15, 1921), 45, 57–58; "Tricks of Traffic Control," *Scientific American* 124 (June 25, 1921), 510, 518–519; Henry Collins Brown, *Fifth Avenue Old and New* (New York: Fifth Avenue Association, 1924), 116.

19. Carlos C. Campbell, "Knoxville Board Installs Street Signs," *American City* 26 (May 1922), 474–475.

20. C. F. Carter, "Chamber Promotes Street Lighting Installation," *American City* 25 (Sept. 1921), 249.

21. George C. Kelcey, "Traffic and Parking Regulations: As They Affect Public Safety and the Business Man," *City Manager Magazine* (International City Managers Association) 8 (Sept. 1926), 22–27, 65 (26).

22. *Efficiency and Uplift: Scientific Management in the Progressive Era* (University of Chicago Press, 1964), esp. x.

23. J. D. Cloud and Company, "The Stockholders of a City Are Its Citizens" (advertisement), *American City* 14 (June 1916), 644.

24. Secretary Perkins of the Sac City Commercial Club, quoted in Fred M. Hansen, "A Town Commercial Club Which Gets Results," *American City* 10 (March 1914), 259–261 (259); "The 'Commercializing' of Civic Movements" (editorial), *American City* 10 (April 1914), 321.

25. Fred W. Powell, "The Problem of Additional Sources of City Revenue," *American City* 16 (Jan. 1917), 31–34 (31).

26. Hays, "The Politics of Reform in Municipal Government in the Progressive Era," *Pacific Northwest Quarterly* 55 (Oct. 1964), 157–169 (159).

27. White, *The City Manager* (Chicago, 1927), ix–x; quoted in Hays, "Politics of Reform," 159.

28. J. D. Cloud and Company, "The Stockholders of a City Are Its Citizens" (advertisement) *American City* 14 (June 1916), 644; Blankenburg quoted in "City Manager Form Urged for Philadelphia," *American City* 14 (Feb. 1916), 194.

29. White, quoted in Hays, "Politics of Reform," 159.

30. H. W. Dodds, "Thumb-Nail Sketches of the Four Principle Types of City Government," *American City* 28 (April 1923), 351–354 (354).

31. Henry G. Barbee, quoted in William H. Jenkins, "What the City Manager Plan Has Meant to Norfolk, Virginia," *American City* 27 (July 1922), 19–21 (21).

32. Harrison G. Otis, "The City Manager Movement—Facts and Figures," *American City* 20 (June 1919), 611, 613 (613).

33. Gantt quoted in Frank Crane, "The Engineer," *American City* 15 (Oct. 1916), 412.

34. Morris L. Cooke, "The Influence of Scientific Management upon Government—Federal, State and Municipal" (paper presented to the Taylor Society, Jan. 26, 1924), *Bulletin of the Taylor Society* 9 (Feb. 1924), 31–38 (33); Cooke echoes the earlier assertion by corporation lawyer William C. Redfield that "the modern spirit in America . . . has set its face to the task of correcting the things that here and now are wrong" by abandoning the "Days of the Rule of Thumb"; Redfield, *The New Industrial Day* (New York, 1912), 16–17; quoted in Morton Keller, *Regulating a New Economy: Public Policy and Economic Change in America, 1900–1933* (Harvard University Press, 1990), 9; see also Taylor, *The Principles of Scientific Management* (New York, 1911; reprint, W. W. Norton, 1967), 16: "The inefficient rule-of-thumb methods . . . are still almost universal in all trades. . . ."

35. Blanchard, "Chambers of Commerce and Public Highways," *American City* 13 (Sept. 1915), 217–220 (217). For a similar call for a publicity effort to overturn "the notion that engineering is merely the 'building of dams and the digging of ditches,' " see "A Cabinet Officer Needs Educating," *Engineering News-Record* 91 (Oct. 18, 1923), 623.

36. Herman G. James, "Building on to Professional Education—What Training Is Needed for the Municipal Service," in *Experts in City Government*, ed. E. Fitzpatrick (Appleton. 1919), 284.

37. Nord, "The Experts Versus the Experts: Conflicting Philosophies of Municipal Utility Regulation in the Progressive Era," *Wisconsin Magazine* of History 58 (Spring 1975), 219–236 (229).

38. See esp. John Duffy, *The Sanitarians: A History of American Public Health* (University of Illinois Press, 1990); Samuel P. Hays, *Conservation and the Gospel of Efficiency: The Progressive Conservation Movement, 1890–1920* (Harvard University Press, 1959); Samuel Haber, *Efficiency and Uplift: Scientific Management in the Progressive Era* (University of Chicago Press, 1964). On public utilities, see below. In conservation the regulators were not engineers, but their methods were analogous.

39. Hausman and Neufeld, "Engineers and Economists: Historical Perspectives on the Pricing of Electricity," *Technology and Culture* 30 (Jan. 1989), 83–104 (102).

40. McCraw, *Prophets of Regulation: Charles Francis Adams, Louis D. Brandeis, James M. Landis, Alfred E. Kahn* (Belknap, 1984), 221. McCraw described the economists as seeking "to upstage the older generation of legally trained regulators"; he emphasized the early importance of lawyers, not engineers (221). In part this is due to McCraw's interest in federal and state-level regulation, in which lawyers were indeed prevalent. Since its founding in 1887, three quarters of the members of the Interstate Commerce Commission have been lawyers (McCraw, 136); a 1929 study found that half of state public utility regulators were lawyers, and only one in six were engineers. (See Morton Keller, *Regulating a New Economy: Public Policy and Economic Change in America, 1900–1933*, Harvard University Press, 1990, 61.) McCraw's findings are also for government regulators of all kinds, not just public utilities regulators, among whom the engineers were concentrated. McCraw's own evidence shows that as late as 1974, on the state of New York's Public Service Commission, engineers were second only to "inspectors and investigators" in number, with lawyers well behind. Wide reading in Progressive Era city government journals will show that at the local level, during the infancy of administrative law, engineers were predominant. See especially *The American City* and *National Municipal Review*. Of course even then judges were the regulators of last resort.

41. Bruce Seely, in *Building the American Highway System: Engineers as Policy Makers* (Temple University Press, 1987), has ably examined the role of engineers as expert administrators in the early federal highway program. In the 1920s, Herbert Hoover's

rapid success was due in part to the assumption that as an engineer, he would be an expert administrator.

42. Taylor, *Principles*, 7, 16. For general treatments of the trend toward formalization and organization in politics, business, and social reform, see especially Robert H. Wiebe, *The Search for Order, 1877–1920* (Hill and Wang, 1967), chapters 5–7 (111–195); Guy Alchon, *The Invisible Hand of Planning: Capitalism, Social Science, and the State in the 1920s* (Princeton University Press, 1985), chapter 1 (8–20); Alfred D. Chandler, *The Visible Hand: The Managerial Revolution in American Business* (Belknap, 1977), 377–500; Stephen Skowronek, *Building a New American State: The Expansion of National Administrative Capacities* (Cambridge University Press, 1982); Morton Keller, *Regulating a New Economy: Public Policy and Economic Change in America, 1900–1933* (Harvard University Press, 1990).

43. For the basic early postulations of scientific management see especially Frederick Winslow Taylor, *The Principles of Scientific Management* (Harper and Row, 1947) and Frank B. Gilbreth, *Motion Study* (D. Van Nostrand, 1911). Of the very extensive historical work on scientific management in American industry, see especially Samuel Haber, *Efficiency and Uplift: Scientific Management in the Progressive Era* (University of Chicago Press, 1964), Daniel Nelson, *Frederick Winslow Taylor and the Rise of Scientific Management* (1980), and Robert Kanigel, *The One Best Way: Frederick Winslow Taylor and the Enigma of Efficiency* (Viking, 1997).

44. *Santa Clara County v. Southern Pacific Railroad* 118 U.S. 394 (1886).

45. See especially Samuel Haber, *Efficiency and Uplift: Scientific Management in the Progressive Era* (University of Chicago Press, 1964).

46. Joel DeWitt Justin, "Selecting an Engineer," *American City* 25 (Oct. 1921), 285.

47. On commission-manager government and budgetary reforms, see below. Several scholarly works examine the example business organization set for public sector reform. The most prominent of the schools holding this view see corporate manipulativeness in this; see for example Gabriel Kolko, *The Triumph of Conservatism: A Reinterpretation of American History, 1900–1916* (Free Press of Glencoe, 1963); James Weinstein, *The Corporate Ideal in the Liberal State, 1900–1918* (Beacon, 1968), and Jackson K. Putnam, "The Persistence of Progressivism in the 1920s: The Case of California," *Pacific Historical Review* 35 (Nov. 1966), 395–411. Many reformers, however, used the business model of government to overcome the obstacles to efficiency they saw in a decentralized state founded upon natural rights liberalism.

48. Straetz, "Scientific Management as a Guide in Traffic Planning," *American City* 32 (May 1925), 579.

49. Two of the most prominent partisans of scientific management made light of their own measurement-fixated domestic lives in an enormously popular book; see Frank and Lillian Gilbreth, *Cheaper by the Dozen* (Crowell, 1948).

50. See chapter 3.

51. John Fairfield and Clay McShane have both argued for a closer connection between the management of city streets and scientific management than is accepted here. Fairfield has related Taylorism to urban transportation and street design in "The Scientific Management of Urban Space," *Journal of Urban History* 20 (Feb. 1994), 179–204; see also his *Mysteries of the Great City: The Politics of Urban Design, 1877–1937* (Ohio State University Press, 1993), chapter 4 (119–157). McShane has traced the "conceptual roots" of traffic engineering to scientific management ("The Origins and Globalization of Traffic Control Signals," *Journal of Urban History* 25, March 1999, 379–404 (390)). For a contemporary comparison of the common principles of traffic control, scientific management, and public utilities regulation, see C. A. Copper, "The Economic Life of the City in Relation to Street Traffic," *AERA* 14 (Sept. 1925), 193–200. Copper was director of research for the Los Angeles Railway and takes the street railways' point of view.

52. *Munn v. Illinois* (1877); Chief Justice Morrison Waite was quoting Lord Chief Justice Hale, *De Portibus Maris* (c. 1670, first published 1776).

53. See Keith Revell, "Beyond Efficiency: Experts, Urban Planning, and Civic Culture in New York City, 1898–1933" (Ph.D. dissertation, University of Virginia, Jan. 1994), chapter 6 (300–377). For a convenient abridgment, see Revell, *Building Gotham: Civic Culture and Public Policy in New York City, 1898–1938* (Johns Hopkins University Press, 2003), chapter 5 (185–226).

54. See especially Robert H. Wiebe, *The Search for Order, 1877–1920* (Hill and Wang, 1967); Alfred D. Chandler, *The Visible Hand: The Managerial Revolution in American Business* (Belknap, 1977); Stephen Skowronek, *Building a New American State: The Expansion of National Administrative Capacities* (Cambridge University Press, 1982); McCraw, *Prophets of Regulation*, chapters 1–4 (1–152); Olivier Zunz, *Making America Corporate, 1870–1920* (University of Chicago Press, 1990).

55. *Second Treatise of Government* (1690), chapter 7, §87.

56. By "positive regulation" I mean regulation that seeks to shape individual social demands, fostering some and discouraging others, for the sake of a social good, as has been a customary ideal in public utility regulation. It is distinct from planning, which shapes ends as well as means. By "negative regulation" I mean regulation that checks individual abuses but leaves social goals and the means of reaching them to "natural law," leaving the state in the role of umpire.

57. See Robert M. Fogelson, *Downtown: Its Rise and Fall, 1880–1950* (Yale University Press, 2001), esp. chapter 3 (112–182); Keith Revell, "Beyond Efficiency: Experts, Urban Planning, and Civic Culture in New York City, 1898–1933" (Ph.D. dissertation, University of Virginia, Jan. 1994), chapter 6 (300–377).

58. Lord Chief Justice Hale introduced the idea of the "public interest" in *De Portibus Maris* (c. 1670, first published 1776). See also Ford P. Hall, *The Concept of a Business Affected with a Public Interest* (Principia, 1940), Walton H. Hamilton, "Affectation with a Public Interest," *Yale Law Journal* 39 (June 1930), 1089–1112, and Breck P. McAllister, "Lord Hale and Business Affected with a Public Interest," *Harvard Law Review* 43 (March 1930) 759–791.

59. 94 U.S. 113. On the adaptation of the idea of the public interest to American regulation, see Ditlev Fredericksen, "The Old Common Law and the New Trusts," 3 *Michigan Law Review* 119 (1904). For excerpts from Waite's majority opinion, see for example Kermit L. Hall, William M. Wiecek, and Paul Finkelman, *American Legal History: Cases and Materials*, second edition (Oxford University Press, 1996), 372–374.

60. *Commonwealth v. Alger* (61 Mass. 53; 1851); see Leonard W. Levy, *The Law of the Commonwealth and Chief Justice Shaw* (Harvard University Press, 1957), 247–254; Morton J. Horwitz, *The Transformation of American Law, 1870–1960: The Crisis of Orthodoxy* (Oxford University Press, 1992), 27; and Revell, *Building Gotham*, 199.

61. These areas have long been recognized as the four realms of the police power. *Black's Law Dictionary* (sixth edition, West, 1990) holds that the police power justifies "restraints on the personal freedom and property rights of persons for the protection of the public safety, health, and morals or the promotion of the public convenience and general prosperity." *Modern Legal Glossary* (Michie, 1980) defines the police power as the "authority of the state [conferred on cities] to regulate health, safety, welfare, and sometimes morality." According to a contemporary engineer, the "health, safety, morals and welfare of the community" were the "four pillars" supporting the police power. See George Herrold, "City Planning and Zoning," *Canadian Engineer* 45 (July 10, 1923), 128–130 (129). See also Morton J. Horwitz, *The Transformation of American Law, 1870–1960: The Crisis of Orthodoxy* (Oxford University Press, 1992), 27–30.

62. Urofsky, "State Courts and Protective Legislation during the Progressive Era: A Reevaluation," *Journal of American History* 72 (June 1985), 63–91 (64).

63. Chief Justice Morrison Waite quoting Lord Chief Justice Hale, *De Portibus Maris*.

64. Waite, quoted in Kermit L. Hall, William M. Wiecek, and Paul Finkelman, *American Legal History: Cases and Materials*, second edition (Oxford University Press, 1996), 374.

65. Charles B. Ball, "Why Zoning Pays" (excerpt from the Chicago City Club Bulletin), *American City* 26 (March 1922), 279.

66. Goodrich, "Zoning and Its Relation to Traffic Congestion," *Annals* 133 (Sept. 1927), 222–233 (229).

67. "To Park or Not to Park," *American City* 35 (Oct. 1926), 461–463 (461).

68. See William J. Hausman and John L. Neufeld, "Engineers and Economists: Historical Perspectives on the Pricing of Electricity," *Technology and Culture* 30 (Jan. 1989), 83–104.

69. Richard L. McCormick, "The Discovery That Business Corrupts Politics: A Reappraisal of the Origins of Progressivism," *American Historical Review* 86 (April 1981), 247–274. These two "discoveries"—that monopolies corrupt free enterprise and that business corrupts politics—are of course related, and McCormick illustrates this relationship (see esp. 271). Alfred D. Chandler has made thorough analyses of the changes in industrial organization and market behavior that underlay the decline of "single-unit enterprises" in favor of "center firms"; see especially *The Visible Hand: The Managerial Revolution in American Business* (Belknap, 1977).

70. See J. H. Hollander, "American School of Political Economy," *Palgrave's Dictionary of Political Economy* 1 (Macmillan, 1926), 804–811.

71. Hollander, "American School of Political Economy."

72. Mill, *Principles of Political Economy* (1848), book 1, chapter 9, §3; book 2, chapter 15, §3. See Toronto University Press edition (1965), volume 1, pp. 140–142, 403–405.

73. *Principles*, book 5, chapter 11, §8–16.

74. On the evolution of early American economics, see especially Dorothy Ross, *The Origins of American Social Science* (Cambridge University Press, 1991).

75. Progressives' expressions of admiration for German economics are abundant. For a classic example, see F. C. Howe, *Socialized Germany* (Charles Scribner's Sons, 1915); see also Herbert Croly, *The Promise of American Life* (Macmillan, 1909), 250–251.

76. J. H. Hollander, "American School of Political Economy," *Palgrave's Dictionary of Political Economy* 1 (Macmillan, 1926), 804–811 (809).

77. Dewey, "The Democratic State," from *The Public and Its Problems* (Henry Holt, 1927); reprinted in Dewey, *The Political Writings* (Indianapolis: Hackett, 1993), 173–183 (183).

78. The leading American neoclassical economist of this time was John Bates Clark, whose *Distribution of Wealth* began, according to American economist Allan G. Gruchy, a "neo-classical revival," the partisans of which included F. A. Fetter, Irving Fisher, and F. W. Taussig (Gruchy, *Modern Economic Thought: The American Contribution*, Prentice-Hall, 1947, 12).

79. Overton H. Taylor, *A History of Economic Thought* (McGraw-Hill, 1960), 337.

80. Marshall, for example, compared the equilibrium price set by supply and demand to "a stone hanging by a string," which, if "displaced from its equilibrium position," would tend by force of gravity to return to it. "But," Marshall adds, "in real life such oscillations are seldom as rhythmical as those of a stone hanging freely from a string" (*Principles of Economics*. Macmillan, 1890, book 5, chapter 3, §5, 405). Such a distinction between theoretical principles and "real life" is characteristic of neoclassical economics.

81. By the 1910 edition of his *Principles*, Marshall argued that "this increased prosperity has made us rich and strong enough to impose new restraints on free enterprise; some material loss being submitted to for the sake of a higher and ultimate greater gain." Marshall, for example, would consider such an intrusion on free enterprise for "the purpose of defending the weak . . . in matters in which they are not able to use the forces of competition in their own defence." Such steps were necessitated by "the quickly changing circumstances of modern industry" (*Principles of Economics*, sixth edition, Macmillan, 1910, 751).

82. Mill, *Principles of Political Economy* (1848), see esp. 143 of the 1909 Longmans edition. Thomas Henry Farrer developed the principle of natural monopoly much more fully, providing the foundations for the modern theory; see *The State in Its Relation to Trade* (1883), esp. 96–98.

83. Little substantive work has been done in the history of the theory of natural monopoly. For a fuller treatment, see esp. William W. Sharkey, *The Theory of Natural Monopoly* (Cambridge University Press, 1982), chapter 2 (12–28), and Edward D. Lowry, "Justification for Regulation: The Case for Natural Monopoly," *Public Utilities Fortnightly*, Nov. 8, 1973, 17–23. In a recent historical comparison of three city services in America (water, electric power, and cable television), Charles David Jacobson has treated some of the peculiar history of natural monopolies but does not examine the history of the idea itself; see *Ties That Bind: Economic and Political Dilemmas of Urban Utility Networks, 1800–1990* (University of Pittsburgh Press, 2000). Perhaps the best historical case study of the peculiar character of city services in America is Maureen Ogle, "Water Supply, Waste Disposal, and the Culture of Privatism in the Mid-Nineteenth-Century American City," *Journal of Urban History* 25 (March 1999), 321–347, though Ogle also does not examine the history of the idea of natural monopoly. Harold L. Platt has briefly examined natural monopoly in the case of electric power service in Chicago; see *The Electric City: Energy and the Growth of the Chicago Area, 1880–1930* (University of Chicago Press, 1991), 74–82, 310 n32. Alfred D. Chandler, especially in *The Visible Hand: The Managerial Revolution in American Business* (Belknap, 1977), has shown the importance of firms which have unusually high economies of scale to the rise of a new economy with new degrees of organization and centralization. Chandler calls such enterprises "center firms," and though none qualify as natural monopolies, their substantial economies of scale gave them monopolistic tendencies, and they ultimately prompted regula-

tion much like the natural monopolies did. McCraw has related Chandler's findings to the rise of state and federal regulation, both negative and positive; see *Prophets of Regulation*, esp. chapters 1–4. See also Morton Keller, *Regulating a New Economy: Public Policy and Economic Change in America, 1900–1933* (Harvard University Press, 1990).

84. *Principles of Political Economy* (1848), book 2, chapter 15, §4 (Longmans, Green and Co., 1923), p. 410. See also p. 143 of Longmans' 1909 edition.

85. Especially Thomas Henry Farrer. See *The State in Its Relation to Trade* (1833) (Macmillan, 1902).

86. McCraw, *Prophets of Regulation*, 9–10.

87. Adams, "Railway Commissions," *Journal of Social Science* 2 (1870), 233–236; quoted in McCraw, *Prophets of Regulation*, 9. In explaining the failure of competition in railroads, Adams was not consistent. He got far more attention for accusing railroads of deliberately subverting competition. See his "A Chapter of Erie," *North American Review* 109 (July 1869), 30–106.

88. Benjamin G. Rader, *The Academic Mind and Reform: The Influence of Richard T. Ely in American Life* (University of Kentucky Press, 1966), 89; Ely, *Problems of To-day* (Crowell, 1888). Ely's *Outlines of Economics* (1893; third edition: Macmillan, 1920) also became a standard text.

89. Adams, "The Relation of the State to Industrial Action," *Publications of the American Economic Association* I (1887), reprinted in Adams, *Two Essays by Henry Carter Adams*, ed. J. Dorfman (Augustus M. Kelly, 1969).

90. Ely, "The Growth of Corporations," *Harper's New Monthly Magazine* 75 (June 1887), 71–79.

91. Bullock, "Monopolies and Trusts" (review), *American Journal of Sociology* 6 (July 1900), 122, cited in Benjamin G. Rader, *The Academic Mind and Reform: The Influence of Richard T. Ely in American Life* (University of Kentucky Press, 1966), 89.

92. Especially British economist Thomas H. Farrer; see Farrer's *The State in Its Relation to Trade* (1883; Macmillan, 1902), esp. 96–98.

93. Ely, *Monopolies and Trusts* (Macmillan, 1900), 61–62.

94. Ely, *Monopolies and Trusts*, 62. Ely used H. C. Adams's "law of increasing returns" without revision; see Adams, "Relation of the State to Industrial Action."

95. Edward D. Lowry, "Justification for Regulation: The Case for Natural Monopoly," *Public Utilities Fortnightly*, Nov. 8, 1973, 17–23 (17).

96. Sharkey, *The Theory of Natural Monopoly* (Cambridge University Press, 1982), 16.

97. "The critical and—if properly defined—all embracing characteristic of natural monopoly is an inherent tendency to decreasing unit costs over the entire extent of the market" (Kahn, *The Economics of Regulation: Principles and Institutions*, volume 2, Wiley, 1971, 119). Sharkey (*Theory of Natural Monopoly*, 20) concludes that by 1982 economists agreed only that natural monopolies have "pervasive economies of scale" (Kahn's "decreasing unit costs," Adams's "law of increasing returns"); thus despite numerous minor modifications to the theory of natural monopoly it remained fundamentally unchanged after 1900.

98. Rader, *Academic Mind and Reform*, 90, 91 n. 15.

99. In general, see Gustavus Robinson, "The Public Utility Concept in American Law," *Harvard Law Review* 41 (1928), 277ff.

100. Mill, *Principles of Political Economy* (1848). Public utilities were, in other words, natural monopolies. Jurists and economists have never quite agreed on this, however; some legal experts have tended to treat public importance as the decisive trait. See for example Henry Rottschaefer, "The Field of Governmental Price Control," *Yale Law Journal* 35 (Feb. 1926), 438, 451–456.

101. William J. Hausman and John L. Neufeld, "Engineers and Economists: Historical Perspectives on the Pricing of Electricity," *Technology and Culture* 30 (Jan. 1989), 83–104 (104).

102. Paul H. Sheldon, "New Electric Light and Power Rates in Houston, Texas," *American City* 12 (Jan. 1915), 27–28; see also Hausman and Neufeld, "Engineers and Economists," esp. 102.

103. Arthur S. Huey, "The Regulation of Public Utilities" (paper presented before the League of American Municipalities, Atlanta, Oct. 1911), quoted in "The Regulation of Public Utilities," *American City* 5 (Dec. 1911), 349–350 (349).

104. Roux, "The Evolution of Public Utilities," *Engineering Magazine* 51 (June 1916), 386–393 (386–387).

105. C. C. Williams, "The Neck of the Highway Transportation Bottle," *Municipal and County Engineering* 68 (Feb. 1925), 93–98 (96).

106. Upson, *Practice of Municipal Administration* (Century, 1926), 524–525.

107. Jones and Bigham, *Principles of Public Utilities* (Macmillan, 1931), 69.

108. J. H. Hanna (vice president, Capital Traction Company, Washington), quoted in the report of the Metropolitan Section of the American Electric Railway Association, Dec. 3, 1926, *AERA* 16 (Jan. 1927), 1070–1084 (1084).

109. Leslie C. Smith, paper presented at the annual convention of the League of American Municipalities, Atlanta, Oct., 1911; quoted in "The Regulation of Public Utilities," *American City* 5 (Dec. 1911), 349–350 (350).

110. Herrold, "The Parking Problem in St. Paul," *Nation's Traffic*, July 1927, 28–30, 47–48 (29).

111. See e.g. Clarence O. Sherrill (engineer and city manager, Cincinnati), "Congested Streets Are Costly," *Electric Railway Journal* 68 (Oct. 9, 1926), 648–650.

112. The twentieth-century crisis of the American political culture stemmed from "the difficulty of reconciling a modern industrial order, necessarily based upon a high degree of collective organization, with democratic postulates, competitive ideals, and liberal individualistic traditions inherited from the nineteenth century" (Ellis Hawley, *The New Deal and the Problem of Monopoly: A Study in Economic Ambivalence*, Princeton University Press, 1966, vii). Good, general histories of water supply in America include Nelson Blake, *Water for the Cities: A History of the Urban Water Supply Problem in the United States* (Syracuse University Press, 1956), and Stanley K. Schultz, *Constructing Urban Culture: American Cities and City Planning, 1800–1920* (Temple University Press, 1989), 162–175.

113. Maureen Ogle, "Water Supply, Waste Disposal, and the Culture of Privatism in the Mid-Nineteenth-Century American City," *Journal of Urban History* 25 (March 1999), 321–347.

114. Tarr, "The Separate vs. Combined Sewer Problem: A Case Study in Urban Technology Design Choice," *Journal of Urban History* 5 (May 1979), 308–339 (334).

115. Ogle, "Water Supply, Waste Disposal, and the Culture of Privatism," 326.

116. George H. Herrold, "The Parking Problem in St. Paul," *Nation's Traffic*, July 1927, 28–30, 47–48 (48).

117. Ogle, "Water Supply, Waste Disposal, and the Culture of Privatism," 329–330.

118. Neptune Meter Company, "Equity—Common Sense" (advertisement), *American City* 10 (Jan. 1914), advertising pp. 14–15 (14). Emphasis in original.

119. Ibid.

120. Bemis, "Water Meters Advantageous or Otherwise?" quoted in Neptune Meter Company, " 'With Universal Metering . . .' " (advertisement), *American City* 10 (March 1914).

121. "Why Meter?" *American City* 20 (June 1919), 522–523.

122. Hodgman, "The Cleaning of City Water Mains," *American City* 10 (March 1914), 303, 305 (303).

123. Clay McShane has also found that "the social and economic roots of traffic engineering lay largely in municipal engineering," see "The Origins and Globaliza-

tion of Traffic Control Signals," *Journal of Urban History* 25 (March 1999), 379–404 (390). The professional biographies of many of these city engineers may be found in *Who's Who in Engineering: A Biographical Dictionary of Contemporaries*, second edition, ed. J. Leonard (Who's Who Publications, 1925).

124. *Who's Who in Engineering*, second edition, 402, 955, 142–143.

125. *Who's Who in Engineering*, second edition, 401, 343, 177. The unnamed Cheyenne engineer was Charles C. Carlisle.

126. Williams, "The Neck of the Highway Transportation Bottle," *Municipal and County Engineering* 68 (Feb. 1925), 93–98 (96). Williams also wrote *Municipal Water Supplies of Colorado*, University of Colorado Bulletin 12, no. 5, 1912.

127. Notice in *American City* 10 (March 1914), advertising p. 94; *Who's Who in Engineering*, second edition, 1180.

128. *Who's Who in Engineering*, second edition, 1180, and Knowles, "Planning and Rebuilding Our Cities for the Motor Age," *Proceedings of the National Safety Council: Thirteenth Annual Safety Congress* (Chicago: the Council, 1925), 809–830. Despite the title of Knowles' paper, he did recommend strict regulation of automobiles; he also recommended "rebuilding" to an extent unusual for traffic control engineers.

129. See early electric railway surveys such as the Massachusetts Public Service Commission's *Transportation Problem of Metropolitan Boston* (Boston: the Commission, 1914–1915). Engineer John A. Beeler conducted several city railway surveys and gradually included other modes of transportation in them. Beeler's major transportation survey of Atlanta in 1924 was ostensibly a street railway survey but was also one of the first extensive general city traffic surveys; see Beeler, *Report to the City of Atlanta on a Plan for Local Transportation* (Atlanta, 1924). One of the first city traffic surveys that examined all traffic modes was the City Plan Commission of Newark, New Jersey's *Comprehensive Plan of Newark* (Newark: the Commission, 1915), but few followed its example. Los Angeles was an early innovator in traffic surveys chiefly concerned with automotive traffic in 1922–24; see chapter 6.

130. *American City* 13 (Sept. 1915), 174.

131. McClintock, *Street Traffic Control* (McGraw-Hill, 1925), 22.

132. Goodrich in B. A. Haldeman et al., "How Can Narrow Streets in Business Districts Be Widened?" *Proceedings of the Fifteenth National Conference on City Planning* (Baltimore, 1923), 145–154 (148).

133. McClintock, *Street Traffic Control*, 9.

134. Bibbins, "Mass-Transportation in the Traffic and City Plan—A Perspective," in National Automobile Chamber of Commerce, *Relief of City Traffic*, Booklet 2 (New York: NACC, 1927), 13.

135. "Traffic for Group Study," *Engineering News-Record* 91 (August 30, 1923), 329. At the next annual meeting, Society members discussed traffic control at length; see *Transactions of the American Society of Civil Engineers* 88 (New York: ASCE, 1925).

136. Halbert P. Gillette, "A New Branch of Highway Engineering—Traffic Control" (editorial), *Roads and Streets* 66 (Nov. 1926), 235. As an engineering field, traffic control owed less to civil engineering than to municipal engineering.

137. "Traffic Control Developments in 1929" (editorial), *Roads and Streets* 70 (Feb. 1930), 62.

138. "Regulation of Street Traffic" (editorial), *American City* 13 (Sept. 1915), 174.

139. *Engineering News-Record*, for example, frequently endorsed major traffic building projects, very few of which were ever built.

## Chapter 5

1. "Take Traffic Control from Police," *Engineering News-Record* 95 (Dec. 24, 1925), 1018–1019 (1019).

2. Bibbins, "Traffic Congestion: The Fundamental Approach," *Electric Railway Journal* 68 (Oct. 16, 1926), 731–732 (731).

3. Fox, in "Metropolitan Section," *AERA* 16 (Jan. 1927), 1070–1084 (1080–1081).

4. Among important works that put modal conflict at the center of their interpretations, see esp. Stephen B. Goddard, *Getting There: The Epic Struggle between Road and Rail in the American Century* (Basic Books, 1994), and David J. St. Clair, *The Motorization of American Cities* (Praeger, 1986).

5. F. H. Caley, "The Parking Problem," *Proceedings of the American Electric Railway Traffic and Transportation Association* (New York: American Electric Railway Association, 1928), 42–50 (50).

6. Frank R. Coates, "Traffic Control Is Essential" (abstract of a paper presented before the U.S. Chamber of Commerce), *Electric Railway Journal* 67 (May 15, 1926), 852–854 (852).

7. See Edwin T. Layton Jr., *The Revolt of the Engineers: Social Responsibility and the American Engineering Profession* (Press of Case Western University, 1971), chapter 3 (53–78).

8. From a synopsis of C. Augustus Vollmer's speech to the Los Angeles Traffic Commission, 1923, in "Engineers Should Control Traffic," *Engineering News-Record* 91 (Oct. 18, 1923), 641.

9. Maxwell Halsey, "What Has Parking Limitation Accomplished?" *Transactions of the National Safety Council: Twentieth Annual Safety Congress* 3 (1931), 83–97 (97).

10. John A. Beeler to Special Traction Committee, Atlanta, Georgia, Dec. 16, 1924, in Beeler, *Report to the City of Atlanta on a Plan for Local Transportation* (1924), iii.

11. Miller McClintock, *Report and Recommendations of the Metropolitan Street Traffic Survey* (Chicago Association of Commerce, 1926), 32.

12. Kelker, "Suggestions for Relief of Street Congestion" (abstract of a paper presented before the American Electric Railway Association, St. Louis, March 4, 1924), *Electric Railway Journal* 63 (March 8, 1924), 373–375 (374). Many traffic engineers identified the automobile as an equal or greater contributor to congestion.

13. Downtown business leaders struggled throughout the twentieth century to protect their central-city investments. See Robert M. Fogelson, *Downtown: Its Rise and Fall, 1880–1950* (Yale University Press, 2001).

14. J. C. Ainsworth, quoted in Charles H. Cheney, "Traffic Street Plan and Boulevard System Adopted for Portland, Oregon," *American City* 25 (July 1921), 47–51 (51).

15. William J. Pedrick, "Street Widenings an Important Factor in Business and Realty Development, but Afford Feeble Traffic Relief in Congested Areas," *American City* 38 (Jan. 1928), 92.

16. Both Mark S. Foster and John D. Fairfield found that most professional city planners advocated urban deconcentration, but neither showed that these sentiments were important in the shaping of the city before 1945. See Foster, *From Streetcar to Superhighway: American City Planners and Urban Transportation, 1900–1940* (Temple University Press, 1981); Fairfield, *The Mysteries of the Great City: The Politics of Urban Design, 1877–1937* (Ohio State University Press, 1993), chapter 4 (119–157); Fairfield, "The Scientific Management of Urban Space: Professional City Planning and the Legacy of Progressive Reform," *Journal of Urban History* 20 (Feb. 1994), 179–204.

17. See esp. Fairfield, *Mysteries of the Great City*, chapter 4 (119–157); Keith Revell, "Beyond Efficiency: Experts, Urban Planning, and Civic Culture in New York City, 1898–1933" (dissertation, University of Virginia, 1994), chapter 5 (235–299); and Revell, *Building Gotham: Civic Culture and Public Policy in New York City, 1898–1938* (Johns Hopkins University Press, 2003), 185–226.

18. Charles B. Ball, Chicago City Club Bulletin; reprinted as "Why Zoning Pays" in *American City* 26 (March 1922), 279.

19. Even amid growing positive regulation, zoning was an uncomfortably collective and intrusive instrument, and particularly vulnerable to the due process protections of the Constitution. Not until 1926 (*Euclid v. Ambler Realty Co.*), after contradictory state court rulings, did the U.S. Supreme Court (with three dissenters) rule zoning constitutional. Besides its tenuous legal foundations, zoning depended upon coalitions of business interests too fragile to back bold measures (as Revell has shown for New York; see "Beyond Efficiency," chapters 5–6). Thus, in practice, zoning's effects were modest, especially in its early years and in its density provisions. Almost invariably, density provisions applied to new construction only; with downtowns already built up before 1920, zoning would have little effect on density for decades. On *Euclid*, see e.g. Revell, "Beyond Efficiency," 351–352, and Revell, *Building Gotham*, 216–217.

20. Charles H. Cheney, "Traffic Street Plan and Boulevard System Adopted for Portland, Oregon," *American City* 25 (July 1921), 47–51 (47). But Cheney avers that there might be a "permanent solution" in "the linking up and widening of a few conveniently located through routes" (47).

21. Herrold, "The Parking Problem in St. Paul," *Nation's Traffic*, July 1927, 28–30, 47–48 (48).

22. Ibid., 28.

23. Miller McClintock, *Report and Recommendations of the Metropolitan Street Traffic Survey* (Chicago Association of Commerce, 1926), 73–74.

24. Ibid., 74.

25. On "social organization," see chapter 4.

26. McClintock, *Street Traffic Control* (McGraw-Hill, 1925), 22.

27. References to the "coordination" of traffic are abundant in urban transportation reports in the early and mid 1920s; thereafter references decline sharply. Important examples include J. Rowland Bibbins (Consulting Engineer, Washington), "Metropolitan Planning for Future Transportation Requirements" (paper presented to the Engineers Club of Philadelphia, April 15, 1924), *Engineers and Engineering* 41 (May 1924), 119–128 (122), and Kelker, De Leuw and Company (engineers), report to the Traffic Survey Commission of Baltimore, abstracted as "Better Car Routing and Traffic Control Proposed for Baltimore" *Electric Railway Journal* 67 (1926): part 1 (May 22), 883–889, part 2 (May 29), 923–927 (see part 1, 884). Instead of traffic *coordination*, McClintock and others preferred the more modest term *segregation*. See esp. *Street Traffic Control*, 94–98, and *Report and Recommendations*, 111–113.

28. Harold M. Gould, "Reserving 'Main Street' for Essential Traffic," *Transactions of the National Safety Council: Eighteenth Annual Safety Congress* 3 (Chicago: the Council, 1929), 75–81 (80).

29. "Traffic Engineering in the City on Seventy Hills," *National Safety News* 16 (August 1927), 15–16, 52–53 (15).

30. Nolen, quoted in Harold Cary, "Will Passenger Cars Be Barred from City Streets? *Motor* 39 (March 1923), 34–35, 66, 68 (66, 68).

31. George Herrold, "City Planning and Zoning," *The Canadian Engineer* 45 (July 10, 1923), 128–130 (129).

32. "Better Car Routing and Traffic Control Proposed for Baltimore" (abstract of a report by Kelker, De Leuw, and Co., to the Traffic Survey Commission of Baltimore), part 1, *Electric Railway Journal* 67 (May 22, 1926), 883–889 (888). For other examples of such recommendations, see City Plan Commission, Newark, New Jersey, *Comprehensive Plan of Newark* (Newark: H. Murphy, 1915), 15, 102; Leo J. Buettner, address to the American Civic Association, Chicago, Nov. 14, 1921, excerpted in "Making the City Plan Effective," *American City* 26 (April 1922), 323–326 (325); George H. Herrold, "The Parking Problem in St. Paul," *Nation's Traffic*, July 1927, 28–30, 47–48 (30).

33. A spate of state court cases settled the question, many in the early 1920s. From *American City*, see "City May Require Removal of Stationary Awnings from Sidewalks," volume 25 (Oct. 1921), 341; "Power to Restrict Use of Sidewalks," volume 11 (Oct. 1914), 288; "Right to Remove Obstructions in Streets Recognized," volume 30 (Jan. 1924), 87; "Obstructions for Private Purposes Upon Public Sidewalks and Streets Cannot Be Licensed by City," volume 31 (Sept. 1924), 263. These trends accorded with the growing police power of the state. See chapter 3 and Mary Kingsbury Simkhovitch, "The City's Care of the Needy" (address to the National Municipal League, Nov. 24, 1916), *National Municipal Review* 6 (March 1917), 255–262 (256).

34. In calling for the removal of sidewalk obstructions, traffic engineers were building upon an earlier trend in the Progressive Era. From *American City*, see Felix Hunt, "A Practical Detail of City Planning," volume 7 (Nov. 1912), 411–415; Warren Wheaton, "Removing Sidewalk Encroachments in Albany," volume 14 (April 1916), 346–347; J. Herbert Kohler, "Up-To-Date Business Streets," volume 20 (Jan. 1919), 56–57. Daniel Bluestone examined the removal of pushcart vendors from New York thoroughfares as city streets were more narrowly defined as traffic ways. See Bluestone, " 'The Pushcart Evil': Peddlers, Merchants, and New York City's Streets, 1890–1940," *Journal of Urban History* 18 (Nov. 1991), 68–92.

35. Maxwell Halsey, "What Has Parking Limitation Accomplished?" *Transactions of the National Safety Council: Twentieth Annual Safety Congress* (1931) 3 (Chicago: the Council, 1932), 83–97 (91).

36. Barber, "The Retailer's Interest in City Motor Traffic," *National Retail Dry Goods Association Year Book* (1928), 207–220 (217).

37. Los Angeles, Pittsburgh, and Rochester were among the cities that used Boy Scouts for traffic survey work. See Paul G. Hoffman, "The Traffic Commission of Los Angeles," *Annals* 116 (Nov. 1924), 246–250 (250); McClintock, *Street Traffic Control*, 13 n. 1; Philip P. Sharples, "Los Angeles Traffic Count by the Boy Scouts," in Frederick Law Olmsted, Harland Bartholomew, and Charles Henry Cheney, *A Major Traffic Street Plan for Los Angeles* (Los Angeles: Traffic Commission of the City and County of Los Angeles, 1924), 67–69 (Los Angeles); "Traffic Engineering in the City on Seventy Hills," *National Safety News* 16 (August 1927), 15–16, 52–53 (15) (Pittsburgh); Leon R. Brown, "A Plan to Relieve Municipalities of the Parking Burden," *Nation's Traffic* 2 (Nov. 1928), 18–21 (18); "How Many Shoppers Park Autos in Streets?" *American City* 38 (Feb. 1928), 163 (Rochester).

38. Frank J. Green, "Educating the Public on Traffic Rules," *American City* 27 (August 1922), 167; T. M. Ford, "Prevention of Street Accidents," *American City* 23 (Dec. 1920), 604–605.

39. McClintock, *Street Traffic Control* (McGraw-Hill, 1925), 47. Speed was still the crucial factor in safety, to McClintock: "Fundamentally all traffic accidents can be reduced to one cause, that is, too great speed under a given set of conditions" (86).

40. These views are represented in Miller McClintock and also, e.g., in a prominent traffic engineer in Washington, J. Rowland Bibbins. See McClintock, *Street Traffic Control* (McGraw-Hill, 1925), 9: "rapidity of movement, properly regulated, is not incompatible with a large degree of safety," and Bibbins, "Mass-Transportation in the Traffic and City Plan—A Perspective," in National Automobile Chamber of Commerce, *Relief of City Traffic*, Booklet 2 (New York: NACC, 1927), 13: "It is not enough to stand behind the slogan 'order and safety.' We must have both safety and speed."

41. J. Rowland Bibbins, "Traffic Congestion: The Fundamental Approach," *Electric Railway Journal* 68 (Oct. 16, 1926), 731–732 (732).

42. William G. Beard, "One Year of Traffic Lights on the 'Boul Mich,'" *National Safety News* 10 (Oct. 1924), 49–50; J. L. Jenkins, "Sees Big Cut in Taxi Rate with Traffic Relief," *Chicago Tribune*, March 29, 1922. On urban taxicabs in general, see Gorman Gilbert and Robert E. Samuels, *The Taxicab: An Urban Transportation Survivor* (University of North Carolina Press, 1982).

43. Barber, "The Retailer's Interest in City Motor Traffic," *National Retail Dry Goods Association Year Book* (1928), 207–220 (212).

44. John A. Dewhurst, "Co-ordinated Lights and Rerouting Speed Up Chicago's Loop Traffic," *Electric Railway Journal* 67 (March 27, 1926), 537–541; John T. Miller, "Chicago's Electric Traffic Signal System," *Electrical World* 87 (March 27, 1926), 658–661; "Dever to Snap On Stop-Go Lights Today," *Chicago Tribune*, Feb. 7, 1926.

45. Dewhurst, "Co-ordinated Lights," 537.

46. J. L. Jenkins, "Maelstrom of Loop Tamed by Traffic Lights," *Chicago Tribune*, Feb. 9, 1926; Dewhurst, "Co-ordinated Lights"; E. J. McIlraith, "Chicago Loop Street Capacity Increased 50 Per cent by New Coordinated Automatic Signal System," *AERA* 15 (May 1926), 555–563.

47. Jenkins, "Maelstrom."

48. "Progressive Light Plan Found Best Traffic Aid," *Automotive Industries* 56 (April 16, 1927), 597.

49. "Slide Rule Chart Determines the Timing of Traffic Lights," *Engineering News-Record* 98 (Feb. 10, 1927), 231.

50. J. L. Jenkins, "Illegal Parking Hinders Work of Stop-Go Lights; Pedestrian Dangers Grow as Loop Speeds Up," *Chicago Tribune*, Feb. 10, 1926.

51. Dewhurst, "Co-ordinated Lights," 537.

52. Jenkins, "Maelstrom."

53. Clarence O. Sherrill, "Congested Streets Are Costly," *Electric Railway Journal* 68 (Oct. 9, 1926), 648–650 (649). For Sherrill's engineering credentials, see John William Leonard, ed., *Who's Who in Engineering*, second edition (Who's Who Publications, 1925), 1890–1891.

54. Kelker, "Suggestions for Relief of Street Congestion" (abstract of a paper presented before the American Electric Railway Association, St. Louis, March 4, 1924), *Electric Railway Journal* 63 (March 8, 1924), 373–375 (374).

55. McClintock, *Street Traffic Control* (McGraw-Hill, 1925), 3.

56. Kelker, De Leuw and Company (engineers), report to the Traffic Survey Commission of Baltimore, abstracted as "Better Car Routing and Traffic Control Proposed for Baltimore" *Electric Railway Journal* 67 (1926): part 1 (May 22), 883–889, part 2 (May 29), 923–927 (see part 1, 884, and part 2, 924).

57. Sherrill, "Congested Streets Are Costly," *Electric Railway Journal* 68 (Oct. 9, 1926), 648–650 (648).

58. Leon R. Brown, "Second Half of Prize Winning Parking Plan," *Nation's Traffic* 2 (Dec. 1928), 18–20 (18).

59. George Edgecombe, "Resolved: That the Parked Car Must Go—But Where?" *Nation's Traffic* 3 (June 1929), 18–21 (18).

60. C. C. Williams, "The Neck of the Highway Transportation Bottle," *Municipal and County Engineering* 68 (Feb. 1925), 93–98 (96).

61. Herrold, "The Parking Problem in St. Paul," *Nation's Traffic*, July 1927, 28–30, 47–48 (47); Leon R. Brown, "A Plan to Relieve Municipalities of the Parking Burden," *Nation's Traffic* 2 (Nov. 1928), 18–21 (18). Herrold's figures were based on the average daytime ridership of each streetcar; Brown evidently assumed full streetcars. See also Morris Knowles, "Planning and Rebuilding Our Cities for the Motor Age," *Proceedings of the National Safety Council* 13 (1924), 809–830 (811–812).

62. Such findings were ubiquitous. For example, an engineering firm's 1924 survey of Baltimore streets found that 60 percent of the vehicles were automobiles, but these were moving only 28 percent of travelers using vehicles; see Kelker, De Leuw and Company (engineers), report to the Traffic Survey Commission of Baltimore, abstracted as "Better Car Routing and Traffic Control Proposed for Baltimore" *Electric Railway Journal* 67 (1926): part 1 (May 22), 883–889, part 2 (May 29), 923–927 (see part 1, 887).

63. Sherrill, "Congested Streets Are Costly," *Electric Railway Journal* 68 (Oct. 9, 1926), 648–650 (649). For a good, recent discussion of the parking problem in the 1920s, see Robert M. Fogelson, *Downtown: Its Rise and Fall, 1880–1950* (Yale University Press, 2001), 282–295. The expression "parking evil" appeared in the 1920s in numerous commentaries on city traffic, by engineers and also by representatives of street railways and merchants' associations. Harold M. Gould, a traffic engineer with a background in public utilities, preferred to call curb parking "the 'poison ivy' of street traffic regulation" ("Reserving 'Main Street' for Essential Traffic," *Transactions of the National Safety Council: Eighteenth Annual Safety Congress* 3, Chicago: the Council, 1929, 75–81 (76)).

64. Herrold, "The Parking Problem in St. Paul," *Nation's Traffic*, July 1927, 28–30, 47–48 (48).

65. Chester T. Crowell drew this conclusion after examining traffic surveys of 16 cities; see Crowell, "Consider the Street Car," *AERA* 14 (Oct. 1925), 412–420 (415).

66. Engineers termed the street surface between curbs the *roadway*. To them, the street included the sidewalks. Perhaps the earliest clear formulation of this distinction is Harland J. Bartholomew et al., *Comprehensive Plan of Newark* (Newark, New Jersey: City Plan Commission, 1915), 28.

67. *Main Street, U.S.A. in Early Photographs*, ed. C. Read-Miller (Dover, 1988) shows vividly the diverse uses of the verges of roadways in cities of all sizes, c. 1890–1915; see esp. 1, 50, 59 (building materials storage); 2 (fire hydrant, telephone pole, trough); 20 (telephone poles, advertisements, barrels); 28 (streetcar poles); 46, 92 (telephone poles); 51 (storage of unidentified objects); 54, 65, 83 (vending); 58, 64 (miscellaneous use); 92 (shoe shining); 105 (storage of crates).

68. Russell Van Nest Black, "The Spectacular in City Building," *Annals* 133 (Sept. 1927), 5–56 (55).

69. Letter to editor quoted in Harold M. Gould, "Reserving 'Main Street' for Essential Traffic," *Transactions of the National Safety Council: Eighteenth Annual Safety Congress* 3 (Chicago: the Council, 1929), 75–81 (78); in this article, Gould, an engineer, also compared parking bans to prohibition under the Eighteenth Amendment.

70. Caley, "The Parking Problem," *Proceedings of the American Electric Railway Transportation and Traffic Association* (New York: the Association, 1928), 42–50 (43). See also William Ullman, "Bootlegging Our Parking: The Car Faces an Eighteenth Amendment of Its Own," *Ohio Motorist* 16 (June 1924), 5, 20–22.

71. Gould, "Reserving 'Main Street' for Essential Traffic," 76. Gould was describing events in Detroit in 1929, following implementation of parking restrictions.

72. Detroit traffic engineer (and former public utilities regulator) Harold M. Gould called for "The return of streets to the purpose for which they were originally intended—that of moving vehicles instead of using them for storage purposes"; see Gould, "Reserving 'Main Street' for Essential Traffic," *Transactions of the National Safety Council: Eighteenth Annual Safety Congress* 3 (Chicago: the Council, 1929), 75–81 (80). Charles Gordon described traffic engineers' work in Chicago as "the conversion of valuable street space from a storage area for private cars to the purpose for which it was intended, namely, the movement of traffic" (Gordon, it should be noted, was an editor of *Electric Railway Journal*); Gordon, "Elimination of Parking Proves Successful in Chicago Loop District," *Electric Railway Journal* 73 (Feb. 23, 1929), 312–318. St. Paul engineer George H. Herrold argued that "the streets are for moving traffic and not for the storage of cars"; Herrold, "The Parking Problem in St. Paul," *Nation's Traffic*, July 1927, 28–30, 47–48 (48). The Chicago engineering firm Kelker, De Leuw recommended to the city of Baltimore that it see that its street space was "kept free for the purpose for which it was intended," that is, "moving traffic"; see report to the Traffic Survey Commission of Baltimore, abstracted as "Better Car Routing and Traffic Control Proposed for Baltimore" *Electric Railway Journal* 67 (1926): part 1 (May 22), 883–889, part 2 (May 29), 923–927 (part 1, 888).

73. Russell Van Nest Black, "The Spectacular in City Building," *Annals* 133 (Sept. 1927), 50–56 (55).

74. McClintock, *Report and Recommendations of the Metropolitan Street Traffic Survey* (Chicago Association of Commerce, 1926), 145.

75. Maxwell Halsey, "What Has Parking Limitation Accomplished?" *Transactions of the National Safety Council: Twentieth Annual Safety Congress* (1931) 3 (Chicago: the Council, 1932), 83–97 (88, see also 91). McClintock also described parking as a "privilege," and one which threatened "the rights of the public" (most of whom traveled primarily by other modes); see *Report and Recommendations*, 166. See also McClintock, *Street Traffic Control* (McGraw-Hill, 1925), 139.

76. See, e.g., the ruling of Judge Sam Hooker et al. in *Ed Butterfield v. City of Oklahoma* (1936), Oklahoma County District Court: "Parking is not a right but a privilege." Hooker cites several cases as precedent. The full opinion is in Public Relations Department, American Automobile Association, Special Information Bulletin 2: "Parking Meter Developments," April 9, 1936, in file "parking," AAA Headquarters Library.

77. Herrold, "Parking Problem in St. Paul," 29.

78. Robert B. Brooks, paper presented to the International Association of Street Sanitation Officials, Toronto, 1928, abstracted as "St. Louis Attacks the Problem of Municipal Parking Space," *American City* 39 (Nov. 1928), 165; see also McClintock, *Street Traffic Control*, 140.

79. Kelker, De Leuw and Company (engineers), report to the Traffic Survey Commission of Baltimore, abstracted as "Better Car Routing and Traffic Control Proposed for Baltimore" *Electric Railway Journal* 67 (1926): part 1 (May 22), 883–889, part 2 (May 29), 923–927 (see part 1, 888).

80. McClintock, *Report and Recommendations of the Metropolitan Street Traffic Survey* (Chicago Association of Commerce, 1926), 146.

81. Kelker, "Suggestions for Relief of Street Congestion" (abstract of a paper presented before the American Electric Railway Association, St. Louis, March 4, 1924), *Electric Railway Journal* 63 (March 8, 1924), 373–375 (375). See also C. C. Williams (a civil engineer expert in city water supply), "The Neck of the Highway Transportation Bottle," *Municipal and County Engineering* 68 (Feb. 1925), 93–98 (94).

82. Kelker, "Suggestions for Relief of Street Congestion," 374.

83. Leon R. Brown, "A Plan to Relieve Municipalities of the Parking Burden," *Nation's Traffic* 2 (Nov. 1928), 18–21 (18).

84. Harold M. Gould, "Reserving 'Main Street' for Essential Traffic," *Transactions of the National Safety Council: Eighteenth Annual Safety Congress* 3 (Chicago: the Council, 1929), 75–81 (76); Herrold, "The Parking Problem in St. Paul," *Nation's Traffic*, July 1927, 28–30, 47–48 (28).

85. Gould, "Reserving 'Main Street,' " 76. In St. Paul in 1927 there was enough downtown curb space for 7 percent of registered autos; see Herrold, "The Parking Problem in St. Paul," 28.

86. Robert B. Brooks, paper presented before the International Association of Street Sanitation Officials, Toronto, 1928, abstracted as "St. Louis Attacks the Problem of Municipal Parking Space" in *American City* 39 (Nov. 1928), 165; Mary A. Burke, "How Street Railways Are Helping to Solve the Traffic Problem," *AERA* 20 (May 1929), 270–276 (272).

87. Clarence O. Sherrill, "Congested Streets Are Costly," *Electric Railway Journal* 68 (Oct. 9, 1926), 648–650 (649).

88. McClintock, *Report and Recommendations of the Metropolitan Street Traffic Survey* (Chicago Association of Commerce, 1926), 146.

89. F. H. Caley, "The Parking Problem," *Proceedings of the American Electric Railway Transportation and Traffic Association* (New York: American Electric Railway Association, 1928), 42–50 (56). Washington's strict building height limitations, broad streets, and vast central business district (if the concentration of federal office buildings may be so called), with its lack of a large industrial population and rapid transit lines, made it a city of automobiles. In nearby Baltimore in 1931, of all persons entering the central business district in vehicles, 34.9 percent were in automobiles; for Washington the figure was 64.5 percent. See Maxwell Halsey, "What Has Parking Limitation Accomplished?" *Transactions of the National Safety Council: Twentieth Annual Safety Congress* (1931) 3 (Chicago: the Council, 1932), 83–97 (88).

90. Simpson, "Are There Too Many Hitching Posts on Main Street?" *AERA* 21 (Dec. 1930), 736–739 (738).

91. Clarence O. Sherrill, "Congested Streets Are Costly," *Electric Railway Journal* 68 (Oct. 9, 1926), 648–650 (649).

92. McClintock, *A Report on the Parking and Garage Problem of the Central Business District of Washington* (Washington: Automobile Parking Committee, 1930), 79.

93. H. H. Hemmings, "Some Economic Considerations Which May Be Applied to Traffic Problems," *Proceedings of the Institute of Traffic Engineers*, Oct. 1931, 72–82 (79).

94. Simpson, "Hitching Posts," 738. Simpson (738) calculated the value of curb space in the center of the largest cities at about $29,000; each space, he thought, was worth $144 a month.

95. See Miller McClintock, *Street Traffic Control* (McGraw-Hill, 1925), 141.

96. McClintock, *Parking and Garage Problem*, 49–51.

97. Herrold, "The Parking Problem in St. Paul," *Nation's Traffic*, July 1927, 28–30, 47–48 (48). A British ruling in 1812 confirmed the place of this tradition in the common law. In a decision which traffic control engineers were fond of citing, Lord Chief Justice Ellenborough ruled that stationary impediments to traffic must not be left in roadways. "Every unauthorized obstruction of a highway to the annoyance of the king's subjects is a nuisance. The king's highway is not to be used as a stable yard" (*Rex v. Cross*). American judges accepted the case as precedent. See " 'The King's Highway Is Not to Be Used as a Stable Yard,' " *American City* 32 (May 1925), 533.

98. Harold M. Gould, "Reserving 'Main Street' for Essential Traffic," *Transactions of the National Safety Council: Eighteenth Annual Safety Congress* 3 (Chicago: the Council, 1929), 75–81 (80).

99. John P. Fox, "Traffic Regulation in Detroit and Toronto," *American City* 13 (Sept. 1915), 175–179 (178). A street railway executive reported a six-hour parking limit in Schenectady, New York, dating from 1880; see Howard F. Fritch, "What Railway Operators Can Do to Improve Traffic Conditions" (paper presented before the New England Street Railway Club, Boston, May 22, 1924), *Electric Railway Journal* 63 (May 31, 1924), 845–847 (846). In his pioneering 1915 survey, Harland Bartholomew recommended half-hour parking limits; see Bartholomew, *Comprehensive Plan of Newark* (Newark: City Plan Commission, 1915), 44.

100. Edwin U. Curtis, "Tagging Traffic Offenders," *American City* 23 (July 1920), 32. This article seems to refer to the earliest regular use of the parking ticket by a large city. Ticketing, however, did not always relieve police of the burden of issuing summons to offenders in person.

101. Chattanooga engineer J. Haslett Bell reported that the curb capacity in the city was 3,969 cars before the 1924 one-hour parking limit, and 13,308 after; see Bell, "Traffic Relief through Parking Regulation," *American City* 31 (August 1924), 140–141 (141).

102. Frank J. Green, "Educating the Public on Traffic Rules," *American City* 27 (August 1922), 167.

103. Coleman (Traffic Division, New York Police Department), address to the Metropolitan Section, American Electric Railway Association, Dec. 3, 1926, in "Metropolitan Section," *AERA* 16 (Jan. 1927), 1074–1078 (1076).

104. Green, "Educating the Public." According to Fox, Detroit police in 1915 towed any illegally parked auto, without informing the owner, with the deliberate intent of making the owner fear the car had been stolen. Police held the car until the owner reported the supposed theft. Fox approved this "touch of humor." See John P. Fox, "Traffic Regulation in Detroit and Toronto," *American City* 13 (Sept. 1915), 175–179 (179).

105. F. C. Lynch, in Maxwell Halsey, "What Has Parking Limitation Accomplished?" *Transactions of the National Safety Council: Twentieth Annual Safety Congress* (1931) 3 (Chicago: the Council, 1932), 83–97 (96).

106. Arnheim, in George Edgecombe et al., "Resolved: That the Parked Car Must Go—But Where?" *Nation's Traffic* 3 (June 1929), 18–21 (21).

107. McClintock, *Street Traffic Control*, 150.

108. Lewis Buddy III, quoted in Edgecombe, "Resolved," 21.

109. Leon R. Brown, "A Plan to Relieve Municipalities of the Parking Burden," *Nation's Traffic* 2 (Nov. 1928), 18–21 (18); Brown, "Second Half of Prize Winning Parking Plan," *Nation's Traffic* 2 (Dec. 1928), 18–20 (19).

110. Herrold, "The Parking Problem in St. Paul," *Nation's Traffic*, July 1927, 28–30, 47–48 (28).

111. Kelker, "Suggestions for Relief of Street Congestion" (abstract of a paper presented before the American Electric Railway Association, St. Louis, March 4, 1924), *Electric Railway Journal* 63 (March 8, 1924), 373–375 (374).

112. Arnheim, in George Edgecombe et al., "Resolved: That the Parked Car Must Go—But Where?" *Nation's Traffic* 3 (June 1929), 18–21 (21).

113. Halsey, "What Has Parking Limitation Accomplished?" 91.

114. Leon R. Brown, "A Plan to Relieve Municipalities of the Parking Burden," *Nation's Traffic* 2 (Nov. 1928), 18–21 (20). Brown estimated a cost of only $500 per space provided by widening, but compared this expense unfavorably to the roughly $75 per year that spaces were worth in private parking lots.

115. Simpson, "Are There Too Many Hitching Posts on Main Street?" *AERA* 21 (Dec. 1930), 736–739 (738).

116. Gerald J. Wagner, quoted in editorial, Grand Rapids *Press*, Oct. 8, 1926, as quoted in "Police Chief for Parking Ban," *AERA* 16 (Dec. 1926), 827–828 (828).

117. George A. Carpenter, "Informing the Autoist Regarding Local Traffic Ordinances," *American City* 27 (Oct. 1922), 341–343 (341). Carpenter was city engineer for Pawtucket, Rhode Island.

118. McClintock, *Street Traffic Control*, 145–146.

119. F. C. Fox, in "Metropolitan Section," *AERA* 16 (Jan. 1927), 1070–1084 (1079–1080).

120. Arnheim in George Edgecombe et al., "Resolved: That the Parked Car Must Go—But Where?" *Nation's Traffic* 3 (June 1929), 18–21 (21).

121. Fox, in "Metropolitan Section," 1080.

122. Charles J. Columbus (secretary, Merchants and Manufacturers Association, Washington), quoted in "Columbus Fights Proposed Ban upon Rush-Hour Parking," *Washington Post*, April 13, 1925.

123. Barber, "The Retailer's Interest in City Motor Traffic," *National Retail Dry Goods Association Year Book* (1928), 207–220 (217–218).

124. John W. Colton, "Traffic Congestion: The Big Problem before American Cities," *AERA* 15 (April 1926), 366–382 (367, 371).

125. Charles Gordon, "Elimination of Parking Proves Successful in Chicago Loop District," *Electric Railway Journal* 73 (Feb. 23, 1929), 312–318 (312, 314). Opponents of no-parking were most often owners of small shops, but these were likely to deal in luxuries. Gordon, for example, admits that "the higher-grade establishment catering to an exclusive clientele, or what was at one time known as a 'carriage trade,' is usually most concerned regarding proposals for the regulation of private automobiles," since its customers were more likely to be motorists (317).

126. Miller McClintock, *A Report on the Parking and Garage Problem of the Central Business District of Washington* (Washington: Automobile Parking Committee, 1930), 47. The May 13, 1930, survey included "seven department stores, and numerous specialty shops."

127. Department of Commerce, Bureau of Foreign and Domestic Commerce, *Vehicular Traffic Congestion and Retail Business* (Washington: Government Printing Office, 1926), 28–30.

128. *Vehicular Traffic Congestion and Retail Business*, 1, 10–16.

129. "Chicago Traffic Ills Attacked by Effective Bureau," *Chicago Tribune*, Feb. 7, 1926; Barber, "The Retailer's Interest in City Motor Traffic," *National Retail Dry Goods Association Year Book* (1928), 207–220 (212).

130. J. L. Jenkins, "Illegal Parking Hinders Work of Stop-Go Lights," *Chicago Tribune*, Feb. 10, 1926; Jenkins, "Lid on Parking Clears the Loop in Record Time," *Chicago Tribune*, Feb. 11, 1926.

131. McClintock, *Report and Recommendations of the Metropolitan Street Traffic Survey* (Chicago Association of Commerce, 1926).

132. McClintock, *Report and Recommendations*, esp. xiii, 3–4.

133. Charles Gordon, "Elimination of Parking Proves Successful in Chicago Loop District," *Electric Railway Journal* 73 (Feb. 23, 1929), 312–318 (314).

134. Miller McClintock's views as summarized in "Traffic Control Developments in 1929" (editorial), *Roads and Streets* 70 (Feb. 1930), 62.

135. "Traffic Engineering on the City of Seventy Hills," *National Safety News* 16 (August 1927), 15–16, 52–53 (15).

136. Mary A. Burke, "How Street Railways Are Helping to Solve the Traffic Problem" *AERA* 20 (May 1929), 270–276 (276).

137. Coleman (Traffic Division, New York Police Department), address to the Metropolitan Section, American Electric Railway Association, Dec. 3, 1926, in "Metropolitan Section," *AERA* 16 (Jan. 1927), 1074–1078 (1075–1076); Burke, "How Street Railways Are Helping to Solve the Traffic Problem," 273.

138. See, for example, the comment by D. E. Evans in "Discussion," *Proceedings of the National Safety Council: Thirteenth Annual Safety Congress* (1924), 831–840 (837).

139. On bans in Baltimore, Dayton, Philadelphia, Pittsburgh, and St. Louis (and Chicago, New York, and Springfield, Massachusetts), see Mary A. Burke, "How Street Railways Are Helping to Solve the Traffic Problem" *AERA* 20 (May 1929), 270–276. On St. Louis, see also Robert E. Lee's comments in "Discussion," *Proceedings of the National Safety Council: Thirteenth Annual Safety Congress* (1924), 831–840 (837). On Cleveland, see A. H. Lintz's comment in "Discussion," *Proceedings of the National Safety Council: Thirteenth Annual Safety Congress* (1924), 831–840 (836).

140. For example, in Pittsburgh; see "Traffic Engineering in the City on Seventy Hills," *National Safety News* 16 (August 1927), 15–16, 52–53 (16).

141. See John Duffy, *The Sanitarians: A History of American Public Health* (University of Illinois Press, 1990).

142. Miller McClintock, *Street Traffic Control* (McGraw-Hill, 1925), 10.

143. Fox, in "Metropolitan Section," *AERA* 16 (Jan. 1927), 1070–1084 (1081).

## Chapter 6

1. Cobb (chairman, Advisory Council, American Electric Railway Association), "Where the Industry Stands Today" (report of the Advisory Council), *AERA* 16 (Nov. 1926), 647–653 (648).

2. Wirth, "Urbanism as a Way of Life," *American Journal of Sociology* 44 (July 1938), 1–24 (16).

3. The engineer Hawley S. Simpson complained of cities installing signals merely "to acquire a metropolitan air"; see Gustave C. Schink, "Traffic Control on City Streets" (paper presented Oct. 7, 1927), *Proceedings of the First Annual Conference on Highway Transport and Sixth Annual Meeting of the Michigan Motor Bus Association* (Ann Arbor: University of Michigan Official Publication 29, no. 25, Dec. 17, 1927), 11–15 (13); Dwight McCracken, *Traffic Regulation in Small Cities* (New York: Municipal Administration Service, 1932), 7.

4. See esp. Clay McShane, *Down the Asphalt Path* (Columbia University Press, 1994), chapter 10 (203–228, 269–276).

5. On engineers' application of the lessons of water supply to city streets, see chapter 4.

6. Goddard, *Getting There*, 282. Goddard's book is the leading recent example of this interpretation; David J. St. Clair makes a more modest and judicious case for this view in *The Motorization of American Cities* (Praeger, 1986).

7. A. J. Brosseau, "Coordination of Motor Vehicle and Electric Railway Service by Electric Railways" (pamphlet; address to the annual convention of the American Electric Railway Association, Oct. 9, 1924). A leading street railway trade journal enthusiastically reported the success of bus operation by streetcar companies; see " 'Four-Track' Operation in Providence," *Electric Railway Journal* 63 (March 15, 1924), 411–413. By the late 1920s, bus manufacturers did express impatience with the railways' "paternalistic attitude" toward them and with the railways' predilection to use the bus as a supplement only; see Norman G. Shidle, "What Is the Future of Bus and Railway Railway Cooperation?" *Automotive Industries* 57 (Nov. 12, 1927), 709–710.

8. American Electric Railway Association, *The Urban Transportation Problem* (New York: the Association, 1932), 5. On the operation of buses by street railways, see also Kenneth L. McKee, "Some Facts about Bus Operation by Electric Railways," *AERA* 14 (Nov. 1925), 718–726, and McKee, "Motor Bus Operation by Railways," *AERA* 16 (Nov. 1926), 583–590. Already in 1924, 115 street railways were operating over 1100 buses; see George M. Graham, "Recent Developments in Highway Transport" (address to the National Chamber of Commerce), in *Cooperation in Transportation* (Washington: the Chamber, 1924), 13–29 (19). The trend persisted: by 1926, almost 300 electric railways supplemented their service with buses; see Frank R. Coates, "Electric Railways Must Represent Car Riders in Discussions of Traffic Congestion" (address before the annual convention of the Chamber of Commerce of the United States, May 11, 1926), *AERA* 15 (June 1926), 738–750 (748).

9. T. C. Powell, "Function of the Motor Truck in Reducing Cost and Preventing Congestion of Freight in Railroad Terminals," *Annals* 116 (Nov. 1924), 87–89; Graham, "Recent Developments in Highway Transport," 14. See also F. W. Fenn, "Transportation–the Keynote of Prosperity," *American City* 23 (Dec. 1920), 598–600).

10. Howard L. Preston, *Automobile Age Atlanta: The Making of a Southern Metropolis, 1900–1935* (University of Georgia Press, 1979), 55–63; Ross D. Eckert and George W. Hilton, "The Jitneys," *Journal of Law and Economics* 15 (Oct. 1972), 293–325.

11. See esp. Barrett, *The Automobile and Urban Transit*, esp. 211–212, and Albro Martin, *Enterprise Denied: Origins of the Decline of American Railroads, 1897–1917* (Columbia University Press, 1971), and *Railroads Triumphant: The Growth, Rejection, and Rebirth of a Vital American Force* (Oxford University Press, 1992). For a concise and skillful analysis of a closely analogous problem, see Christopher J. Castaneda and Clarence M. Smith, *Gas Pipelines and the Emergence of America's Regulatory State: A History of Panhandle Eastern Corporation, 1928–1993* (Cambridge University Press, 1996).

12. See esp. American Electric Railway Association, *The Urban Transportation Problem*, 38–43.

13. Ibid., 13.

14. Ibid., 38.

15. Gerit Fort, "Interests of Railroad and Automotive Industries Identical, Fort Says," *Automotive Industries* 49 (Sept. 20, 1923), 571–572 (572). See also D. W. Pontius, "The Traffic Congestion Problem, Traffic Control, and Rapid Transit," *Proceedings of the American Electric Railway Association* 44 (1925), 177–185 (183).

16. C. C. Williams, "The Neck of the Highway Transportation Bottle," *Municipal and County Engineering* 68 (Feb. 1925), 93–98 (95).

17. Anderson, "The Business of Street Management," *AERA* 16 (Nov. 1926), 689–695 (693).

18. Cobb, "Where the Industry Stands Today," *AERA* 16 (Nov. 1926), 647–653 (648).

19. Halbert O. Crews, "Concentration in Publicity," *AERA* 14 (Oct. 1925), 374–382 (380).

20. Copper, "The Economic Life of the City in Relation to Street Traffic," *AERA* 14 (Sept. 1925), 193–200 (200).

21. Department of Commerce, Bureau of the Census, *Historical Statistics of the United States* (Washington: Government Printing Office, 1975), 1:8, 2:716.

22. Philip H. Brockman, "No Auto Saturation Likely to Be Reached," *St. Louis Star*, Oct. 22, 1922.

23. Historians have tended to see the significance of the saturation crisis chiefly in the new marketing methods it helped inspire and in the industry's consolidation, but its importance is much broader. Bradford Snell rightly argued that the saturation crisis led auto interests to adopt a more aggressive urban strategy, yet he dated their first coordinated response to it only to 1932, when GM began to buy street railways. By then the auto industry was suspending manufacturing for lack of demand anywhere. Action really came years earlier. See Snell, "American Ground Transport," in Senate, Committee on the Judiciary, *The Industrial Reorganization Act: Hearings Before a Subcommittee on S. 1167*, 93rd Cong., 2d sess. (1974), 26–49 (28–30). James Flink found in the saturation crisis a chronic annoyance culminating in the Great Depression. He noted trade journalists' denials of the crisis to maintain "business confidence." Flink took such denials as evidence that the industry "failed to face the problem squarely." On the contrary, the saturation crisis transformed it. Before the end of 1924, the industry was mobilized to reverse it. See Flink, *The Automobile Age* (MIT Press, 1988), 189–193. David J. St. Clair has sketched a connection between the industry's saturation fears and its backing of urban highways, but he considered only the industry's role in federal policy, and that only after 1930. Yet the auto

industry stepped into city traffic matters before the federal government had any urban transportation policy; the first critical link between the industry and traffic was forged from 1923 to 1924. See St. Clair, *The Motorization of American Cities* (Praeger, 1986), esp. 124–126.

24. Herrold, "City Planning and Zoning" (paper read before the Minnesota Federation of Architectural Societies), *Canadian Engineer* 45 (July 10, 1923), 128–130 (129). In 1925, McClintock found Herrold's ratio "the most widely accepted estimate of the saturation point" (*Street Traffic Control*, McGraw-Hill, 1925, 4). For another saturation forecast by an engineer, see T. Glenn Phillips, "The Traffic Problems in Detroit and How They Are Met," *Annals* 116 (Nov. 1924), 241–243 (243).

25. M. Warren Baker, "Putting More Cars on the Road," *Motor Age* 48 (Sept. 24, 1925), 10–12, 43 (10).

26. Baker, "Putting More Cars on the Road," 10.

27. McClintock, *Street Traffic Control*, 8; see also Herrold, "City Planning and Zoning," 129.

28. Fisher quoted in Edward S. Jordan, "The Future of the Automobile," in National Motorists Association, "Report on the National Convention of Automobile Club Officials," Cleveland, Sept. 20–22, 1923 (mimeographed typescript), AAA Headquarters Library.

29. Jordan, "Future of the Automobile," 8.

30. M. Warren Baker, "Putting More Cars on the Road," *Motor Age* 48 (Sept. 24, 1925), 10–12, 43.

31. Kelker, De Leuw and Company, "Better Car Routing and Traffic Control Proposed for Baltimore" (abstract of a report to the Traffic Survey Commission of Baltimore), part 1, *Electric Railway Journal* 67 (May 22, 1926), 883–889 (884); Bureau of the Census, *Historical Statistics of the United States, Colonial Times to 1970* (Washington: Government Printing Office, 1975), 2:716.

32. Kelker, De Leuw and Company, "Better Car Routing and Traffic Control Proposed for Baltimore," part 2, *Electric Railway Journal* 67 (May 29, 1926), 923–927 (927).

33. Harold Cary, "Will Passenger Cars Be Barred from City Streets?" *Motor* 39 (March 1923), 34–35, 66, 68 (35).

34. "Sales of Automotive Vehicles Limited by Traffic Conditions" *Automotive Industries* 50 (May 29, 1924), 1184–1185 (1185).

35. M. Warren Baker, "Putting More Cars on the Road," *Motor Age* 48 (Sept. 24, 1925), 10–12, 43 (10).

36. Paul G. Hoffman, "Congestion of Traffic Retards Motor Buying," *New York Times*, Jan. 7, 1934. See also McClintock, *Street Traffic Control*, 4.

37. Edward S. Jordan, "The Future of the Automobile," in National Motorists Association, "Report on the National Convention of Automobile Club Officials," Cleveland, Sept. 20–22, 1923 (mimeographed typescript), AAA Headquarters Library. Jordan's "floor space" analogy gained some currency within the industry in the mid 1920s; see for example J. Borton Weeks, "Philadelphia's Traffic Problems and Their Solution," *Annals* 116 (Nov. 1924), 235–240 (236), and Alfred Reeves, "Have We Enough Highway Floor Space?" *Nation's Business* 12 (Dec. 1924), 30–31.

38. Alfred Reeves, address to the Rotary Club of Detroit, July 18, 1923, quoted in "Use of Cars Limited by City Congestion," *Automotive Industries* 49 (July 19, 1923), 150.

39. Harold Cary, "Will Passenger Cars Be Barred from City Streets?" *Motor* 39 (March 1923), 34–35, 66, 68 (35).

40. Russell Van Nest Black, "The Spectacular in City Building," *Annals* 133 (Sept. 1927), 50–56.

41. Frank R. Coates, "Electric Railways Must Represent Car Riders in Discussions of Traffic Congestion" (address, U.S. Chamber of Commerce, Washington, May 11, 1926), *AERA* 15 (June 1926), 738–750 (747).

42. Macauley in *Automotive Industries*, Jan. 22, 1925, reprinted as "Active City Planning Body Needed in Every Town," in Macauley, *City Planning and Automobile Traffic Problems* (Detroit: Packard Motor Car Company, 1925), 40–45 (41).

43. Macauley in *Public Works*, Sept. 1924, reprinted as "Adapting the City to the Automobile," in Macauley, *City Planning and Automobile Traffic Problems* (Detroit: Packard Motor Car Company, 1925), 5–10 (5–7).

44. "Graphical Presentation: Traffic Accident Records," March 1926, file 02770, Commerce Papers, Hoover Presidential Library.

45. Alfred Reeves, address to the monthly meeting of directors, NACC, Buffalo, July 1925, quoted in "Makers See Decline Lasting Two Months," *Automotive Industries* 49 (July 26, 1923), 200.

46. Two-car homes: see H. B. Peabody, "Answering the Parking Question" (paper presented to the City Traffic Conference of the National Automobile Chamber of Commerce, Chicago, April 13–14, 1927), *Relief of City Traffic*, booklet 4 (New York: the Chamber, 1927), and M. Warren Baker, "Putting More Cars on the Road," *Motor Age* 48 (Sept. 24, 1925), 10–12, 43 (10). Annual model change: Robert Paul Thomas, "Style Change and the Automobile Industry During the Roaring Twenties," in Louis P. Cain and Paul J. Uselding, eds., *Business Enterprise and Economic Change: Essays in*

*Honor of Harold F. Williamson* (Kent State University Press, 1973), 118–138; David A. Hounshell, *From the American System to Mass Production: The Development of Manufacturing Technology in the United States, 1800–1932* (Johns Hopkins University Press, 1984), 263–266; and James J. Flink, *The Automobile Age* (MIT Press, 1988), 239–244. "Used car evil": Alvan Macauley, "And What of the Motor Car Industry?" *Nation's Business* 13 (Jan. 1925), 46, 48, 50 (48, 50); see also Flink, *Automobile Age*, 131, 193, 212, 229, "Many Dealers Favor Scrapping Old Cars," *Motor Age* 51 (April 21, 1927), 13, 20; and K. P. Albridge, "Radical Used Car Plan Suggested" (letter to editor), *Automotive Industries* 51 (July 31, 1924), 248–249.

47. See chapters 3 and 8.

48. Macauley, "Adapting the City to the Automobile," 5–7.

49. "Industry's Spotlight Is Turned on Traffic Problems," *Automotive Industries* 56 (April 23, 1927), 607–610 (609–610). The engineer-Cassandra was Arthur S. Tuttle. His warning was a staple of traffic control engineering in the 1920s, but it was independently rediscovered and popularized in the late 1950s and the 1960s.

50. On curb parking, see chapter 5. See also John A. Miller Jr., "Lax Enforcement of Traffic Rules: A Prevalent Cause of Congestion," *Electric Railway Journal* 75 (August 1931), 400–404.

51. Frank R. Coates, "Electric Railways Must Represent Car Riders in Discussions of Traffic Congestion," *AERA* 15 (June 1926), 738–750 (743); "Local Chambers and Traffic," *Nation's Business*, extra edition: "Self-Government in Business" (June 5, 1926), 52–54 (53). The electric railways' growing direct role in traffic matters helped to subvert the authority of experts (traffic engineers) who were generally sympathetic to the railways' interests.

52. Anderson, "The Business of Street Management," *AERA* 16 (Nov. 1926), 689–695 (693).

53. Emmons, " 'The Intellectual Revolution' Hits the Street Cars," *AERA* 16 (Oct. 1926), 358–362 (362).

54. "An Appeal to Courtesy," *American City* 30 (March 1924), 289.

55. Vincent O. Law, "Advertising Will Aid, but It Cannot Do Everything," *AERA* 16 (August 1926), 67–73.

56. Labert St. Clair, "A Little Journey to the Home of the Toonerville Trolley," *AERA* 16 (August 1926), 1–12. For a large collection of Fox's "Toonerville Trolley" cartoons, see Fox, *Fontaine Fox's Toonerville Trolley*, ed. H. Galewitz and D. Winslow (Weathervane Books, 1972).

57. Editor's introduction to Edmund J. Murphy, "Why Some Electric Railway Lines Have Been Discontinued," *AERA* 16 (Sept. 1926), 183–196 (183).

58. Kansas City (Missouri) Railways Company, "How Many People in These Pictures? You'd Be Surprised!" (advertisement), reproduced in E. B. Sanders, "We Know the Car Riders Read Our Advertisements," *AERA* 16 (August 1926), 22–30 (26). Chicago Surface Lines used the same visual method to bring home the "superior efficiency of street cars"; see "Street Car Efficiency" (newspaper advertisement), reproduced in *Popularizing Public Transportation, 1927*, ed. H. Norris (New York: American Electric Railway Association, 1928), plate XII.

59. Henry H. Norris, ed., *Popularizing Public Transportation, 1927* (New York: American Electric Railway Association, 1928), 95.

60. Westinghouse Electric and Manufacturing Company, "Street Cars Relieve Street Congestion" (advertisement) in, e.g., *Nation's Business* 14 (July 1926), 75; Lesley C. Paul, "Manufacturers Can Help Industry in Various Ways," *AERA* 16 (August 1926), 88–94; " 'Reach the Individual,' G. E. Motto in Advertising," *AERA* 16 (August 1926), 81–87.

61. According to James J. Flink, between 1920 and 1929 the number of automobile manufacturers fell from 108 to 44; see Flink, *The Automobile Age* (MIT Press, 1988), 70. In 1922, according to the president of a small manufacturer, there were still "about one hundred and twenty-nine manufacturers of automobiles today"; see Philip H. Brockman, "No Auto Saturation Likely to Be Reached," *St. Louis Star*, Oct. 22, 1922.

62. Alvan Macauley and B. F. Everitt, "Some Causes and Effects of Automotive Prosperity," *Industrial Management* 71 (June 1926), 329–333.

63. Paul G. Hoffman, "The Traffic Commission of Los Angeles: Its Work on the Traffic Problem," *Annals* 116 (Nov. 1924), 246–250 (246).

64. Automobile Club of Southern California, *The Los Angeles Traffic Problem* (Los Angeles: the Club, 1922), 6.

65. On Los Angeles's "parking crisis" of 1920, see Bottles, *Los Angeles and the Automobile*, 63–89; Virginia Scharff, "Of Parking Spaces and Women's Places: The Los Angeles Parking Ban of 1920," *NWSA Journal* 1 (Sept. 1988), 37–51; Robert M. Fogelson, *Downtown: Its Rise and Fall, 1880–1950* (Yale University Press, 2001), 290–291; G. Gordon Whitnall, " 'No Parking of Autos,' " *American City* 22 (May 1920), 484.

66. Automobile Club of Southern California, *Los Angeles Traffic Problem*, esp. 3, 6.

67. Los Angeles Traffic Commission, *The Los Angeles Plan* (Los Angeles: the Commission, 1922).

68. Frederick Law Olmsted, Harland Bartholomew, and Charles Henry Cheney, *A Major Traffic Street Plan for Los Angeles* (Los Angeles: the Traffic Commission, May, 1924), esp. 5, 7; Bottles, *Los Angeles and the Automobile*, chapter 4.

69. Paul G. Hoffman, "The Traffic Commission of Los Angeles: Its Work on the Traffic Problem," *Annals* 116 (Nov. 1924), 246–250 (247); *Major Traffic Street Plan*, 16.

70. Hoffman in 1938 recalling his plans in 1922, in "Free Enterprise—Can It Survive?" (speech before the Bond Club of New York, Dec. 8, 1938), *Vital Speeches of the Day* 5 (Jan. 15, 1939, 205–208 (205).

71. "Paul G. Hoffman in Charge Sales Studebaker Car," *Brenham* (Texas) *Banner-Press*, April 10, 1925, 1, box: newspaper clippings, 1924–1927, Studebaker National Museum, South Bend, Indiana. (I thank Robert Denham of the museum for kindly photocopying this article and sending it to me).

72. "Traffic Expert Retained," *Los Angeles Times*, July 2, 1924; Thomas Sugrue, "Miller McClintock," *Scribner's Magazine* 102 (Dec. 1937), 9–13, 99 (12). Sugrue's article remains the best account of McClintock's career in traffic. See also *The National Cyclopaedia of American Biography*, volume 44 (James T. White, 1962), 14–15. On Hoffman, see Alan R. Raucher, "Paul G. Hoffman, Studebaker and the Car Culture," *Indiana Magazine of History* 79 (Sept. 1983), 209–230.

73. Published in New York by McGraw-Hill. Intimations of Hoffman's and McClintock's personalities come through their writings and through the articles by Raucher and Sugrue.

74. McClintock's dissertation is "The Street Traffic Problem" (Harvard, 1924); the manuscript is at the Harvard University Archives.

75. "Council Scorns Free Help," *Los Angeles Times*, July 3, 1924.

76. Hoffman, "The Traffic Commission of Los Angeles: Its Work on the Traffic Problem," *Annals* 116 (Nov. 1924), 246–250 (247).

77. Hoffman, "The Traffic Commission of Los Angeles: Its Work on the Traffic Problem," *Annals* 116 (Nov. 1924), 246–250 (247); McClintock, "Interesting Features of Los Angeles' New Traffic Ordinance," *American City* 32 (March 1925), 333, 335; McClintock, *Street Traffic Control* (McGraw-Hill, 1925), 18–184.

78. McClintock, "Simple Rule Code Needed for Traffic," *Los Angeles Times*, Nov. 16, 1924; see also "Rights of Autoist Upheld," *Los Angeles Times*, Oct. 15, 1924.

79. "Traffic Law Explained," *Los Angeles Times*, Jan. 23 and 24, 1925. On the implementation of the code, see chapter 8.

80. McClintock, "Interesting Features of Los Angeles' New Traffic Ordinance"; D. W. Pontius, "Congestion, Traffic Control and Rapid Transit" (abstract of a paper presented before the American Electric Railway Association, Oct., 1925), *Electric Railway Journal* 66 (Oct. 10, 1925), 618–619 (618).

81. " 'Jaywalker' Law Reduces Accidents throughout City," *Los Angeles Times*, Feb. 8, 1925; "Success of New Traffic Rule Causes Other Cities to Gasp, 'How Did You Do It?' " *Los Angeles Times*, Sept. 27, 1925; Pontius, "Congestion, Traffic Control and Rapid Transit," 618; McClintock, *Street Traffic Control*, 98–101, 145–146.

82. "California Cities Adopt Traffic Code," *New York Times*, May 2, 1926.

83. "Paul G. Hoffman in Charge"; Sugrue, "Miller McClintock," 12–13.

84. Sugrue, "Miller McClintock," 12–13.

85. "Research in Traffic for University," *Los Angeles Times*, Oct. 14, 1925; "Establish First Traffic Research Bureau," *Automotive Industries* 53 (Oct. 22, 1925), 725; "Endows Fellowships for Traffic Study," *New York Times*, Oct. 25, 1925; "Bureau for Street Traffic Research Created," *American City* 33 (Dec. 1925), 689; Raucher, "Paul G. Hoffman, Studebaker, and the Car Culture," 215; "Four Frictions," *Time*, August 3, 1936, 41–43 (41). Donald T. Critchlow's definitive corporate history of Studebaker (*Studebaker: The Life and Death of an American Corporation*, Indiana University Press, 1996) unfortunately has nothing to say about McClintock or the Erskine Bureau, although it may have been through them that Studebaker made its most important mark on America. Albert Russel Erskine had a penchant for creating namesakes; months after he established the Erskine Bureau he unveiled a new (and unsuccessful) small car, the "Erskine" (Critchlow, *Studebaker*, 91; Alan R. Raucher, "Albert Russel Erskine," in *The Automobile Industry, 1920–1980*, ed. G. May (Bruccoli Clark Layman and Facts on File, 1989), 136).

86. "The Classical and Medieval Origins of the Character of Diomedes as He Appears in Chaucer's *Troilus and Cressida*"; see Sugrue, "Miller McClintock," 9.

87. Sugrue, "Miller McClintock," 9–10, 12.

88. Mary A. Burke, "How Street Railways Are Helping to Solve the Traffic Problem," *AERA* 20 (May 1929), 270–276 (270).

89. McClintock, *Report and Recommendations of the Metropolitan Street Traffic Survey* (Chicago Association of Commerce, 1926), 4, 10. Paul Barrett agrees that "McClintock's 1925 book on street traffic control contains most of the recommendations which the engineer made one year later for Chicago"; see Barrett, *The Automobile and Urban Transit: The Formation of Public Policy in Chicago, 1900–1930* (Temple University Press, 1983), 155.

90. McClintock, "Diagnosing the Ills of San Francisco's Traffic," *Motor Land* 21 (August 1927), 20–21, 48–50 (20).

91. McClintock, *Report and Recommendations*, 12. McClintock reported here that the Street Traffic Committee of the Chicago Association of Commerce shared this view.

92. McClintock, "Diagnosing the Ills," 21.

93. McClintock, "Can We Solve the Traffic Problem by Guesswork?" (abstract of a paper presented at the Central States Safety Congress, Kansas City, Mo., April 15, 1927), *Good Roads* 70 (1927), 269–271 (269).

94. *Street Traffic Control*, 4, 5–6, 110, 6–7. Already in this book McClintock countenanced more ambitious efforts to accommodate automobiles, but considered them steps for an unspecified and distant future; see esp. 106–107, 113.

95. McClintock, "How the City Traffic Problem Will Be Solved," *The Automobilist*, May 1927, 5–7 (5); see also McClintock, "Remedies for Traffic Congestion," *S.A.E. Journal* 23 (Nov. 1928), 443–446 (443).

96. *Street Traffic Control*, 113.

97. "City Traffic Problem," 5; see also McClintock, "Remedies," 443.

98. *Street Traffic Control*, 56; "City Traffic Problem," 5.

99. "City Traffic Problem," 7.

100. "City Traffic Problem," 6. See also McClintock, "Highways and Streets Inadequate for Cars," *New York Times*, Jan. 6, 1929.

101. "City Traffic Problem," 5, 7.

102. "Just Among Ourselves," *Automotive Industries* 49 (Nov. 8, 1923), 940–941 (940).

103. Edward S. Jordan, "Signal Lights Cannot Solve Peak Traffic," *Good Roads*, Sept. 1927, 399.

104. "N. A. C. C. Conference Gets Traffic Views," *Automotive Industries* 56 (April 16, 1927), 597.

105. Sam Shelton, "Automobile Makers and City Officials Discuss Traffic Problems," *Motor Age* 51 (April 21, 1927), 9, 20 (9).

106. "Sessions on Traffic Held in New York and Chicago," *Engineering News-Record* 98 (April 21, 1927), 666.

107. Russell Van nest Black et al., "Committee Recommendations for Reorganization of the American City Planning Institute," Oct. 21, 1934, Walter Blucher Papers, Cornell University, quoted in Eugenie Ladner Birch, "Advancing the Art and Science of Planning: Planners and Their Organizations, 1909–1980," *APA Journal*, Jan. 1980, 22–49 (28). Harold Buttenheim, editor of *The American City*, gave wide publicity to practical city planning projects and took a keen personal interest in the place of the automobile in the city. When the American City Planning Institute refused to admit him, Buttenheim helped to found the American Society of Planning Officials (Birch, "Art and Science of Planning," 29).

108. Planning Foundation of America, *New Cities for the New Age* (New York: the Foundation, 1929), 25–31; Birch, "Advancing the Art and Science of Planning," 28.

109. *New Cities for the New Age*, 6, 10, 25, 27, 29.

110. "Born to the U.S. A.—A New and Gigantic A.A.A.," *American Motorist* 16 (August 1924), 24–25, 34, 36 (24). Under Henry, the statement of purpose quoted (and variations of it) became a slogan at annual meetings of the AAA.

111. "What? No Traffic Jam" (untraceable newspaper clipping, c. 1955), and "Professional Data—Burton W. Marsh," June 14, 1962 (typescript), both in vertical file, Biographies, Marsh, AAA Headquarters Library; *Annual Meeting of the Councillors of the American Automobile Association* (Washington: the Association, 1934), 44; "Burton Marsh Receives NSC Safety Award," *AAA News Review* 34 (Nov. 1974), 3; " 'Mr. ITE': Burton Wallace Marsh (1898–1988)," *ITE Journal* 58 (March 1988), 11–14 (11–12).

## Chapter 7

1. McClintock, "Preventive and Palliative Measures for Street Traffic Relief," *City Planning* 6 (April 1930), 99–105 (99).

2. Miller McClintock, *Street Traffic Control* (McGraw-Hill, 1925); Maxwell Halsey, *Traffic Accidents and Congestion* (Wiley, 1941), 16.

3. Carter H. Harrison, "The Regulation of Public Utilities" (address, Nov. 13, 1914), *Annals* 57 (Jan. 1915), 54–61 (54).

4. Delos F. Wilcox, "Public Utility Advice from a Public Point of View," *American City* 8 (March 1913), 264, 266 (266).

5. American Electric Railway Association, *The Urban Transportation Problem* (New York: the Association, 1932), 30–31. See also Eliot Jones and Truman C. Bigham, *Principles of Public Utilities* (Macmillan, 1931, 1939), 53–58; G. Lloyd Wilson, James M. Herring, and Roland B. Eutsler, *Public Utility Industries* (McGraw-Hill, 1936), esp. 181.

6. Charles E. Merriam and Harold Lasswell, "Current Public Opinion and the Public Service Commissions," in *Public Utility Regulation*, ed. M. Cooke (Ronald, 1924), 286–289, 294–295.

7. Farley Gannett, "The Change in Attitude of the Public and Public Service Companies Toward State Regulation," *American City* 24 (March 1921), 280–281 (281).

8. In the years 1917–1920, 117 street railways went into receivership. At the end of 1917, fares averaged 5.05 cents; and at the end of 1927 fares averaged 7.75 cents. Eliot Jones and Truman C. Bigham, *Principles of Public Utilities* (Macmillan, 1931, 1939), 57, 378 n. 2.

9. On the street railway crisis of 1917–1921, see e.g. Jones and Bigham, *Principles of Public Utilities*, 56–58, and G. Lloyd Wilson, James M. Herring, and Roland B. Eutsler, *Public Utility Industries* (McGraw-Hill, 1936), 181.

10. Editorial, *Public Works*, June 1924, reprinted as "Traffic as an Engineering Problem," in Alvan Macauley, *City Planning and Automobile Traffic Problems* (Detroit: Packard Motor Car Company, 1925), 11–12 (12).

11. Cooke, "Public Engineering and Human Progress," *Journal of the Cleveland Engineering Society* 9 (Jan. 1917), 252, quoted in Edwin T. Layton Jr., *The Revolt of the Engineers: Social Responsibility and the American Engineering Profession* (Press of Case Western Reserve University, 1971), 179.

12. Layton, *Revolt of the Engineers*, 192.

13. Historian Guy Alchon has examined this trend carefully, and called it "techno-corporatist legitimation"; see *The Invisible Hand of Planning: Capitalism, Social Science, and the State in the 1920s* (Princeton University Press, 1985), chapter 7 (112–128), esp. 112.

14. John A. Beeler to Special Traction Committee, Atlanta, Dec. 16, 1924, in Beeler, *Report to the City of Atlanta on a Plan for Local Transportation* (Atlanta: Foote and Davies, 1924), iii.

15. P. G. Agnew, "How Business Is Policing Itself," *Nation's Business* 13 (Dec. 1925), 41–43 (43).

16. Hoffman, "The Traffic Commission of Los Angeles: Its Work on the Traffic Problem," *Annals* 116 (Nov. 1924), 246–250 (246, 250).

17. Macauley in *Automotive Industries*, Jan. 22, 1925, reproduced as "Active City Planning Body Needed in Every Town," in Macauley, *City Planning and Automobile Traffic Problems* (Detroit: Packard Motor Car Company, 1925), 40–45 (41).

18. The term is Ellis Hawley's, derived from Hoover's frequent use of the word "associative." See esp. Hawley, "Herbert Hoover, the Commerce Secretariat, and the Vision of an 'Associative State,' 1921–1928," *Journal of American History* 61 (1974), 116–140; Hawley, *The Great War and the Search for a Modern Order: A History of the American People and Their Institutions, 1917–1933* (St. Martin's Press, 1979); Hawley, "Three Facets of Hooverian Associationalism: Lumber, Aviation, and the Movies, 1921–1930," in *Regulation in Perspective*, ed. T. McCraw (Harvard University Press, 1981), 95–123, 221–233; Himmelberg, *Origins of the National Recovery Administration*, chapters 3–5; McCraw, *Prophets of Regulation*, 147–152.

19. And thus avoid positive regulation and remain true to the ideals of classical liberalism. The quotation is from an untitled collection of Hoover quotations (n.d.), in box 487, Commerce Papers, Hoover Library. On the state as umpire, see also Locke, *Second Treatise on Government*, chapter 7, §88.

20. William J. Donovan, address to the Association of Attorneys General, Sept. 25, 1925, *Proceedings of the Academy of Political Science* 11 (Jan. 1926), 19–26, quoted in Himmelberg, *Origins of the National Recovery Administration*, 48. On the Supreme Court decision (*Maple Flooring Manufacturers Association v. United States*, 1925), see esp. Himmelberg, *Origins of the National Recovery Administration*, 46–48, 52 n. 9; McCraw, *Prophets of Regulation*, 146–147; Morton J. Horwitz, *The Transformation of American Law, 1870–1960: The Crisis of Legal Orthodoxy* (Oxford University Press, 1992), 207, 326 n. 77. On the Federal Trade Commission's change of course, see esp. McCraw, *Prophets of Regulation*, 149–152.

21. Among important the scholarship on associationism, only Himmelberg (*Origins of the National Recovery Administration*) ascribes to the National Chamber its due importance in the mid 1920s; see esp. 44–45.

22. Hoover, *The Memoirs of Herbert Hoover: The Cabinet and the Presidency, 1920–1933* (Macmillan, 1952), 173.

23. P. G. Agnew, "How Business Is Policing Itself," *Nation's Business* 13 (Dec. 1925), 41–43; "Self-Government for Industry," *Nation's Business* 14 (Feb. 1926), 40; Merle Thorpe, "Home Rule for Business," 9–10, and Julius H. Barnes, "Self-Government in Business," 16–18, both in *Nation's Business* 14 (extra edition of June 5, 1926). The general subject of the fourteenth annual meeting of the National Chamber in 1926 was "Self-Government in Business"; in part the participants considered how far "self-imposed standards of practice" could "do away with the need of regulation by the state or municipality." See *Nation's Business* 14: " 'Self-Government in Business' " (April 1926), 75, and "Self-Government in Business" (May 1926), 40. It is ironic that just when local chambers of commerce were backing regulatory traffic control, their national body (the National Chamber) was working to reduce regulation. Local chambers turned to traffic control, in part, for relief from chaotic police traffic regulation, which was more intrusive and less effective (to those concerned above all with traffic efficiency). The National Chamber, in practice, never championed individual liberty. It stood for business cooperation and fought "wasteful" competition.

24. On the ubiquity of the conference's byname, see Ernest Greenwood to Hoover, Dec. 11, 1926, file 02772, box 161, Commerce Papers, Hoover Library.

25. Members convened again in 1934, but by then automotive interest groups conducted little of their work behind the Commerce Department aegis, and the conference was a shadow of its former self.

26. Hoover, "Opening Address of the Chairman," March 23, 1926, in *Second National Conference on Street and Highway Safety* (Washington: the Conference, 1926), 1–12 (11).

27. George M. Graham, "Safeguarding Traffic," *Annals* 116 (Nov. 1924), 174–185 (177).

28. Hoover, address, in *First National Conference on Street and Highway Safety* (Washington: the Conference, 1924), 7–11 (7, 11).

29. Hoover, address, in *First National Conference*, 37–39 (38).

30. Hoover, address, Washington, Dec. 16, 1924, in *First National Conference on Street and Highway Safety* (Washington: the Conference, 1924), 37–38. The proportion of engineers among conference participants is estimated from a comparison of *First National Conference*, 40–51, and John William Leonard, ed., *Who's Who in Engineering 1925* (Who's Who Publications, 1925).

31. The lone traffic engineer was W. J. Cox of New York. Excluded from this total of "professional engineers" are delegates of the Society of Automotive Engineers (whose work was confined to automobile design) and engineers who worked for the steam railroads.

32. Hoover, "We Can Cooperate and Yet Compete," *Nation's Business* 14 (June 5, 1926), 11–14 (13).

33. Coolidge, address, in *First National Conference on Street and Highway Safety* (Washington: the Conference, 1924), 5–6 (5).

34. F. H. Caley, "The Parking Problem," *Proceedings of the American Electric Railway Transportation and Traffic Association* (New York: American Electric Railway Association, 1928), 42–50 (43). Caley served on the Hoover Conference committee that drafted the Uniform Vehicle Code for adoption by states and the Model Municipal Traffic Ordinance for cities (see below).

35. See press releases in file 02781, box 160, Commerce Papers, Hoover Library; e.g. "Accident Statistics Show Danger Spots; Permit Remedy," Sept. 20, 1924.

36. Editor's introduction to Ernest Greenwood, "Make the Streets Safe," *The Survey* 53 (Dec. 15, 1924), 317–319 (317). Greenwood was Hoover's secretary for the conference.

37. "Remedial and Other Measures for Unravelling Traffic," *Electric Railway Journal* 62 (Dec. 22, 1923), 1035.

38. "Report of the Legislative Committee," *Proceedings of the Twenty-Seventh Annual Meeting of the American Automobile Association*, Buffalo, July 1–2, 1929 (Washington, the Association: 1929), 12–15 (15).

39. Hoover, address, in *First National Conference on Street and Highway Safety* (Washington: the Conference, 1924), 37–38.

40. Frank R. Coates, "Electric Railways Must Represent Car Riders in Discussions of Traffic Congestion" (address at the annual convention of the U.S. Chamber of Commerce, Washington, May 11, 1926), *AERA* 15 (June 1926), 738–750 (739). For an

earlier, similar claim, see "Remedial and Other Measures for Unravelling Traffic," *Electric Railway Journal* 62 (Dec. 22, 1923), 1035.

41. W. G. Strait, address to the Metropolitan Section, American Electric Railway Association, New York, Dec. 3, 1926, in "Metropolitan Section," *AERA* 16 (Jan. 1927), 1070–1084 (1078).

42. E. J. McIlraith, "Promoting the Best Use of City Streets" (abstract of an address before the annual meeting of the New York Electric Railway Association, Bluff Point, New York, June 24–25, 1927), *Electric Railway Journal* 70 (July 2, 1927), 23–25 (23).

43. Mary A. Burke, "How Street Railways Are Helping to Solve the Traffic Problem," *AERA* 20 (May 1929), 270–276 (274). Street railway advertisements frequently noted that the railways served the majority.

44. A. E. Mittendorf, "Cooperation of Motor Clubs with Dealers' Associations and Other Civic Groups," *Second Conference of Club and Association Secretaries*, Washington, March 25–27, 1926, 87–90 (87).

45. Stanley H. Horner, "Dealer Co-operation with Motor Clubs," in American Automobile Association, *Third Conference of Club and Association Secretaries*, Washington, March 23–25, 1927, 69–72 (71).

46. Press release, Automobile Club of the District of Columbia, for Nov. 7, 1915. See clippings, *Washington Post* and *Washington Star*, Nov. 7, 1915, in Seiler Scrap Book, AAA Headquarters Library.

47. Fred H. Caley, "The Parking Problem," in *Proceedings of the American Electric Railway Transportation and Traffic Association* (New York: American Electric Railway Association, 1928), 42–50 (43).

48. J. Clyde Myton, "Legislative Activities of Motor Clubs," *Conference of Club and Association Secretaries* (Washington: American Automobile Association, 1925), 29–33 (30); "Hog-Tying the Automobile," *Ohio Motorist*, Oct. 1924, 8. The earliest example of an auto club attack on restrictive traffic regulation that the author has found is Herbert Buckman, "Traffic Problems" (address to the National Motorists Association), in "Report on the National Convention of Automobile Club Officials," Cleveland, Sept. 20–22, 1923 (typescript), AAA Headquarters Library.

49. Fred H. Caley, "The Parking Problem," in *Proceedings of the American Electric Railway Transportation and Traffic Association* (New York: American Electric Railway Association, 1928), 42–50 (43); Charles C. Janes, address, *Conference of Club and Association Secretaries* (Washington: American Automobile Association, 1925), 34–40 (36).

50. Edwin T. Layton Jr., *The Revolt of the Engineers: Social Responsibility and the American Engineering Profession* (Press of Case Western Reserve University, 1971), 193.

51. Hoover, "The Problem of Distribution" (address before the National Distribution Conference, Jan. 14, 1925), p. 3, box 153, Commerce Papers, Hoover Library; Hoover repeated the policeman metaphor in "Yes, We Can Cut Marketing Costs," *Nation's Business* 13 (March 1925), 48, 50, 74, 76–77 (48) and in "We Can Cooperate and Yet Compete," *Nation's Business* 14 (June 5, 1926), 11–14 (12).

52. Editorial, *Nation's Business* 14 (Jan. 1926), 6. *Nation's Business*, the National Chamber's journal, formed its "Fewer Laws Club" in March 1925; see Merle Thorpe, "Through the Editor's Spectacles," *Nation's Business* 13 (May 1925), 7–8, 10–12 (7). The superiority of "natural law" unencumbered by human lawmakers, the surfeit of legislation, and the ability of business to regulate itself are constant themes in *Nation's Business* in the mid 1920s.

53. Since the first history of America in the 1920s, Frederick Lewis Allen's *Only Yesterday: An Informal History of the Nineteen-Twenties* (Harper and Row, 1931), the decade has been characterized as a time when business "was regarded with a new veneration" (146).

54. Slauson, "The Municipal Traffic Problem" (address to the Motor and Accessory Manufacturers Association, Sept. 21, 1923), *Roads and Streets*, Nov. 1923, 976–978 (978).

55. Macauley in *Automotive Industries*, Jan. 22, 1925, reprinted as "Active City Planning Body Needed in Every Town," in Macauley, *City Planning and Automobile Traffic Problems* (Detroit: Packard Motor Car Company, 1925), 40–45 (41).

56. "Highways Transport Committee's General Traffic Regulations," *Good Roads*, Sept. 3, 1919, 122–123; "Size and Weight Restrictions in State Motor Vehicle Laws in Effect July 1, 1920," *American City* 23 (Oct. 1920), 410–411 (411).

57. For an early instance of the participation of the American Electric Railway Association in efforts to achieve uniformity, see "American Association Suggestions," *Electric Railway Journal* 60 (Oct. 7, 1922), 585.

58. J. J. Cavanagh, "Legal Work for Motor Clubs" (address, May 1, 1925) *Conference of Club and Association Secretaries* (Washington: American Automobile Association, 1925), 176–183 (177).

59. " 'There Ought to Be a Law' " (advertisement, Dec. 1, 1924, in *New York Times*, Chicago *Daily News*, Cleveland *Plain Dealer*, and Kansas City *Star*), reproduced in *Nation's Business* 13 (Feb. 1925), 111.

60. Editor's introduction to Agnes C. Laut, "Lawmaking Still Runs Wild," *Nation's Business* 14 (July 1926), 23–25 (23).

61. " 'There Ought to Be a Law' " (poem by Strickland Gillilan, cartoon by Charles Dunn), *Nation's Business* 13 (August 1925), 33.

62. "The Clear Road for Motor Legislation" (editorial), *Nation's Business* 14 (Dec. 1926), 27.

63. Craig, "What Did You Go and Adopt More For?" (cartoon), Rochester, New York, *Democrat and Chronicle*, reproduced in *The Outlook* 139 (Feb. 18, 1925), 247; Fitzpatrick, "The Laws of Moses and the Laws of Today" (cartoon), *St. Louis Post-Dispatch*, reproduced in *Nation's Business* 14 (July 1926), 24.

64. Hoover, "Opening Address of the Chairman," March 23, 1926, *Second National Conference on Street and Highway Safety* (Washington: the Conference, 1926), 7–12 (8).

65. By "positive regulation" I mean regulation that seeks to shape individual social demands, fostering some and discouraging others, for the sake of a social good, as has been a customary ideal in public utility regulation. It is distinct from *planning*, which shapes ends as well as means. By "negative regulation" I mean regulation that checks individual abuses but leaves social goals and the means of reaching them to "natural law," leaving the state in the role of umpire.

66. Barber, "Need for Unified Action by All Public Utilities in Approach to Traffic Congestion Solution," *Proceedings of the American Electric Railway Association* 44 (New York: the Association, 1925), 171–177 (174–175); National Conference on Street and Highway Safety, *Report of the Committee on Uniformity of Laws and Regulations* (Washington: the Conference, 1926), esp. 315; *Second National Conference on Street and Highway Safety* (Washington, the Conference, 1926), 4, 8, 14–17, 29–31, esp. 30; National Conference on Street and Highway Safety, *Final Text of Uniform Vehicle Code* (Washington: the Conference, 1926), esp. 76–77. The Committee on Uniformity of Laws and Regulations consisted of 34 men and women from a broad spectrum of interests, including five representatives of motordom and two of street railways.

67. National Conference on Street and Highway Safety, *Report of the Committee on Uniformity of Laws and Regulations* (Washington: the Conference, Jan. 29, 1926), § 23, p. 315; National Conference on Street and Highway Safety, *Final Text of Uniform Vehicle Code* (Washington: the Conference, August 20, 1926), § 23, p. 76.

68. "The Uniform Code" (editorial), *AERA* 16 (Jan. 1927), 1057–1058.

69. See, e.g., the exchange between Barber and Fred H. Caley of the Cleveland Automobile Club on Ohio's code, March 23, 1927, in American Automobile Association, *Third Conference of Club and Association Secretaries*, Washington, March 23–25, 1927, 65–66.

70. Barber to Hoover, May 23, 1927, file 02773, box 161, Commerce Papers, Hoover Library.

71. Hoover, in National Conference on Street and Highway Safety, *Ways and Means to Traffic Safety* (Washington: the Conference, 1930).

72. The phrase "organized motordom" was a favorite at AAA; see e.g. *Twenty-Fourth Annual Meeting, American Automobile Association: Summary of Proceedings*, Chicago, June 7–8, 1926 (Washington: AAA, 1926), 5.

73. M. Warren Baker, "Putting More Cars on the Road," *Motor Age* 48 (Sept. 24, 1925), 10–12, 43 (11).

74. A. E. Mittendorf, "Good Roads Work by Motor Clubs," 44–48 (46), and Charles C. Janes, "Address," 34–39 (38), both in *Conference of Club and Association Secretaries* (Washington: American Automobile Association, 1925).

75. William A. Jensen, "Address," *Conference of Club and Association Secretaries* (Washington: American Automobile Association, 1925), 16–20 (18). On the value of such information to legislators, and the consequent ability of interest groups that can supply it to gain access to legislators, see John Mark Hansen, *Gaining Access: Congress and the Farm Lobby, 1919–1981* (University of Chicago Press, 1991, esp. 12–17, 103–106.

76. "Report of Legislative Committee," June 16, 1927, *Proceedings of the Twenty-Fifth Annual Meeting of the American Automobile Association*, Philadelphia, June 15–17, 1927, 13–17 (16).

77. James Lightbody, "Knowing What You Want Necessary to Advertising Success," *AERA* 16 (August 1926), 74–80 (78).

78. A. Leroy Hodges, "Five Causes of Traffic Congestion" (abstract of a paper presented before the New York Electric Railway Association, Bluff Point, New York, June 24, 1927), *Electric Railway Journal* 70 (July 1927), 22–23 (23).

79. R. F. Kelker Jr., "Suggestions for Relief of Street Congestion" (abstract of a paper presented to the American Electric Railway Association, St. Louis, March 4, 1924), *Electric Railway Journal* 63 (March 8, 1924), 373–375 (374).

80. A decade later a New York clergyman noted "the dual personalities with which many people are possessed and which frequently assert themselves as one passes from the rôle of driver to that of pedestrian." Ralph W. Sockman quoted in "Safety Officials Urged for Cities," *New York Times*, March 4, 1936.

81. "California Cities Adopt Traffic Code," *New York Times*, May 2, 1926; "Traffic Code Gains Favor," *Los Angeles Times*, Jan. 8, 1928; Barber, "How the States Are Working Towards a Uniform Motor Vehicle Code," *American City* 35 (Dec. 1926), 851–854 (851–852); Barber to Hoover, May 23, 1927, file 02773, box 161, Commerce Papers, Hoover Library; Barber, "Making Our Traffic Laws Uniform," *Annals* 133 (Sept. 1927), 128–133 (133); Barber, "Street and Highway Safety Conferences," *Roads and Streets* 67 (Sept. 1927), 405–406 (406).

82. "William Metzger, Auto Pioneer, Dies," *New York Times*, April 12, 1933.

83. Barber to Hoover, May 23, and Hoover to Barber, May 29, 1927, file 02773, box 161, Commerce Papers, Hoover Library.

84. National Conference on Street and Highway Safety, *Tentative Draft of Model Municipal Traffic Ordinance* (Washington: the Conference, May 1, 1928); Barber, "Making Our Traffic Laws Uniform," *Annals* 133 (Sept. 1927), 128–133 (133); "Model Traffic Laws Complete," *Los Angeles Times*, August 26, 1928.

85. National Conference on Street and Highway Safety, *Tentative Draft of Model Municipal Traffic Ordinance* (Washington: the Conference, 1928), 54.

86. "Traffic Code Gains Favor," *Los Angeles Times*, Jan. 8, 1928; "Model Traffic Laws Complete," *Los Angeles Times*, August 26, 1928; *Tentative Draft of Model Municipal Traffic Ordinance*, §§ 18, 23, pp. 17–19, 57–60; C. W. Stark, "The Model Municipal Traffic Ordinance," *National Municipal Review* 17 (Nov. 1928), 684–689 (687).

87. Stark, "Model Municipal Traffic Ordinance," 686.

88. Draft of Committee Report, in "This document contains . . . ," 1927, pp. 2, 34, file 02778, box 161, Commerce Papers, Hoover Library.

89. McIlraith in "Committee Comments in Field of Subcommittee V," p. 1, in untitled collection of comments on subcommittee work on the Model Municipal Traffic Ordinance (1927), file 02777, Street and Highway Safety Papers, box 161, Commerce Papers, Hoover Library.

90. Frank A. Goodwin in "Committee Comments in Field of Subcommittee V," p. 2.

91. Hayes in "Committee Comments in Field of Subcommittee VI," p. 5, file 02777.

92. National Conference on Street and Highway Safety, *Tentative Draft of Model Municipal Traffic Ordinance* (Washington: the Conference, 1928), esp. 4, 17–19, 41, 57–60, 63, 81–84.

93. A practical guide appeared as Robert O'Brien, *A Handbook of the Theory and Practice of Safety First* (St. Louis: Robert O'Brien, 1929).

94. "Model Traffic Code Indorsed," *Los Angeles Times*, Oct. 7, 1928.

95. Miller McClintock, "Highways and Streets Inadequate for Cars," *New York Times*, Jan. 6, 1929.

96. National Conference on Street and Highway Safety, *Ways and Means to Traffic Safety* (Washington: the Conference, 1930), 3, 10–11, 43. See also Barber, "How the States Are Working Towards a Uniform Motor Vehicle Code," *American City* 35 (Dec.

1926), 851–854. On the absence of calls for stringent regulation by traffic engineers, see esp. *Proceedings of the Institute of Traffic Engineers* 1–3 (1930–1932), bound mimeograph volume, University of Michigan Library.

97. Hoover, "Masonic R[e]view and Radio Broadcast," n.d. (c. August 1922), box 489, Commerce Papers, Hoover Library.

98. Hoover, "Statement by the Secretary of Commerce at the Opening of the Radio Conference," Feb. 27, 1922, 2, box 489, Commerce Papers, Hoover Library.

99. Hoover to Gene Brown, Jan. 30, 1923, box 489, Commerce Papers, Hoover Library.

100. Hoover, "We Can Cooperate and Yet Compete," *Nation's Business* 14 (June 5, 1926), 11–14 (12).

101. Hoover to Brown, Jan. 30, 1923.

102. Hoover, "Statement by the Secretary of Commerce," 3.

103. Hoover, "Statement by the Secretary of Commerce," 1.

104. Anderson, "The Business of Street Management," *AERA* 16 (Nov. 1926), 689–695 (692). See also Deane S. Kintner, "Traffic Troubles in the Air," *Ohio Motorist* 16 (Dec. 1924), 16–18, 28.

105. Hoover, "Statement by the Secretary of Commerce," 3; Hoover, statement, Nov. 9, 1925, reprinted in Hoover, *The Memoirs of Herbert Hoover: The Cabinet and the Presidency, 1920–1933* (Macmillan, 1952), 144.

106. Hoover, "Statement by the Secretary of Commerce at the Opening of the Radio Conference," Feb. 27, 1922, p. 1, box 489, Commerce Papers, Hoover Library.

107. See Susan Smulyan, *Selling Radio: The Commercialization of American Broadcasting, 1920–1934* (Smithsonian Institution Press, 1994).

108. Hoover to C. E. Skinner, March 14, 1927, file 02758, box 160, Commerce Papers, Hoover Library; Hoover in National Conference on Street and Highway Safety, *Ways and Means to Traffic Safety* (Washington: the Conference, 1930). Readers may wonder why Hoover, if he truly sought to avoid private disputes, would so roundly condemn the "advertising chatter" that crowded the ether. Susan Smulyan has shown, however, that advertisers saw little promise in radio in the early 1920s, and so took little offense at such criticism; see Smulyan, *Selling Radio*, 68–71.

109. Hoover, address, in *First National Conference on Street and Highway Safety* (Washington: the Conference, 1924), 7–11 (38). On the sudden improvement of the prospects of associationism in 1924–1925, see Robert F. Himmelberg, *The Origins of the National Recovery Administration: Business, Government, and the Trade Association*

*Issue, 1921–1933*, second edition (Fordham University Press, 1993), chapters 3–4 (43–74).

110. On the origins and early history of the gasoline tax, see esp. John Chynoweth Burnham, "The Gasoline Tax and the Automobile Revolution," *Mississippi Valley Historical Review* 48 (Dec. 1961), 435–459, and (especially for California) Jeffrey Richard Brown, "Trapped in the Past: The Gas Tax and Highway Finance" (M.A. thesis, University of California, Los Angeles, 1998), 70–108.

111. New York levied the first auto registration fee in 1901; by 1913, all 48 states and the District of Columbia (and numerous local jurisdictions) collected fees from motorists; see James W. Follin, "Taxation of Motor Vehicles in the United States," *Annals* 116 (Nov., 1924), 141–159, esp. 144, 150.

112. John Chynoweth Burnham, "The Gasoline Tax and the Automobile Revolution," *Mississippi Valley Historical Review* 48 (Dec. 1961), 435–459 (443).

113. Reeves, "Why No More Taxes Should Be Placed on Motor Cars and Motor Trucks" (address, Oct. 22, 1920), p. 13, in bound collection marked "Pamphlets" (New York: National Automobile Chamber of Commerce, 1919–1933), Library, United States Department of Transportation, Washington.

114. Quoted in "Cost $375 to Collect $318,000 Gasoline Tax," *Automotive Industries* 49 (July 26, 1923), 197.

115. H. R. Trumbower, "The Economic Aspects of Passenger Transportation over the Highways," *Proceedings of the Twenty-Fifth Annual Meeting of the American Automobile Association*, Philadelphia, June 15–17, 1927, 23–31 (24), AAA Headquarters Library. Before the close of the decade, all 48 states collected gasoline taxes (Burnham, "Gasoline Tax," 446).

116. Burnham, "Gasoline Tax," 446.

117. Burnham, "Gasoline Tax," 448; Austin E. Heiss, "Motorists Goosestep for Tax Collector," *National Petroleum News* 17 (August 19, 1925), 84, 88 (84).

118. California auto clubs were early champions of the gasoline tax. See Jeffrey Richard Brown, "Trapped in the Past: The Gas Tax and Highway Finance" (M.A. Thesis, University of California, Los Angeles, 1998), 81–83.

119. Douglas Shelor, "Address of Mr. Douglas Shelor," American Automobile Association, *Conference of Club and Association Secretaries* (Washington, April 30–May 1, 1925), 40–42 (42). After helping to pass the state's gas tax, the Automobile Club of Washington lobbied to raise the tax rate (Shelor, "Address," 41–42). The last four states to pass a gasoline tax were among the most urbanized: Illinois, Massachusetts, New Jersey, and New York; see H. R. Trumbower, "The Economic Aspects of Passenger Transportation over the Highways," *Proceedings of the Twenty-Fifth Annual*

*Meeting of the American Automobile Association*, Philadelphia, June 15–17, 1927, 23–31 (24).

120. J. Clyde Myton, "Legislative Activities of Motor Clubs," American Automobile Association, *Conference of Club and Association Secretaries* (Washington, April 30–May 1, 1925), 29–33 (31–32). The gasoline tax bill passed despite auto club opposition (Myton, "Legislative Activities," 31). On the Pennsylvania Motor Federation's opposition to the tax, see also C. H. Hites, "Thomas P. Henry of Detroit Elected President of A.A.A.," *American Motorist* 15 (June 1, 1923), 1–3, 13 (2).

121. Robert P. Hooper, quoted in Hites, "Thomas P. Henry of Detroit Elected President of A.A.A." (3).

122. C. M. Talbert, "A City Tax on Gasoline," *American City* 23 (July 1920), 35–36; W. W. Horner, "How Modern Traffic and the City Plan Affect Distribution of Paving Costs," *National Municipal Review* 13 (Nov. 1924), 626–631 (628–629).

123. Thomas P. Henry, "The Motorist Pays—and Pays," *The Independent* 115 (Oct. 17, 1925), 437–439 (437).

124. "Report of the Resolutions Committee," *Proceedings of the Twenty-Fifth Annual Meeting of the American Automobile Association*, Philadelphia, June 15–17, 1927, 34–38 (36), AAA Headquarters Library. An earlier resolution pledged a similar defense of the revenues of "all special taxes levied against motor vehicles" ("Resolutions Adopted by the Annual Meeting," *Twenty-fourth Annual Meeting, American Automobile Association*, Chicago, June 7–8, 1926, 11–14 (13)).

125. Legislatures, under the scrutiny of motordom, honored this commitment in the 1920s. Late in the decade, about 98 percent of revenues were spent on roads and streets (Burnham, "Gasoline Tax," 455).

126. Reeves, address to the Cincinnati Automobile Dealers Association, Cincinnati, Oct., 1923, quoted in "Accidents Can Be Prevented," *Cincinnati Enquirer*, Oct. 21, 1923 (Automobile Section).

127. In 1920, 28.7 percent of state highway revenues came from imposts on motorists; by 1930, 65.6 percent were so derived. See U.S. Bureau of the Census, *Historical Statistics of the United States, Colonial Times to 1970* (Washington: Government Printing Office) part 1, 712.

128. Jeffrey Richard Brown, "Trapped in the Past: The Gas Tax and Highway Finance" (M.A. Thesis, University of California, Los Angeles, 1998), 70.

129. Maxwell Halsey, "What Has Parking Limitation Accomplished?" (paper presented at the annual meeting of the National Safety Council, Chicago, Oct. 13, 1931), *Transactions of the National Safety Council: Twentieth Annual Safety Congress* 3 (1931), 83–97 (83).

130. While city planners might help implement the motor age city, the initiative lay with the funding mechanism and the new view of roads and streets that it implied. Thus explanations of the motor age city cannot begin with city planners' visions. The automotive city did not begin on the drafting boards of prophetic designers such as Clarence Stein and Henry Wright (the designers of the automotive new town of Radburn, New Jersey) or Norman Bel Geddes (designer of General Motors' "Futurama": its vision of the city of 1960 presented at the New York World's Fair of 1939–1940). The most effective proponents of the motor age city saw more possibilities in highway engineers, who were represented as mere technicians responding almost automatically to the demands of consumers (gasoline-tax-paying motorists).

131. The reality was somewhat different. Though highway engineers (especially c. 1930–1970) represented themselves as technicians working to fulfill others' demands, in practice they did much more. For example, in locating highways they sometimes involved themselves in sensitive policy questions (e.g. urban renewal), and in meeting motorists' demands they promoted motor transportation over other modes. See Peter Norton, "Fighting Traffic: U.S. Transportation Policy and Urban Traffic Congestion, 1950–1970," *Essays in History* 38 (1996), available at http://etext. lib.virginia.edu.

132. "The Story of the 'Snitching Post,' " *Highway User*, Nov. 1965, 19–21 (19).

133. Public Relations Department, AAA, Special Information Bulletin 2, "Parking Meter Developments," in Vertical File: "Parking," AAA Headquarters Library.

134. AAA referred to the national anti-meter camaign as a "war" and to each city anti-meter effort as a "fight." See Department of Public Relations, AAA, Special Information Bulletin 5: "The 'Parking Tax' War: Motor Clubs Score Signal Victories in Many Sectors, Business Groups Join Fight Against Meters," Sept. 29, 1936, in Vertical File: "Parking," AAA Headquarters Library. In one case (Mobile, Alabama), the introduction of meters precipitated a threat of a "protest march" (presumably orchestrated by the auto club), which was called off only when the city agreed to remove the meters. See "The 'Parking Tax' War," p. 4.

135. Though many cities considered meters, only 90 had adopted them by c. 1939–40; see Norman Bel Geddes, *Magic Motorways* (Random House, 1940), 230. For details on the spread of meters, 1935–1941, see Vertical File: Parking, AAA Headquarters Library.

136. Department of Public Relations, AAA, Special Information Bulletin: "The Park-O-Meter: A Survey of Its Operation, Cost and Public Reception," March 17, 1936, p. 4, in Vertical File: "Parking," AAA Headquarters Library.

137. "Parking Meters Operate in 1,209 Cities," *American City* 62 (Sept. 1947), 133.

138. Editorial, *Public Works*, reprinted as "Traffic as an Engineering Problem," in Alvan Macauley, *City Planning and Automobile Traffic Problems* (Detroit: Packard Motor Car Company, 1925), 11–12. The change in the industry position is striking. In 1920, Alfred Reeves of the National Automobile Chamber of Commerce argued against automotive excise taxes for roads on the grounds that roads are for everybody, not just motorists; see Reeves, "Why No More Taxes Should Be Placed on Motor Cars and Motor Trucks" (address, Oct. 22, 1920), p. 15, in bound collection marked "Pamphlets" (New York: National Automobile Chamber of Commerce, 1919–1933), Library, U.S. Department of Transportation, Washington.

139. Harold M. Gould, "Reserving 'Main Street' for Essential Traffic," *Transactions of the National Safety Council: Eighteenth Annual Safety Congress* 3 (Public Safety Division, 1929), 75–81 (80).

140. The traffic control engineer George Herrold of St. Paul remained in traffic work but resisted the new trends. In 1958 the 90-year-old Herrold fought the routing of Interstate 94 through the Twin Cities. See "Official Fought Freeway Route Near Capitol," *Session Weekly* (Public Information Office, Minnesota House of Representatives) 16 (March 26, 1999), 4, 17; available online at http://www.house.leg.state. mn.us/hinfo/swkly3.htm. Other traffic control professionals, most of them much younger than Herrold, left their original paradigm for new opportunities planning the motor age city. Examples include Miller McClintock of Los Angeles and Burton Marsh of Pittsburgh; on their conversion, see chapter 6.

141. Alvan D. Macauley, "There's No Need to Clog the Streets," *Nation's Business* 17 (March 1929), 49–50, 52, 208 (50).

142. McClintock in Los Angeles Board of City Planning Commissioners, *Second Conference on Mass Transportation* (Los Angeles: City Planning Commission, 1930), 5–18.

143. In practice the distinction was not so neat. Many municipal and traffic control engineers were trained as civil engineers. For example, C. C. Williams, a civil engineer, was a builder, an expert in city water supply, and an adherent of public utility principles in street traffic; see his *Municipal Water Supplies of Colorado*, University of Colorado Bulletin 12, no. 5, 1912, and "The Neck of the Highway Transportation Bottle," *Municipal and County Engineering* 68 (Feb. 1925), 93–98. Yet traffic control engineers were not builders. If a civil engineer is a builder, the civil engineering solution to congested traffic was not important until the late 1920s.

144. Macauley in *Public Works*, Sept. 1924, reprinted in Macauley, *City Planning and Automobile Traffic Problems* (Detroit: Packard Motor Car Company, 1925), 5–10 (5, 7).

145. Editorials: "Motor Killings and the Engineer," volume 89 (Nov. 9, 1922), 775; "Capitalizing Traffic Congestion Cost," volume 91 (Nov. 8, 1923), 747; "For a New

View of Traffic Control," volume 93 (Oct. 16, 1924), 615; "The Leaven Is Working," volume 92 (Feb. 21, 1924), 307.

146. Corbett, in "Discussion," *Transactions of the American Society of Civil Engineers* 88 (New York: the Society, 1925), 223–230 (223); see also Corbett, "City Replanning," *Public Works* 55 (August 1924), 258–260. Corbett's case is a good example of the ineffectiveness of visionary planning in the absence of interest group support.

147. Ernest F. Ayres, "Opportunities in City Planning for the Highway Engineer," *Roads and Streets Monthly Issue of Engineering and Contracting* 59 (March 1923), 571–574.

148. On Wacker Drive, see esp. Hugh E. Young, "The South Water Street Improvement," *Journal of the Western Society of Engineers* 30 (March 1925), 73–80. See also T. A. Evans, "Design and Construction Features," *Journal of the Western Society of Engineers* 30 (March 1925), 80–95; "Design and Structure of Double-Deck Street, Chicago," *Engineering News-Record* 95 (Oct. 15, 1925), 632–635, and "Construction Methods on Double-Deck Street, Chicago," *Engineering News-Record* 95 (Oct. 22, 1925), 662–665.

149. H. P. Gillette, "A New Branch of Highway Engineering—Traffic Control" (editorial), *Roads and Streets* 66 (Nov. 1926), 235. Gillette made no distinction between highway engineers and traffic control engineers.

150. American Automobile Association, *Society's Responsibilities* (Washington: AAA, 1937), 40.

151. MacDonald quoted in Eugene S. Taylor, "Arterial Street Planning," *Proceedings of the American Road Builders Association* 23 (1926), 69–72 (72), cited in Bruce E. Seely, *Building the American Highway System: Engineers as Policy Makers* (Temple University Press, 1987), 157. Seely's book is by far the best work of scholarship on MacDonald and the Bureau of Public Roads.

152. "Memorandum of Minutes of the First and Organization Meeting of the Institute of Traffic Engineers," Oct. 2, 1930, in *Proceedings of the Institute of Traffic Engineers* 1–3 (1930–1932), bound mimeograph volume, University of Michigan Library, Ann Arbor.

153. "Regional Plan Committee Discusses Traffic," *Electric Railway Journal* 63 (May 24, 1924), 819.

154. In greater New York there were 3.5 miles of express highways in 1928, but 107 early in 1933. The trend antedated the Great Depression by more than a year. See "Extraordinary Development of Express Highways in the New York Region," *American City* 48 (July 1933), 40.

155. L. H. Robbins, "New Jersey Weaves a Vast Fabric of Super-Highways," *New York Times*, Oct. 19, 1930; Paul G. Hoffman ("with" Neil M. Clark), "The White Line

Isn't Enough," *Saturday Evening Post* 210 (March 26, 1938), 12–13, 32, 37, 39, 41 (12).

156. Earle Duffy, "This Motor Ache," *Outlook and Independent* 158 (August 19, 1931), 491–493.

157. Committee on Traffic and Public Safety, City of Chicago, *Limited Ways for the Greater Chicago Traffic Area* (City of Chicago, 1932); see also Eugene S. Taylor, "Chicago's Superhighway Plan," *National Municipal Review* 18 (June 1929), 371–376; Robert Kingery, "Grade Separation Structures Untangle Traffic Jams," *Concrete* 37 (Nov. 1930), 17–19; Hugh E. Young, "Chicago Needs Superhighways," in Glenn A. Bishop and Paul T. Gilbert, eds., *Chicago's Accomplishments and Leaders* (Bishop Publishing Co., 1932), 532, 534, 536, 538; and "160-Mile Elevated Super-Highway System Proposed for Chicago," *Roads and Streets* 76 (Dec. 1933), 433–437.

158. "Extraordinary Development of Express Highways in the New York Region," *American City* 48 (July 1933), 40.

159. Paul Hoffman of Studebaker defined a "limited way" as "a city street planned and built exclusively for automobiles in a motor age." See Hoffman ("with" Neil M. Clark), "The White Line Isn't Enough," *Saturday Evening Post* 210 (March 26, 1938), 12–13, 32, 37, 39, 41 (32). Hoffman developed this article into a book in which he used the same definition but substituted "highway" for "street"; see Hoffman, *Seven Roads to Safety: A Program to Reduce Automobile Accidents* (Harper and Brothers, 1939), 48.

160. George Baker Anderson, "The Business of Street Management," *AERA* 16 (Nov. 1926), 689–695 (691, 694); see also W. Bruce Cobb (magistrate, New York City Traffic Court), in "Nation Roused against Motor Killings," *New York Times*, Nov. 23, 1924: "This is a Government based on the idea of the greatest good for the greater number."

161. J. Rowland Bibbins, "What an Adequate Traffic Signal System Must Do," *Roads and Streets* 70 (Nov. 1930), 397–400 (399).

162. McClintock, "Dat Ole Debbil Speed," *American City* 51 (March 1936), 97.

163. Roy D. Chapin, "The Motor's Part in Transportation," *Annals* 116 (Nov. 1924), 1–8 (4–5).

164. Reeves, address to the Cincinnati Automobile Dealers Association, Cincinnati, Oct., 1923, quoted in "Accidents Can Be Prevented," *Cincinnati Enquirer*, Oct. 21, 1923 (Automobile Section).

165. Reeves, "Have We Enough Highway Floor Space?" *Nation's Business* 12 (Dec. 1924), 30–31 (31).

166. Hoffman, "Free Enterprise—Can It Survive?" (speech before the Bond Club of New York, Dec. 8, 1938), *Vital Speeches of the Day* 5 (Jan. 15, 1939), 205–208.

## Chapter 8

1. Graham, address to the Society of Automotive Engineers, Washington, Dec. 15, 1924, published as "Cause and Prevention of Accidents," *Journal of the Society of Automotive Engineers* 16 (Jan. 1925), 12–13 (12).

2. Hertz, quoted in "Agree on Code of Sane Speed for Speedy U.S.," *Chicago Tribune*, March 25, 1926.

3. Hayes, quoted in J. E. Bulger, "Developing the Safety Manner," *American Motorist*, Feb. 1926, 30.

4. "Personnel of Safety Committee Appointed," *Automotive Industries* 49 (Nov. 29, 1923), 1122. The group was officially known as the Traffic Planning and Safety Committee.

5. Price, "Automotive Industry Should Lead in Safety Movement," *Automotive Industries* 49 (Dec. 13, 1923), 1187–1190.

6. Graham, address to the Society of Automotive Engineers, Washington, Dec. 15, 1924, published as "Cause and Prevention of Accidents," *Journal of the Society of Automotive Engineers* 16 (Jan. 1925), 12–13 (12).

7. Graham, "Education, Punishment and Traffic Safety" (address, Feb. 14, 1924), pp. 2–3, in bound collection marked "Pamphlets" (New York: National Automobile Chamber of Commerce, 1919–1933), Library, United States Department of Transportation, Washington; reprinted as Graham, "Stern Action Is Urged to Banish Death from Highways," *Automotive Industries* 50 (Feb. 28, 1924), 495–498, 511 (496, 497). See also Graham, "Safeguarding Traffic: A Nation's Problem—A Nation's Duty" (address, Sept. 23, 1924), *Annals* 116 (Nov. 1924), 179.

8. Chapin to Hugh Chalmers, quoted in J. C. Long, *Roy D. Chapin* (privately published presentation copy, 1945), 95.

9. The *Cincinnati Enquirer* and the *Los Angeles Times* are cases in point. To the *Los Angeles Times*, for example, a jaywalker was merely a person lacking "thrift enough to own a car" (editorial, "Jaywalkers," April 1, 1926). In neither case, however, has a definitive relationship between sponsorship by auto interests and editorial policy been proved, and in Los Angeles pro-automobile coverage might plausibly have been a fair reflection of readers' sympathies.

10. See Alfred G. Seiler, "Scrap Book," catalogue no. A200 Am35s 1920–29, AAA Headquarters Library.

11. In *Chicago Tribune*, see "Two Dead, 8 Hurt in Another Day of Auto Mishaps," Jan. 24, 1923; "Child Is Killed As Wolff Plans New Speed Curb," March 21, 1923; "Two Boys Die As Coroner Again Wars on Speed," July 17, 1923; "Auto Deaths Now Total 702," Dec. 21, 1923; "Two More Die; Courts Speed Up War on Speed," Sept. 24, 1924.

12. Smith, " 'Be Careful' Campaigns" (address), in National Motorists Association, "Report on the National Convention of Automobile Club Officials," Cleveland, Sept. 20–22, 1923, AAA Headquarters Library. Smith did not cite a source for his "statistics"; for a like example, see "Motor Vehicle Fatalities Can Be Eliminated by Scientific Traffic Regulation," *Automotive Industries* 48 (June 14, 1923), 1297.

13. Chicago Motor Club, "Who Is to Blame for Accidents?" ("Traffic Talks no. 2"), *Chicago Tribune*, Sept. 27, 1923, 18.

14. See *Chicago Tribune*: "Two More Die; Courts Speed Up War on Speed," Sept. 24, 1924, 7; Chicago Motor Club, "Why Does Coroner Wolff Attack the Chicago Motor Club?" ("Traffic Talks no. 52"), Oct. 3 1924, 14; Chicago Motor Club, " 'The Passing of Jay Walking' " ("Traffic Talks no. 50"), Sept. 18 1924, 22.

15. Graham, "Education, Punishment and Traffic Safety," 5; Graham, "Stern Action Is Urged," 497.

16. "Trying to Get Newspapers to Tell Accident Causes," *Automotive Industries* 50 (Jan. 17, 1924), 125 (emphasis added); Graham, "Education, Punishment and Traffic Safety," 6. According to *Automotive Industries*, "the real purpose . . . is to get the press to treat automobile accidents more fairly than in the past."

17. Graham, "Education, Punishment and Traffic Safety" (address, Feb. 14, 1924), pp. 2–3, in bound collection marked "Pamphlets" (New York: National Automobile Chamber of Commerce, 1919–1933), Library, United States Department of Transportation, Washington; reprinted as Graham, "Stern Action Is Urged to Banish Death from Highways," *Automotive Industries* 50 (Feb. 28, 1924), 495–498, 511 (496, 497). See also Graham, "Safeguarding Traffic: A Nation's Problem—A Nation's Duty" (address, Sept. 23, 1924), *Annals* 116 (Nov. 1924), 179.

18. Cobb, in "Nation Roused against Motor Killings," *New York Times*, Nov. 23, 1924.

19. See, e.g., press releases in file 02781, box 160, Commerce Papers, Hoover Library; e.g., "Accident Statistics Show Danger Spots; Permit Remedy," Sept. 20, 1924.

20. National Conference on Street and highway Safety, "Announce Program of Traffic Safety Conference" (press release), Dec. 9, 1924, in file 02792, box 163, Commerce Papers, Hoover Library.

21. J. E. Bulger, "Developing the Safety Manner," *American Motorist*, Feb. 1926, 30 (including quotation by Charles M. Hayes of the Chicago Motor Club). Hayes' appar-

ent linking of "emotional" safety reform to women was probably deliberate. Women were prominent in safety councils (see chapter 1) and safety patrols organized through local councils often included girls (e.g. in Newark; see Fred M. Rosseland, "Nine Years Without a Serious Injury to Children on Their Way to or from School," *American City* 35 (Nov. 1926), 684). The AAA and its member clubs were almost exclusively male, and its "schoolboy patrols" were for boys only.

22. Elkins, "Safety Work for Motor Clubs," May 1, 1925, American Automobile Association, *Conference of Club and Association Secretaries*, Washington, April 30–May 1, 1925, 150–156 (151–152).

23. American Automobile Association, Twenty-Fourth Annual Meeting, *Summary of Proceedings*, Chicago, June 7–8, 1926, 11.

24. Hertz quoted in "Agree on Code of Sane Speed for Speedy U.S.," *Chicago Tribune*, March 25, 1926, 6.

25. "Fourth Annual Good Roads Essay Contest," *American City* 28 (March 1923), 247.

26. "13-Year-Old Girl Wins Prize for Safety Essay," *Automotive Industries* 49 (Oct. 18, 1923), 820; "Prizes of the Highway Education Board," *School and Society* 27 (April 7, 1928), 416–417; Roy D. Chapin, in "Industry Ready to Co-Operate," *Cincinnati Enquirer*, Oct. 14, 1923 (Automobile Section; J. C. Long, *Roy D. Chapin* (privately published presentation copy, 1945), 206.

27. *Detroit Free Press*, August 23, 1927.

28. See chapter 6.

29. *First National Conference on Street and Highway Safety*, Washington, Dec. 15–16, 1924, 40–51. Representatives of automotive interests outnumbered street-railway representatives 80 to 29. Pedestrians in general were not represented, except that educators were invited to speak for the interests of school children. Best represented of all were the steam railroads (86 representatives), but in city traffic matters their interests were confined to the safety of grade crossings.

30. M. Warren Baker, "Putting More Cars on the Road," *Motor Age* 48 (Sept. 24, 1925), 10–12, 43 (11).

31. "What's Your Hurry?" *Milwaukee Journal*, Sept. 12, 1920; Carl M. Saunders, "What's Your Hurry?" *Milwaukee Journal*, Sept. 26, 1920.

32. Blanchard, in *Motor*, 1922, excerpted in "Two Cures for Motor Accidents," *Literary Digest*, July 1, 1922, 60–63 (60–61). Blanchard argued that governors were ineffective because trucks were involved in a disproportionate number of accidents and because many (perhaps most) operators of truck fleets equipped their rigs with speed governors. Blanchard presumably resorted to this flimsy case because to challenge

governors only on practical grounds would leave the premise of the inherent danger of speed unquestioned. Blanchard understood that this premise threatened the future of the automobile in the city.

33. "Make It Unanimous!" (editorial), *Cincinnati Enquirer*, Oct. 24, 1923. Similarly, a 1925 editorial in *Ohio Motorist* claimed that "speed in itself is a minor cause" of traffic accidents; see "Nailing Misfacts," *Ohio Motorist* 17 (Oct. 1925), 3, 42–43 (42).

34. "Speed Limit Here 10 Miles an Hour Still By Old Law," *St. Louis Star*, Nov. 7, 1923. This speed limit was a dead letter before 1920, but remained on the books some years longer. A judge in the municipal court of St. Louis explained that judges arranged with city police "to permit automobilists in St. Louis to travel 25 miles an hour without being arrested for speeding." See George E. Mix, "Law Enforcement and Its Application," *Proceedings of the National Safety Council*, Chicago, 1920, 337, cited in McClintock, *Street Traffic Control* (McGraw-Hill, 1925), 88–89n. 1. For more on Britton's campaign, see "Auto Club Leader Urges Speed Limit Be Abolished Here," *St. Louis Star*, Nov. 8, 1923; "Brockman Warns Against Dropping Auto Speed Laws," *Star*, Nov. 9, 1923; "Britton's Traffic Plan 'Too Vague,' Declares Brockman, Arguing for Two fixed Speed Limits in City," *Star*, Nov. 12, 1923; No Speed Limit Plea Amplified by Roy Britton," *Star*, Nov. 17, 1923.

35. Miller McClintock, *Report and Recommendations of the Metropolitan Street Traffic Survey* (Chicago Association of Commerce, 1926), 105. McClintock added: "In placing limits upon speed, legislators have almost always fallen short of current demands for rapidity of movement" (107). McClintock's drastic revision of his principles, described in chapter 6, was not complete until 1927; his work for the Chicago Association of Commerce was more consistent with the prevailing traffic control principles of the early-to-mid 1920s.

36. Bob Beiser, "Ridicule Would Come to City," *Cincinnati Enquirer*, Oct. 28, 1923 (Automobile Section).

37. "More Jail Sentences Needed," *American Motorist* 16 (Jan. 1924), 11.

38. M. O. Eldridge (secretary, Washington branch, AAA) quoted in "What's an Auto 'Jay-Walker?' " *Washington Post*, May 21, 1922. In 1925 Eldridge was appointed Washington's traffic director. He quickly became controversial for loose use of the term "jaywalker" against pedestrians (see chapter 3).

39. "The Motor 'Flivverboob,' " *New York Times*, July 9, 1922; "Consider the 'Flivverboob,' a Horrible Example," *Chicago Tribune*, Sept. 17, 1922. The word was submitted by F. B. Simpson of Cedar Rapids, Iowa. According to the *Times*, it described "reckless and careless" (but not necessarily fast) motorists. David Blanke of Texas A&M University and the author independently rediscovered this term in 2004.

40. "Make It Unanimous!" (editorial), *Cincinnati Enquirer*, Oct. 24, 1923.

41. "Drivers Should Be Licensed," *Cincinnati Enquirer*, Sept. 23, 1923 (Automobile Section).

42. "Crush That Ordinance!" (editorial), *Cincinnati Enquirer*, Nov. 5, 1923.

43. "Accidents Can Be Prevented," *Cincinnati Enquirer*, Oct. 21, 1923 (Automobile Section). The words quoted are from the newspaper's synopsis of Reeves' statement.

44. The earliest of such rare examples that this author can find dates from 1915. The auto club in Washington urged that "something ought to be done quickly to corral a large number of dangerous drivers and break them of their wild ways." See press release of Washington auto club, reprinted in *Washington Post* (under "Stop Careless Driving") and *Washington Star* (under "Auto Club News: District of Columbia") for Nov. 7, 1915, or see clippings in Alfred G. Seiler, "Scrap Book," catalogue no. A200 Am35s, AAA Headquarters Library. In 1919 William Ullman of AAA made a stronger case for going after the reckless: "in automobiling as well as in everything else . . . a [whole] class receives its bad reputation from its few bad members." This is from an untraced clipping of a newspaper article by Ullman dated March 8, 1919; the title ends " . . . Speed Regulations." The clipping is in the Seiler Scrap Book, AAA Library. After 1923 warnings of reckless drivers' threat were suddenly ubiquitous. The article probably was published in the *Washington Star*, which often printed material written at the AAA.

45. "Crush That Ordinance!" (editorial), *Cincinnati Enquirer*, Nov. 5, 1923.

46. "Drivers Should Be Licensed," *Cincinnati Enquirer*, Sept. 23, 1923 (Automobile Section).

47. "Nailing Misfacts," *Ohio Motorist* 17 (Oct. 1925), 3, 42–43 (42).

48. George M. Graham, "Recent Developments in Highway Transport" (address), Transportation and Communication Group Session, Chamber of Commerce of the United States, Twelfth Annual Meeting, 1924, reprinted in "Cooperation in Transportation," U.S. Chamber of Commerce, 1924, 13–29 (27).

49. Macauley, "Banish the Automobile Nuisance," *Collier's* 78 (July 17, 1926), 27.

50. The tag was titled "Ten Commandments of Safety for Motorists"; its full text is duplicated in "N.A.C.C. Seeks Aid in Safety Effort," *Automotive Industries* 49 (August 2, 1923), 243.

51. Graham, "Education, Punishment and Traffic Safety" (address, Feb. 14, 1924), p. 10, in bound collection marked "Pamphlets" (New York: National Automobile Chamber of Commerce, 1919–1933), Library, United States Department of Transportation, Washington.

52. Traffic Planning and Safety Committee, National Automobile Chamber of Commerce, "Getting the Most from Your Car" (New York: NACC, 1924), 12. Despite repeated attempts, I have not found a copy of this pamphlet, but see George M. Graham (NACC), "Recent Developments in Highway Transport" (address), Transportation and Communication Group Session, Chamber of Commerce of the United States, Twelfth Annual Meeting, 1924, reprinted in "Cooperation in Transportation," U.S. Chamber of Commerce, 1924, 13–29 (27): "We believe it is not enough to take the license. Take away the car too. Let the punishment follow the car." See also Graham, "Stern Action Is Urged to Banish Death from Highways," *Automotive Industries* 50 (Feb. 28, 1924), 495–498, 511; "Impound Motorist's Car for Reckless Driving," *New York Times*, March 9, 1924; Graham, "Careless Drivers = Poor Citizens," *Ohio Motorist* 16 (Sept. 1924), 4, 19.

53. George M. Graham, "Recent Developments in Highway Transport" (address), Transportation and Communication Group Session, Chamber of Commerce of the United States, Twelfth Annual Meeting, 1924, reprinted in "Cooperation in Transportation," U.S. Chamber of Commerce, 1924, 13–29 (27–28).

54. "Power of Industry Behind Traffic Booklet," *Automotive Industries* 51 (August 14, 1924), 319.

55. William Ullman, *Star*, Dec. 1924, quoted in "Making the Automobile Safe for Everybody," *Literary Digest* 84 (Jan. 1925), 56–58 (57).

56. "Getting the Most from Your Car"; "Each New Purchaser to Get Safety Book," *Automotive Industries* 51 (August 7, 1924), 304.

57. Editorial, including Vane's comments, in "Hog-Tying the Automobile," *Ohio Motorist* 16 (Oct. 1924), 8.

58. Vane, quoted in "Hog-Tying the Automobile."

59. Kettering's address to the National Safety Council, Louisville, quoted in Sam Shelton, "Lack of Human Intelligence Is Cause of Most Traffic Accidents," *Automotive Industries* 51 (Oct. 9, 1924), 646–647 (646).

60. See Joel W. Eastman, *Styling vs. Safety: The American Automobile Industry and the Development of Automotive Safety, 1900–1966* (University Press of America, 1984).

61. Ernest N. Smith (general manager, AAA), Oct. 27, 1926, in "Drivers' License Laws," *Transactions of the National Safety Council, Fifteenth Annual Safety Congress*, Detroit, Oct. 25–29, 1926, volume 3: Public Safety and Education Sections, 67; see also "Note to Sec. 2," in National Conference on Street and Highway Safety, *Final Text of Uniform Vehicle Code* (Washington: the Conference, August 20, 1926), 60.

62. "Uniform Motor Vehicle Operators' and Chauffeurs' License Act," in National Conference on Street and Highway Safety, *Final Text of Uniform Vehicle Code* (Washington: the Conference, August 20, 1926), 43–61 (see Title II, pp. 46–57).

63. Robert P. Lamont, address, in National Conference on Street and Highway Safety, *Ways and Means to Traffic Safety* (Washington: the Conference, May, 1930), 7–13 (10).

64. "Swat It!" *Cincinnati Enquirer*, Sept. 30, 1923 (Automobile Section).

65. Letter to editor ("Those Reckless Pedestrians," signed "A Driver"), *St. Louis Star*, Nov. 14, 1923.

66. *Post-Dispatch*, "Safety and the Motors" (editorial), Nov. 19, 1923.

67. J. E. Bulger, "Developing the Safety Manner," *American Motorist*, Feb. 1926, 30.

68. Frederick J. Haynes, "Safety—A Necessity of the Motor Age," Oct. 26, 1926, *Transactions of the National Safety Council, Fifteenth Annual Safety Congress*, Detroit, Oct. 25–29, 1926, volume 3: Public Safety and Education Sections, 4–12 (9).

69. Until the middle 1920s, "pleasure car" was synonymous with "passenger car." See, e.g., Ernest Greenwood to Harold Phelps Stokes, May 8, 1925, box 39, Commerce Papers, Hoover Library.

70. "Auto Worm Turns; Blames Walkers" *New York Times*, May 13, 1923.

71. From Harding's message to Congress, April 12, 1921. Six days after Harding delivered his message, the quotation appeared hand typed on NACC stationery (see, e.g., Alfred Reeves to Hoover, April 18, 1921, box 428, Commerce Papers, Hoover Library). A week later, NACC stationery included the quotation as an integral, professionally printed epigraph (see, e.g., Pyke Johnson to Hoover, April 25, 1921, box 39, Commerce Papers, Hoover Library. To attack automotive excise taxes, Reeves challenged the notion that automobiles were luxuries as early as fall, 1920; see his address of Oct. 22: "Why No More Taxes Should Be Placed on Motor Cars and Motor Trucks," p. 7, in bound collection marked "Pamphlets" (New York: National Automobile Chamber of Commerce, 1919–1933), Library, United States Department of Transportation, Washington.

72. R. H. Harper, quoted in "Auto Overcomes Expensive Toy Idea," *Washington Post*, Dec. 3, 1922. See also Philip H. Brockman, "No Auto Saturation Likely to Be Reached," *St. Louis Star*, Oct. 22, 1922.

73. Herbert Hoover, "Your Automotive Industry," 1–5 (3), typescript of an article for *Collier's* of Jan. 7, 1922, in "Public Statements," Hoover Library. My thanks to Brian Balogh of the University of Virginia, who gave me a photocopy of this document.

74. "Tell It to Congress!" *Cincinnati Enquirer*, Oct. 14, 1923 (Automobile Section). The words quoted are from the *Enquirer*. See also "The Assured Future of the Automobile" (editorial), *Scientific American* 123 (Dec. 18, 1920), 606.

75. "Jaywalkers Given Grace," *Los Angeles Times*, Jan. 24, 1925; E. B. Lefferts, "Regulation of Pedestrians" (address to the National Safety Council, Sept. 28, 1927), *Transactions of the National Safety Council*, Sixteenth Annual Safety Congress, volume 3 (Chicago, 1927), 19–24. See also Lefferts, "Effective Regulation of Pedestrians," *American City* 37 (Oct. 1927), 434–436. The most defiant pedestrians were arrested immediately, however; see "New Law on Jaywalking Is in Force," *Los Angeles Times*, Jan. 25, 1925.

76. Lefferts, "Regulation of Pedestrians" (address to the National Safety Council, Sept. 28, 1927), *Transactions of the National Safety Council*, Sixteenth Annual Safety Congress, volume 3 (Chicago, 1927), 19–24 (20–21); Lefferts, "Effective Regulation of Pedestrians," *American City* 37 (Oct. 1927): 434–436 (434–435).

77. Lefferts, "Regulation of Pedestrians," 21; Lefferts, "Effective Regulation," 436. In *Domesticating the Street: The Reform of Public Space in Hartford, 1850–1930* (Ohio State University Press, 1999), Peter C. Baldwin reports similar use of unwelcome public attention directed at violators arrested in Hartford in 1929 (p. 224).

78. "Pedestrian Is King between the White Lines," *Los Angeles Times*, Sept. 27, 1925.

79. "Jaywalker to Face New Foe," *Los Angeles Times*, April 19, 1925.

80. "Jaywalking Leads to Jail," *Los Angeles Times*, Jan. 27, 1925.

81. Lefferts, "Regulation of Pedestrians," 21; Lefferts, "Effective Regulation," 436.

82. Katherine Lipke, "The Jaywalker," *Los Angeles Times*, Jan. 28, 1925.

83. " 'Jaywalker' Law Reduces Accidents Throughout City," *Los Angeles Times*, Feb. 8, 1925.

84. "Officer Carries Fractious Woman Jaywalker to Curb," *Los Angeles Times*, Feb. 8, 1925.

85. As had happened elsewhere—for example in Joliet, Illinois, in 1921; see *Benison v. Dembinsky* (1926), 241 Illinois Court of Appeals 530.

86. "Jaywalking Leads to Jail," *Los Angeles Times*, Jan. 27, 1925; "Jaywalker to Face New Foe," *Los Angeles Times*, April 19, 1925.

87. Oscar Hewitt, "Motor Toll Cut, Traffic Speeded—in Los Angeles," *Chicago Tribune*, Jan. 10, 1926.

88. "Success of New Traffic Rule Causes Other Cities to Gasp, 'How Did You Do It?' " *Los Angeles Times*, Sept. 27, 1925.

89. "School Tunnel Action Urged," *Los Angeles Times*, Oct. 25, 1925.

90. "Praises Los Angeles Traffic Methods," *New York Times*, Sept. 19, 1927.

91. Lefferts, "Regulation of Pedestrians"; Lefferts, "Effective Regulation"; "Jaywalker to Face New Foe," *Los Angeles Times*, April 19, 1925.

92. "Melbourne, Chicago Adopt Los Angeles Traffic Plan," *Los Angeles Times*, Sept. 19, 1926.

93. Herbert L. Towle, "The Motor Menace," *Atlantic Monthly* 136 (July 1925), 98–107 (99).

94. Patrick F. Bridgeman, in "The Inquiring Reporter," *Chicago Tribune*, Sept. 14, 1926. See also letter to editor ("More about Motorists and Pedestrians," signed "A Pedestrian"), *Chicago Tribune*, Sept. 25, 1923; "Pedestrian Rights Recognized—but Try to Exercise Them," *Chicago Tribune*, Dec. 12, 1926.

95. Letter to editor ("Pedestrians' Rights," signed "Pedestrian"), *Chicago Tribune*, March 17, 1926; F. E. Latta, letter to editor ("In Defense of Jay Walking"), *Chicago Tribune*, Oct. 18, 1926. See also A. R., letter to editor ("Protection for the Pedestrian"), Jan. 3, 1926, *Chicago Tribune*, Jan. 10, 1926; Ben Leavitt, in "The Inquiring Reporter," *Chicago Tribune*, Feb. 12, 1926; Lora Byrne, in "The Inquiring Reporter," *Chicago Tribune*, Sept. 14, 1926; Leonard Porges, letter to editor ("Lights Are Too Fast"), *Chicago Tribune*, June 17, 1928; Letter to editor ("Jay Parking," signed "Pedestrian"), Dec. 10, 1928, *Chicago Tribune*, Dec. 30, 1928; Jessie L. Sterling, letter to editor ("Pedestrian's Complaint"), Oct. 8, 1929, *Chicago Tribune*, Oct. 13, 1929.

96. Harvey H. Wells, letter to editor ("Pedestrians' Rights"), *Chicago Tribune*, Oct. 19, 1926.

97. R. W. P., letter to editor ("Manners for Motorists"), *Chicago Tribune*, July 1, 1928.

98. F. R. D., letter to editor ("A Much Needed Safeguard"), *Chicago Tribune*, July 1, 1928.

99. A. R., letter to editor ("Protection for the Pedestrian"), Jan. 3, 1926, *Chicago Tribune*, Jan. 10, 1926.

100. "Better Car Routing and Traffic Control Proposed for Baltimore" (abstract of a report by Kelker, De Leuw, and Co., to the Traffic Survey Commission of Baltimore), part 1, *Electric Railway Journal* 67 (May 22, 1926), 883–889 (888–889).

101. "Traffic Control Developments in 1930," *Roads and Streets* 70 (Feb. 1930), 62.

102. Cobb, in "Nation Roused against Motor Killings," *New York Times*, Nov. 23, 1924.

103. Philip Vyle to editor ("Pedestrians' Rights Ignored"), *New York Times*, July 6, 1925.

104. Stoeckel, "Pedestrians Are Responsible for Many Motor Vehicle Accidents," 108–109.

105. *Ohio Motorist* 17 (Oct. 1925), 3, 42–43 (42).

106. "Accident Prevention on a Business Basis," *Los Angeles Times*, Oct. 2, 1921.

107. William E. Metzger, "Report of Traffic and Safety Committee," July 1, 1929, *Proceedings of the Twenty-Seventh Annual Meeting of the American Automobile Association*, Buffalo, July 1–2, 1929, 17–18 (17).

108. William E. Metzger, "Report of Traffic and Safety Committee," July 1, 1929, *Proceedings of the Twenty-Seventh Annual Meeting of the American Automobile Association*, Buffalo, July 1–2, 1929, 17–18 (17).

109. See collection, "AAA: The Traffic Safety Posters," volume 1, nos. 1–81 (1928–1936), AAA Headquarters Library.

110. "AAA: The Traffic Safety Posters," volume 1, nos. 1–81 (1928–1936). Posters quoted were drafted 1928–1929 and issued 1929–1930. The other posters of 1928–1929: "The left side is the right side for hiking on the road"; "Hitch-riding is dangerous—play safe!; "Don't dart across—be careful!"; "Dodging between parked cars is—dodging into danger"; no caption (don't play in the street); no caption (Thanksgiving: injured or healthy?); no caption (look before crossing—Christmas traffic); "Ride hopping is dangerous—play safely!" The other posters of 1929–1930: "Look left, then right before crossing streets"; "Merry Christmas and a safe new year"; "Ever alert—never hurt!"; "We always stop, look, listen at each corner"; "Ride begging is unsafe and unfair."

111. "10,000 Children Join Safety March," *New York Times*, Oct. 10, 1922; "The Nation's Needless Martyrdom," *Literary Digest* 75 (Oct. 28, 1922), 29–31 (29); "Tragedies to 25 Children Are Recalled," *Louisville Herald*, June 5, 1923.

112. "School Safety Patrols Chronology," Nov. 6, 1984, file "AAA Traffic Safety," AAA Headquarters Library; "Pupils Are Protected in Traffic," *Los Angeles Times*, Feb. 23, 1923. Detroit's patrols were organized in 1920, those of Los Angeles in 1922 or 1923. On the school patrol of the Detroit Automobile Club see also H. O. Rounds, "Traffic Accidents, the Subconscious Mind, and Law Enforcement" *American City* 37 (Nov. 1927), 670.

113. G. C. Smith, in discussion following Smith, "Factors That Affect the Size of Traffic Squads," Oct. 2, 1929, *Transactions of the National Safety Council, Eighteenth Annual Safety Congress*, Chicago, Sept. 30–Oct. 4, 1929, volume 3, 23–30 (28).

114. William E. Metzger, "Report of Traffic and Safety Committee," July 1, 1929, *Proceedings of the Twenty-Seventh Annual Meeting of the American Automobile Association*, Buffalo, July 1–2, 1929, 17–18 (18).

115. "Schoolboy Patrol Saves Many Lives," *Washington Post*, Feb. 18, 1927 (this article is apparently from a local auto club's press release; see file "school safety patrols," AAA Headquarters Library); American Automobile Association, "They Approve the School Boy Patrol" (pamphlet), Washington, 1928 and "Did You Know?" (column for club publications), AAA Club Editorial Service, Dec. 1985, file "safety patrols," AAA Library. See also "Highlights of Nearly Four Decades of Service," (c. 1959), in file "AAA Traffic Safety"; "Announcer's Fact Sheet," School Safety Patrol Parade, 1977, and "The School Safety Patrol: Selected Highlights" (c. 1977), both in file "safety patrols"; all in AAA Headquarters Library.

116. "School Kids Sit in Judgment on Jay Walking Companions," *Chicago Tribune*, March 8, 1925, A11.

117. A. L. Morgan, "Pupil Participation in School Control," *Peabody Journal of Education* 7 (March 1930), 268.

118. Hoover in each of several letters to interest groups, March 26, 1924, in file 02765, box 160, Commerce Papers, Hoover Library; see also "National Automobile Chamber of Commerce," box 428, Commerce Papers.

119. Hoover, untitled draft, stamped "B.F.D.C. RECD MAR 31 1924," in file 02765, box 160, Commerce Papers, Hoover Library.

120. Owen to Hoover, April 3, 1924, and attached speech drafts, in file 02765, box 160, Commerce Papers, Hoover Library.

121. See undated draft address (1924), marked "For the Secretary—From Ernest Greenwood," in file 02764, box 160, Commerce Papers, Hoover Library.

122. Hoover, address, Dec. 15, 1924, in *First National Conference on Street and Highway Safety* (U.S. Department of Commerce, 1924), 7–11.

123. Stokes to Hoover ("Memo for Chief"), Nov. 25, 1924, in file 02764, box 160, Commerce Papers, Hoover Library.

124. National Conference on Street and Highway Safety, press release, "Summary of Report of Committee on Traffic Control," for Nov. 25, 1924, in file 02764, box 160, Commerce Papers, Hoover Library. The committee included eight representatives of motordom, but only one street-railway representative. Many representatives, however, were local and state government officials without obvious interests.

125. A. B. Barber (director, National Conference on Street and Highway Safety), "Conference on Street and Highway Safety," Oct. 2, 1924, Proceedings of the National Safety Council, Thirteenth Annual Safety Congress, Louisville, Sept. 29–Oct. 3, 1924, 119–122. The original typescript of Barber's address is in file 02764, box 160, Commerce Papers, Hoover Library; see p. 3.

126. National Conference on Street and Highway Safety, press release, "Summary of Report of Committee on Traffic Control," for Nov. 25, 1924, in file 02764, box 160, Commerce Papers, Hoover Library.

127. According to President Hoover's Secretary of Commerce, Robert P. Lamont, by 1930 "some twenty-three states have adopted one or more of the acts" of the Uniform Vehicle Code, "or have modified their laws to bring them into greater conformity with it." See Lamont, address, in *Ways and Means to Traffic Safety* (Washington: National Conference on Street and Highway Safety, 1930), 7–13 (10).

128. Stokes to Hoover ("Memo for Chief"), Nov. 25, 1924, in file 02764, box 160, Commerce Papers, Hoover Library.

129. Stokes to Hoover ("Memo for Chief"), Nov. 25, 1924.

130. Hoover to C. E. Skinner (chairman, American Engineering Standards Committee), March 14, 1927, in file 02758, box 160, Commerce Papers, Hoover Library.

131. Ray W. Sherman, *If You're Going to Drive Fast* (Crowell, 1935), esp. 11–12, 17–18, 138. The composition of the 27-member Accidents Committee was diverse; motordom's representation was lower on it than on other committees. The largest contingents were state and local officials (7) and representatives of insurers (5). Motordom was represented by just 4 members, or about a seventh of the committee. Street railways had one representative.

132. Press release, "Clinics Needed for Reckless Drivers," for Sept. 29, 1925, National Conference on Street and Highway Safety, Washington, in file 02768, box 160, Commerce Papers, Hoover Library. The committee's source was apparently records of the state of Connecticut, which were reported to the conference's Statistics Committee. While strictly true, the Accident Committee's reference to the Connecticut figures obscures the fact that of those accidents ascribed to recklessness or carelessness, Connecticut found the motorist at fault in 83.6 percent of cases, pedestrians (child and adult) in 16.4 percent. See Statistics Committee's draft report, Oct. 17, 1924, in file 02764, box 160, Commerce Papers, Hoover Library. See also Department of Civil Engineering, Yale University, "Connecticut Motor Vehicle Accident Statistics," April 9, 1924, in *Proceedings of a Conference on Motor Vehicle Traffic*, New Haven, April 9–11, 1924, ed. R Kirby (Yale University Press, 1924), 3–31.

133. Undated press release, "Study Insurance Measures to Make Traffic Safe," U.S. Department of Commerce, 1924, in file 02781, box 162, Commerce Papers, Hoover Library.

134. "Report of the National Conference on Street and Highway Safety," March 25, 1926, 5, in file 02792, box 163, Commerce Papers, Hoover Library.

135. "Making the Automobile Safe for Everybody," *Literary Digest* 84 (Jan. 10, 1925), 56–58 (56).

136. Hoffman, "Speed and Safety" (address), Sept. 30, 1929, *Transactions of the National Safety Council, Eighteenth Annual Safety Congress*, Chicago, Sept. 30–Oct. 4, 1929, 4–9 (5).

137. McClintock, *Street Traffic Control* (McGraw-Hill, 1925), 86.

138. Ibid., 87–89.

139. McClintock, "Speed Control Without a 'Prima Facie' Rule," *American City* 44 (Jan. 1931), 137–138.

140. See above, under "Speed Can Be Safe"; for another argument against speed limits, see H. W. Slauson, "A New Plan for Traffic Laws," *Scientific American* 132 (May 1925), 296–298.

141. McClintock, "Speed Control Without a 'Prima Facie' Rule," 137–138. The accompanying cartoon (p. 137) attacking prima facie speed laws is captioned "A Strain on the Mentality of Officer and Motorist," and appeared originally in the *Detroit News*. In it, a policeman addresses a motorist he has stopped: "If you go 30 and bust something you're unconstitutional because of prima facie and that stuff." The motorist replies "How do I prime my face?"

142. McClintock, address to the annual meeting of the American Standards Association, reprinted as "Dat Ole Debbil Speed," *American City* 51 (March 1936), 97. McClintock may not have chosen this title.

143. See the narration of GM's Futurama, its 1939–1940 World's Fair exhibit, in General Motors, *Futurama* (1939), a presentation edition of 1,000 issued Oct. 16, 1939. A copy is available at the Special Collections Library, University of Virginia, Charlottesville.

144. H. P. Gillette, "How to Reduce Highway Accidents," *Roads and Streets* 70 (March 1930), 123–124.

145. "Unfit for Modern Traffic," *Fortune* 14 (August 1936), 85–92, 94, 96, 99 (87). The words quoted are *Fortune*'s summary of the prevailing view of traffic safety experts.

146. Richard Shelton Kirby, "A Study of Motor Vehicle Accidents in the State of Connecticut, 1924 and 1925"; see "Accident Statistics Graphically Shown," *American City* 35 (Oct. 1926), 530.

147. *Ohio Motorist* 17 (Oct. 1925), 3, 42–43 (42). One of the earliest clear connections of motor highways, traffic flow, and traffic safety is Norman G. Shidle, "Traffic Accidents Can Be Made Impossible," *Automotive Industries* 51 (Nov. 27, 1924), 913–914. Motordom did little with this idea for about three years.

148. Edward S. Jordan (president, Jordan Motor Car Co.), "Signal Lights Cannot Solve Peak Traffic," *Good Roads*, Sept. 1927, 399; Alvan Macauley (president, Packard

Motor Car Co.), "Over-Passes Overcome Traffic Congestion in Cities," *Automotive Industries* 52 (Sept. 29, 1927), 39, 43.

149. Cox, paper presented at a conference of the American Society of Civil Engineers, New York, May, 1927, reprinted as "Population Density as a Factor in Traffic-Accident Rates," in *American City* 37 (August 1927), 207–209 (209). Cox also predicted (209) that "the bulk of residential streets will be . . . given over more and more to pedestrian use."

150. Norman G. Shidle, "Traffic Accidents Can Be Made Impossible," *Automotive Industries* 51 (Nov. 27, 1924), 913–914 (913).

151. Kettering's words according to a synopsis by Sam Shelton, in Shelton, "Lack of Human Intelligence Is Cause of Most Traffic Accidents," *Automotive Industries* 51 (Oct. 9, 1924), 646–647 (646).

152. Automobile Manufacturers Association, *You in Your Car on City Streets* (Detroit: Automobile Manufacturers Association, 1936), 60.

153. American Automobile Association, *Driver and Pedestrian Responsibilities* (third printing, Washington: AAA, 1936), 50. The quoted passage is similar to one in a 1923 pamphlet issued by the Automobile Club of America (a New York organization, founded in 1899): "Pedestrians often appear stupid or careless, and lots of them are." See W. Bruce Cobb, "Making the Roads Safe" (pamphlet issued by *The Evening Telegram*, the Automobile Club of America, and the National Automobile Chamber of Commerce, New York, 1923), 7.

154. Norman G. Shidle, "Traffic Accidents Can Be Made Impossible," *Automotive Industries* 51 (Nov. 27, 1924), 913–914 (913).

155. "Pedestrian Problem Seen Uppermost in Accident Situation," *Automotive Industries* 59 (Oct. 27, 1928), 601.

156. C. H. Claudy, "Building the Road to Fit the Car," *Illustrated World* 35 (April 1921), 299–301 (301).

157. McClintock, "Four Frictions," *Time* 28 (August 3, 1936), 41–43.

158. Arnold H. Vey (traffic engineer, New Jersey Department of Motor Vehicles, Trenton), "The Safety of Divided Roadways," *American City* 48 (Sept. 1933), 69–70 (69).

159. See, e.g., L. H. Robbins, "New Jersey Weaves a Vast Fabric of Super-Highways," *New York Times*, Oct. 19, 1930; "Extraordinary Development of Express Highways in the New York Region," *American City* 48 (July 1933), 40. The latter article dates the beginning of the highway construction "surge" to 1928–1929.

160. Macauley, "There's No Need to Clog the Streets," *Nation's Business* 17 (March 1929), 49–50, 52, 208 (208).

161. Hoffman, "Speed and Safety" (address), Sept. 30, 1929, *Transactions of the National Safety Council, Eighteenth Annual Safety Congress,* Chicago, Sept. 30–Oct. 4, 1929, volume 3, 4–9 (5–6).

162. L. H. Robbins, "New Jersey Weaves a Vast Fabric of Super-Highways," *New York Times,* Oct. 19, 1930.

163. "Speed and Safety," 6.

164. Hugh E. Young, "Chicago Needs Superhighways," in Glenn A. Bishop and Paul T. Gilbert, eds., *Chicago's Accomplishments and Leaders* (Bishop Publishing Co., 1932), 532, 534, 536, 538 (536).

165. Miller McClintock, "Pedestrian Tunnels for School Children," *American City* 34 (Jan. 1926), 81–82; "Pedestrian Tunnels or Signals or Schoolboy Patrols for School Children's Safety?" *American City* 38 (April 1928), 113.

166. Henry M. Propper, "A New Town Planned for the Motor Age," *American City* 38 (Feb. 1928), 152–154.

167. American Society for Municipal Improvements, resolution, Philadelphia, Oct., 1929, quoted in "Highway Sidewalks Urged by A.S.M.I.," *American City* 41 (Nov., 1929), 21.

168. Leslie J. Sorenson (president, Institute of Traffic Engineers), quoted in *Pedestrian Protection* (Washington: American Automobile Association, 1939), inside back cover.

169. Hoffman (president, Automotive Safety Foundation), quoted in *Pedestrian Protection* (Washington: American Automobile Association, 1939), inside front cover.

170. L. G. Holleran, "Grade Crossing Elimination in Westchester Co.," *Public Works* 59 (May 1928), 184–187 (186); Holleran, "Development of Parks and Parkways in Westchester County, New York," *Journal of the Boston Society of Civil Engineers* 16 (Sept. 1929), 381–391 (386–388).

171. Alvan Macauley, "Over-Passes Overcome Traffic Congestion in Cities," *Motor Age* 52 (Sept. 29, 1927), 39, 43; Robert Kingery, "Grade Separation Structures Untangle Traffic Jams," *Concrete* 37 (Nov. 1930), 17–19.

172. L. H. Robbins, "New Jersey Weaves a Vast Fabric of Super-Highways," *New York Times,* Oct. 19, 1930.

173. Hoffman ("with" Neil M. Clark), "The White Line Isn't Enough," *Saturday Evening Post* 210 (March 26, 1938), 12–13, 32, 37, 39, 41 (12).

174. L. H. Robbins, "New Jersey Weaves a Vast Fabric of Super-Highways," *New York Times,* Oct. 19, 1930.

175. "The White Line Isn't Enough," 12.

176. McClintock, "The Fool-Proof Highway of the Future," *Safety Engineering* 68 (July 1934), 22.

177. "160-Mile Elevated Super-Highway System Proposed for Chicago," *Roads and Streets* 76 (Dec. 1933), 433–437.

178. "The White Line Isn't Enough," 32. Hoffman developed this article into a book, in which he used the same definition but substituted "highway" for "street"; see Hoffman, *Seven Roads to Safety: A Program to Reduce Automobile Accidents* (Harper and Brothers, 1939), 48.

179. McClintock, "Fool-Proof Highway."

180. "Four Frictions," *Time* 28 (August 3, 1936), 41–43. At about the same time, *Fortune* quoted McClintock a little differently: "If it were possible to apply everything we know about traffic control, we could eliminate 98 per cent of all accidents and practically all congestion." *Fortune* called this statement "one of Miller McClintock's favorite pronouncements." See "Unfit for Motor Traffic," *Fortune* 14 (August 1936), 85–92, 94, 96, 99 (99).

181. McClintock, "Fool-Proof Highway."

182. Committee on Traffic and Public Safety, City of Chicago, *Limited Ways for the Greater Chicago Traffic Area* (City of Chicago, 1932), 12. While it cannot be known if McClintock drafted the words quoted, he was the chief traffic consultant for the project and signed his name to this report.

183. McClintock, "Fool-Proof Highway."

184. Hoover accepted the new role for the Bureau of Public Roads Sept. 19, 1931. See correspondence, June 10 to Sept. 19, 1931 (Robert P. Lamont to Hoover, June 10; Lawrence Richey to Arthur M. Hyde, June 19; Hyde to Richey, undated, received June 24; Hyde to Richey, undated, received Sept. 18; Hoover to Hyde, Sept. 19; Richey to Lamont, Sept. 19), all in file "Agriculture, Public Roads, Correspondence," box 7 (Presidential Papers), Hoover Library. Ten days after Hoover agreed to it, Chief of the Bureau of Public Roads Thomas MacDonald announced the new policy (without pledging specific commitments) in a speech to state highway departments; see "Current Phases of Highway Building and Maintenance" (typescript of an address to the American Association of State Highway Officials, Salt Lake City, Sept. 29, 1931), 14–15, in file "Agriculture, Public Roads," Correspondence, box 7 (Presidential Papers), Hoover Library. For an excellent study of the Bureau of Public Roads, see Bruce E. Seely, *Building the American Highway System: Engineers as Policy Makers* (Temple University Press, 1987). Seely's work does not consider the role of safety in the Bureau of Public Roads' entry into the realm of urban highways, concentrating instead on congestion and work relief. Seely also dates the beginning of the Bureau's role in urban highways to the New Deal (see esp. 154–155, 157–158). The Bureau's budgets did indeed grow much faster after 1933, but the crucial precedent was

established in Hoover's 1931 acceptance, on grounds of safety, of a federal role in urban highways.

185. Hoffman ("with" Neil M. Clark), "America Goes to Town," *Saturday Evening Post* 211 (April 29, 1939), 8–9, 30, 32, 35 (35). "No city . . . " is Hoffman's direct quotation of MacDonald; "we must dream . . . " is Hoffman's paraphrase of MacDonald.

186. "10,000 Children Join Safety March," *New York Times*, Oct. 10, 1922; "The Nation's Needless Martyrdom," *Literary Digest* 75 (Oct. 28, 1922), 29–31 (29); AAA traffic safety poster no. 13, in collection, "AAA: The Traffic Safety Posters," volume 1, nos. 1–81 (1928–1936), AAA Headquarters Library.

## Chapter 9

1. Macauley, "There's No Need to Clog the Streets," *Nation's Business* 17 (March 1929), 49–50, 52, 208 (208).

2. McClintock quoted in "Cities of Tomorrow," *Safety Engineering* 74 (July 1937), p. 35. See also McClintock, " 'Of Things to Come' " (address to the National Planning Conference, Detroit, June 1937), *American Civic and Planning Annual* 10 (1938), 383–387 (384).

3. On the cloverleaf interchange at Woodbridge, N. J., see L. H. Robbins, "New Jersey Weaves a Vast Fabric of Super-Highways," *New York Times*, Oct. 19, 1930; C. S. Hill, "Intersection Design a Primary Highway Problem in New Jersey," *Engineering News-Record* 107 (Nov. 26, 1931), 834–838 (836–837); "Job and Office Note," *Engineering News-Record* 104 (March 27, 1930), 535; and Sigvald Johannesson, *Highway Economics* (McGraw-Hill, 1931), 110–111. The relatively suburban Woodbridge cloverleaf is widely recognized as the first of its kind, but similar grade-separation structures appeared within large cities earlier. See examples listed in Alvan Macauley, "Over-Passes Overcome Traffic Congestion in Cities," *Motor Age* 52 (Sept. 29, 1927), 39, 43, and in Robert Kingery, "Grade Separation Structures Untangle Traffic Jams," *Concrete* 37 (Nov. 1930), 17–19.

4. Peter Norton, "Fighting Traffic: U.S. Transportation Policy and Urban Traffic Congestion, 1950–1970," *Essays in History* 38 (1996), available at http://etext.lib. virginia.edu.

5. Ray W. Sherman, *If You're Going to Drive Fast* (Crowell, 1935), esp. 138. Sherman was a retired auto manufacturer and former editor of the auto industry trade journal *Motor*. In 1924 he served on the National Conference on Street and Highway Safety's Committee on the Causes of Accidents.

6. Furnas, "—And Sudden Death," *Reader's Digest* 27 (August 1935), 21–26. Anedith Jo Bond Nash has documented the article's extraordinarily wide notice in her

dissertation, "Death on the Highway: The Automobile in American Culture, 1920–40" (University of Minnesota, 1983), 38–42, 59–60n. 15; see also Daniel M. Albert, "Order out of Chaos: Automobile Safety, Technology and Society, 1925 to 1965" (Ph.D. dissertation, University of Michigan, 1997), 10–12.

7. A. J. Bracken, "The Aftermath of Sudden Death," *Reader's Digest* 27 (Dec. 1935), 52–54; quotation from editor's preface, p. 52.

8. Hadley Cantril and Mildred Strunk, eds., *Public Opinion, 1935–1946* (Princeton University Press, 1951), 35–36.

9. Henry, address, Nov. 21, 1936, *Annual Meeting of the Councillors of the American Automobile Association*, 34th annual meeting, Detroit, Nov. 20–21, 1936, 19–23 (23).

10. Furnas, "—And Sudden Death," 22, 26.

11. Ibid., 21.

12. Hoffman in address to graduates, Traffic Institute of Northwestern University, 1935; quoted in James Playsted Wood, *Of Lasting Interest: The Story of The Reader's Digest* (Doubleday, 1958), 62.

13. "A Statement by General Motors," *New York Times*, Nov. 3, 1935.

14. Maxwell Halsey, "Accident Prevention Is Now a Work of Plan and System," *New York Times*, Nov. 3, 1935.

15. "Unfit for Modern Traffic," *Fortune* 14 (August 1936), 85–92, 94, 96, 99 (88, 87).

16. Automobile Manufacturers Association, *You in Your Car on City Streets* (Detroit: AMA, 1936), 44.

17. "Auto Makers Open Wide Safety Drive," *New York Times*, Jan. 22, 1936. The words quoted were spoken by Alvan Macauley. ASF funding soon changed the Erskine Bureau's name to the Harvard Bureau for Street Traffic Research. In 1938 the bureau moved to Yale, where a grant from GM's Alfred P. Sloan Jr. supplemented ASF funds; see "Yale to Open Bureau on Traffic Research," *New York Times*, April 12, 1938. McClintock left the bureau and traffic work in 1942.

18. "Safety Foundation Formed," *Automotive Industries* 76 (June 5, 1937), 825, 829. Among manufacturers Ford was the only conspicuous absentee. AAA operated a like-minded safety effort outside of ASF.

19. Hoffman, *Seven Roads to Safety: A Program to Reduce Automobile Accidents* (Harper, 1939), 58 (here Hoffman quotes an unspecified ASF document), 60–61.

20. Hoffman ("with" Neil M. Clark), "The White Line Isn't Enough," *Saturday Evening Post* 210 (March 26, 1938), 12–13, 32, 37, 39, 41. See also the book into

which this article grew: Hoffman, *Seven Roads to Safety: A Program to Reduce Automobile Accidents* (Harper, 1939). Hoffman ("with" Neil M. Clark), "America Goes to Town," *Saturday Evening Post* 211 (April 29, 1939), 8–9, 30, 32, 35; Hoffman, "She Shall Have Safety," *Good Housekeeping* 108 (June 1939), 48ff.; Hoffman, "Winning the War on Traffic Accidents," *Popular Mechanics* 72 (Oct. 1939), 568–571, 116A.

21. The Bill of Rights was unanimously adopted by AAA, Nov. 21, 1936; see "Bill of Rights for Motorists," in "Report of the Resolutions Committee," *Annual Meeting of the Councillors of the American Automobile Association*, 34th annual meeting, Detroit, Nov. 20–21, 1936, 12–16 (12–13).

22. Sheets, "New Roads to Safety," *Safety Engineering* 74 (Oct. 1937), 44–45.

23. "Pedestrian Heaven in City of Future Is Exhibited Here," *New York Times*, July 20, 1937; McClintock, " 'Of Things to Come' " (address to the National Planning Conference, Detroit, June 1937), *American Civic and Planning Annual* 10 (1938), 383–387 (384). According to the *Times*, "The model city was designed by Dr. Miller McClintock . . . and Norman Bel Geddes." McClintock seems to have supplied traffic expertise to Bel Geddes, who built the model.

24. Norman Bel Geddes, "City 1960," *Architectural Forum* 67 (July 1937), 57–62 (58).

25. "Pedestrian Heaven."

26. Shell Oil Company, "In the City of Tomorrow" (advertisement), *Saturday Evening Post* 210 (Nov. 13, 1937), 35.

27. "Road Safety Seen as Engineers' Job," *New York Times*, June 25, 1938.

28. "Horse of Old Held Faster Than Autos," *New York Times*, Sept. 22, 1937; "Making a Motor Market," *Wall Street Journal*, Sept. 24, 1937.

29. McClintock, "Road Design Is Aid to Safety," *New York Times*, Oct. 24, 1937.

30. "Exposition to Portray City of 1999," *National Safety News* 39 (Jan. 1939), 74.

31. For a transcript of the "Voice," see General Motors, *Futurama* (1939), a presentation edition of 1,000 issued Oct. 16, 1939. A copy is available in the Special Collections Library of the University of Virginia at Charlottesville.

32. Norman Bel Geddes, *Magic Motorways* (Random House, 1940), 57.

33. Norman Bel Geddes ("as told to" Maxwell Hamilton), "Traffic and Transit in the World of Tomorrow," *Transit Journal* 83 (August 1939), 272–273, 290–291; Robert Coombs, "Norman Bel Geddes: Highways and Horizons," *Perspecta* 13 (1971), 11–27.

34. AAA Safety Division; Insurance Institute for Highway Safety.

35. Hoffman ("with" Neil M. Clark), "America Goes to Town," *Saturday Evening Post* 211 (April 29, 1939), 8–9, 30, 32, 35 (8); Hoffman, "Winning the War on Traffic Accidents," *Popular Mechanics* 72 (Oct. 1939), 568–571, 116A (568).

36. Eisenhower, "Special Message to the Congress from President Eisenhower on a National Highway Program," Feb. 22, 1955, in *Public Papers of the Presidents of the United States: Dwight D. Eisenhower*, volume 3 (1954; Washington: National Archives and Records Service, 1960), 275.

37. In *Public Papers of the Presidents of the United States: Dwight D. Eisenhower*, volume 8 (1960–61; Washington: National Archives and Records Service, 1961), see "Remarks at the Dedication of the Hiawatha Bridge," Red Wing, Minn., Oct. 18, 1960, 780–781 (781: " . . . will save 4,000 lives every year"), and "Address in Philadelphia at a Rally of the Nixon for President Committee of Pennsylvania," Oct. 28, 1960, 815–816 (815: " . . . will save four thousand American lives a year").

38. Barber to Hoover, May 23, 1927, file 02773, box 161, Commerce Papers, Hoover Library.

39. Danielian to Eisenhower, August 16, 1955, in Robert L. Branyan and Lawrence H. Larsen, eds., *The Eisenhower Administration, 1953–1961: A Documentary History* (Random House, 1971), volume 1, pp. 550–551.

40. President's Advisory Committee on a National Highway Program, *A Ten-Year National Highway Program: A Report to the President* (Government Printing Office, 1955).

41. Hoffman ("with" Neil M. Clark), "The White Line Isn't Enough," *Saturday Evening Post* 210 (March 26, 1938), 12–13, 32, 37, 39, 41 (12).

42. Department of Commerce, Bureau of the Census, *Historical Statistics of the United States, Colonial Times to 1970* (Government Printing Office, 1975), part 2, 719.

43. In this development, Ralph Nader played the role of a latter-day J. C. Furnas, with greater long-term success. See Nader, *Unsafe at Any Speed: The Designed-In Dangers of the American Automobile* (Grossman, 1965).

44. For a study of MADD, including a brief history of its origins, see Craig Reinarman, "The Social Construction of an Alcohol Problem: The Case of Mothers Against Drunk Drivers and Social Control in the 1980s," *Theory and Society* 17 (Jan. 1988), 91–120. MADD was influential in the revival of roadside memorials to accident victims; see Holly Everett, "Roadside Crosses and Memorial Complexes in Texas," *Folklore* 111 (April 2000), 91–103, esp. 92–93.

## Conclusion

1. Ludwik Fleck, *Genesis and Development of a Scientific Fact* (1935) (University of Chicago Press, 1979), 37.

2. Trevor Pinch, "The Social Construction of Technology: A Review," in *Technological Change: Methods and Themes in the History of Technology*, ed. R. Fox (Overseas Publishers Association, 1996), 35.

3. The frequent misunderstandings between empirical historians (on the one hand) and theorists of technology and society (on the other) are perhaps best exemplified in a 2002 exchange between historian Nick Clayton, and theorists Wiebe E. Bijker and Trevor J. Pinch, in *Technology and Culture* 43 (April). See Clayton, "SCOT: Does It Answer?," 351–360; Bijker and Pinch, "SCOT Answers, Other Questions," 361–369; and Clayton, "Rejoinder," 369–370.

4. Thomas P. Hughes, *Networks of Power: Electrification in Western Society, 1880–1930* (Johns Hopkins University Press, 1983).

5. Mark H. Rose, "Urban Environments and Technological Innovation: Energy Choices in Denver and Kansas City, 1900–1940," *Technology and Culture* 25 (July, 1984), 503–539 (esp. 523–24).

6. Historians of these networks have been attentive to such questions. On the persistence of private cisterns in American cities, long after the practical advantages of large, public water systems were established, see Maureen Ogle, "Water Supply, Waste Disposal, and the Culture of Privatism in the Mid-Nineteenth-Century American City," *Journal of Urban History* 25 (March 1999), 321–347.

7. Wiebe E. Bijker, *Of Bicycles, Bakelites, and Bulbs: Toward a Theory of Sociotechnical Change* (MIT Press, 1995), 262.

8. Ronald Kline and Trevor Pinch, "Users as Agents of Technological Change: The Social Construction of the Automobile in the Rural United States," *Technology and Culture* 37 (Oct. 1996), 763–795; Nelly Oudshoorn and Trevor Pinch, eds., *How Users Matter: The Co-Construction of Users and Technologies* (MIT Press, 2003).

9. Efforts to control definitions of use and misuse, as a "closure mechanism" (or means of controlling the construction of a system), are implicit in much social constructivist research. In "Technology, Stability, and Social Theory," John Law and Wiebe E. Bijker briefly consider misuse explicitly; see Bijker and Law, *Shaping Technology/Building Society* (MIT Press, 1992), 290–308 (295).

10. Kline and Pinch, "Users as Agents of Technological Change: The Social Construction of the Automobile in the Rural United States," *Technology and Culture* 37 (Oct. 1996), 763–795.

11. Paul Rosen, "The Social Construction of Mountain Bikes: Technology and Post-modernity in the Cycle Industry," *Social Studies of Science* 23 (August 1993), 479–513.

12. Dale Rose and Stuart Blume, "Citizens as Users of Technology: An Exploratory Study of Vaccines and Vaccination," in Oudshoorn and Pinch, *How Users Matter*.

13. Bessie E. Buckley to editor, in "Where Can the Kiddies Play? Send Views to *Journal*," *Milwaukee Journal*, Sept. 29, 1920.

## Inside Technology

*Ham Radio's Technical Culture*
Kristen Haring

*The Radiance of France: Nuclear Power and National Identity after World War II*
Gabrielle Hecht

*On Line and On Paper: Visual Representations, Visual Culture, and Computer Graphics in Design Engineering*
Kathryn Henderson

*Unbuilding Cities: Obduracy in Urban Sociotechnical Change*
Anique Hommels

*Pedagogy and the Practice of Science: Historical and Contemporary Perspectives*
David Kaiser, editor

*Biomedical Platforms: Reproducing the Normal and the Pathological in Late-Twentieth-Century Medicine*
Peter Keating and Alberto Cambrosio

*Constructing a Bridge: An Exploration of Engineering Culture, Design, and Research in Nineteenth-Century France and America*
Eda Kranakis

*Making Silicon Valley: Innovation and the Growth of High Tech, 1930–1970*
Christophe Lécuyer

*Viewing the Earth: The Social Construction of the Landsat Satellite System*
Pamela E. Mack

*Inventing Accuracy: A Historical Sociology of Nuclear Missile Guidance*
Donald MacKenzie

*Knowing Machines: Essays on Technical Change*
Donald MacKenzie

*Mechanizing Proof: Computing, Risk, and Trust*
Donald MacKenzie

*An Engine, Not a Camera: How Financial Models Shape Markets*
Donald MacKenzie

*Building the Trident Network: A Study of the Enrollment of People, Knowledge, and Machines*
Maggie Mort

*How Users Matter: The Co-Construction of Users and Technology*
Nelly Oudshoorn and Trevor Pinch, editors

*Building Genetic Medicine: Breast Cancer, Technology, and the Comparative Politics of Health Care*
Shobita Parthasarathy

*Framing Production: Technology, Culture, and Change in the British Bicycle Industry*
Paul Rosen

*Coordinating Technology: Studies in the International Standardization of Telecommunications*
Susanne K. Schmidt and Raymund Werle

*Structures of Scientific Collaboration*
Wesley Shrum, Joel Genuth, and Ivan Chompalov

*Making Parents: The Ontological Choreography of Reproductive Technology*
Charis Thompson

*Everyday Engineering: An Ethnography of Design and Innovation*
Dominique Vinck, editor

# Index